Problem Books in Mathematics

Edited by P.R. Halmos

Problem Books in Mathematics

Series Editor: P.R. Halmos

Polynomials
by *Edward J. Barbeau*

Problems in Geometry
by *Marcel Berger, Pierre Pansu, Jean-Pic Berry and Xavier Saint-Raymond*

Problem Book for First Year Calculus
by *George W. Bluman*

Exercises in Probability
by *T. Cacoullos*

An Introduction to Hilbert Space and Quantum Logic
by *David W. Cohen*

Problems in Analysis
by *Bernard Gelbaum*

Theorems and Counterexamples in Mathematics
by *Bernard R. Gelbaum and John M.H. Olmsted*

Exercises in Integration
by *Claude George*

Algebraic Logic
by *S.G. Gindikin*

Unsolved Problems in Number Theory
by *Richard K. Guy*

An Outline of Set Theory
by *James M. Henle*

Demography Through Problems
by *Nathan Keyfitz and John A. Beekman*

Theorems and Problems in Functional Analysis
by *A.A. Kirillov and A.D. Gvishiani*

(continued after index)

Bernard R. Gelbaum   John M.H. Olmsted

Theorems and Counterexamples in Mathematics

With 24 Illustrations

Springer-Verlag
New York Berlin Heidelberg
London Paris Tokyo Hong Kong

Bernard R. Gelbaum
Department of Mathematics
State University of New York
at Buffalo
Buffalo, New York 14214-3093
USA

John M.H. Olmsted
Department of Mathematics
Southern Illinois University
Carbondale, Illinois 62901
USA

Editor

Paul R. Halmos
Department of Mathematics
Santa Clara University
Santa Clara, California 95053, USA

Mathematical Subject Classifications: 00A07

Library of Congress Cataloging-in-Publication Data
Gelbaum, Bernard R.
   Theorems and counterexamples in mathematics / Bernard R. Gelbaum,
John M.H. Olmsted.
     p.   cm — (Problem books in mathematics)
   Includes bibliographical references and index.
   1. Mathematics. I. Olmsted, John Meigs Hubbell, 1911–
II. Title. III. Series.
QA36.G45   1990
510 – dc20
                           90-9899
                           CIP

Printed on acid-free paper

Photocomposed copy prepared by the authors using T_EX.
Printed and bound by R.R. Donnelly & Sons, Harrisonburg, Virginia.
Printed in the United States of America.

9 8 7 6 5 4 3 2 1

ISBN 0-387-97342-7 Springer-Verlag New York Berlin Heidelberg
ISBN 3-540-97342-7 Springer-Verlag Berlin Heidelberg New York

PREFACE

The gratifying response to *Counterexamples in analysis* (CEA) was followed, when the book went out of print, by expressions of dismay from those who were unable to acquire it.

The connection of the present volume with CEA is clear, although the sights here are set higher. In the quarter-century since the appearance of CEA, mathematical education has taken some large steps reflected in both the undergraduate and graduate curricula. What was once taken as very new, remote, or arcane is now a well-established part of mathematical study and discourse. Consequently the approach here is designed to match the observed progress. The contents are intended to provide graduate and advanced undergraduate students as well as the general mathematical public with a modern treatment of some theorems and examples that constitute a rounding out and elaboration of the standard parts of algebra, analysis, geometry, logic, probability, set theory, and topology.

The items included are presented in the spirit of a conversation among mathematicians who know the language but are interested in some of the ramifications of the subjects with which they routinely deal. Although such an approach might be construed as demanding, there is an extensive GLOSSARY/INDEX where all but the most familiar notions are clearly defined and explained. The object of the body of the text is more to enhance what the reader already knows than to review definitions and notations that have become part of every mathematician's working context.

Thus terms such as *complete metric space, σ-ring, Hamel basis, linear programming, [logical] consistency, undecidability, Cauchy net, stochastic independence, etc.* are often used without further comment, in which case they are italicized to indicate that they are carefully defined and explained in the GLOSSARY/INDEX.

The presentation of the material in the book follows the pattern below:

i A *definition* is provided either in the text proper or in the GLOSSARY/INDEX. The term or concept defined is usually *italicized* at some point in the text.

ii. A THEOREM for which proofs can be found in most textbooks and monographs is stated often without proof and always with at least one reference.

iii A result that has not yet been expounded in a textbook or monograph is given with at least one reference and, as space permits, with a proof, an outline of a proof, or with no proof at all.

iv Validation of a counterexample is provided in one of three ways:

    a. As an **Exercise** (with a *Hint* if more than a routine calculation is involved).

    b. As an **Example** and, as space permits, with a proof, an outline of a proof, or with no proof at all. Wherever full details are not given at least one reference is provided.

c. As a simple statement and/or description together with at least one reference.

Preceding the contents there is a **GUIDE** to the principal items treated.

We hope this book will offer at least as much information and pleasure as CEA seems to have done to (the previous generation of) its readers.

State University of New York at Buffalo   B. R. G.

Carbondale, Illinois   J. M. H. O.

Contents

The list below provides the sequence in which the essential items in the book are presented.

In this GUIDE and in the text proper, the boldface numbers **a.b.c.d. e** following an [Item] indicate [Item] **d** on **page e** in **Chapter a, Section b, Subsection c**; similarly boldface numbers **a.b.c. d** following an [Item] indicate [Item] **c** on **page d** in **Chapter a, Section b**; e.g., **Example 1.3.2.7. 35.** refers to the **seventh Example** on **page 35** in **Subsection 2** of **Section 3** of **Chapter 1**; LEMMA **4.2.1. 218.** refers to the **first** LEMMA on **page 218** in **Section 2** in **Chapter 4**.

Group Theory

1. Faulty group axioms.

Example 1.1.1.1. 2, Remark 1.1.1.1. 2.

2. Lagrange's theorem and the failure of its converse.

THEOREM 1.1.2.1 3, **Exercise 1.1.2.1. 3.**

3. Cosets as equivalence classes.

Exercise 1.1.2.2. 3.

4. A symmetric and transitive relation need not be reflexive.

Exercise 1.1.2.3. 3.

5. A subgroup H of a group G is normal iff every left (right) coset of H is a right (left) coset of H.

Exercise 1.1.2.4. 3

6. If $G : H$ is the smallest prime divisor p of $\#(G)$ then H is a normal subgroup.

THEOREM **1.1.2.2. 4.**

7. An exact sequence that fails to split.

Example 1.1.3.1. 5.

8. If the topological group H contains a countable dense set and if the homomorphism $h : G \mapsto H$ of the locally compact group G is measurable on some set P of positive measure then h is continuous (everywhere).

THEOREM **1.1.4.1. 5.**

9. If A is a set of positive (Haar) measure in a locally compact group then AA^{-1} contains a neighborhood of the identity.

pages 5–6.

10. The existence of a Hamel basis for $\mathbb{R}$ over $\mathbb{Q}$ implies the existence in $\mathbb{R}$ of a set that is not Lebesgue measurable.

page 6.

11. If f (in $\mathbb{R}^{\mathbb{R}}$) is a nonmeasurable function that is a solution of the functional equation $f(x + y) = f(x) + f(y)$ then a) f is unbounded both above and below in every nonempty open interval and b) if R is one of the relations $<, \leq, >, \geq$ and $E_{R,\alpha} \overset{\text{def}}{=} \{\, x \,:\, f(x)\, R\, \alpha \,\}$, then for all α in $\mathbb{R}$ and for every open set U, $E_{R,\alpha} \cap U$ is dense in U.

Exercise 1.1.4.1. 6.

12. There are nonmeasurable midpoint-convex functions.

Exercise 1.1.4.2. 7.

13. There exists a Hamel basis B for $\mathbb{R}$ over $\mathbb{Q}$ and $\lambda(B) = 0$.

THEOREM **1.1.4.2. 7.**

14. For the Cantor set C_0: $C_0 + C_0 = [0, 2]$.

Exercise 1.1.4.3. 7, Note 1.1.4.1. 7.

15. The Cantor set C_0 contains a Hamel basis for $\mathbb{R}$ over $\mathbb{Q}$.

Exercise 1.1.4.4. 7.

16. *Finiteness* is a Quotient Lifting (QL) property of groups.

Example 1.1.4.1. 8.

17. *Abelianity* is not a QL property of groups.

Example 1.1.4.2. 8.

18. *Solvability* is a QL property of groups.

Exercise 1.1.4.5. 8.

19. *Compactness* is a QL property of locally compact topological groups.

Example 1.1.4.3. 9.

20. If X is a set there is a free group on X.

Exercise 1.1.5.1. 9.

21. **The** free group on X.

Note 1.1.5.1. 10.

22. Every group G is the quotient group of some free group $F(X)$.

Exercise 1.1.5.2. 10.

23. A group G can be the quotient group of different free groups.

Note 1.1.5.2. 11.

24. The undecidability of the word problem for groups.

Note 1.1.5.2. 11.

25. There is a finitely presented group containing a finitely generated subgroup for which there is no finite presentation.

Note 1.1.5.2. 11.

26. An infinite group G presented by a finite set $\{x_1, \ldots, x_n\}$ of generators and a finite set

$$\{x_i^{k_i} \sim \tilde{\emptyset}\},\ k_i \in \mathbb{N},\ 1 \leq i \leq n,$$

of identities.

Note 1.1.5.2. 11.

27. The Morse-Hedlund nonnilpotent semigroup Σ generated by three nilpotent elements.

pages 11–12.

28. Every quaternion $\mathbf{q}$ is a square.

29. Two pure quaternions commute iff they are linearly dependent over $\mathbb{R}$.

30. If X is a completely regular topological space there is a free topological group $F_{top}(X)$ on X.

31. A quaternion $\mathbf{q}$ is of norm 1: $|\mathbf{q}| = 1$ iff $\mathbf{q}$ is a commutator.

32. The commutator subgroup of $\mathbb{H}^*$ is the set of quaternions of norm 1: $Q(\mathbb{H}^*) = \{\mathbf{q} : \mathbf{q} \in \mathbb{H}, |\mathbf{q}| = 1\}$.

33. In $\mathbb{H}^*$ there is a free subset T such that $\#(T) = \#(\mathbb{R})$.

34. A faulty commutative diagram.

35. The square root function is not continuous on $\mathbb{T}$.

36. The classification of finite simple groups.

37. For two (different) primes p and q, are the natural numbers

$$\frac{p^q - 1}{p - 1} \text{ and } \frac{q^p - 1}{q - 1}$$

relatively prime?

Algebras

38. Over $\mathbb{H}$, a polynomial of degree two and for which there are infinitely many zeros.

39. There are infinitely many different quaternions of the form $\mathbf{qiq}^{-1}$.

40. If the quaternion $\mathbf{r}$ is such that $\mathbf{r}^2 + 1 = 0$ then for some quaternion $\mathbf{q}, \mathbf{r} = \mathbf{r_q} \overset{\text{def}}{=} \mathbf{qiq}^{-1}$.

41. A nonassociative algebra.

42. The Jacobi identity.

43. Lie algebras and groups of Lie type.

Linear Algebra

Measure Theory

Topological Vector Spaces

Complex Variable Theory

The Euclidean Plane

Topological spaces

Exotica in Differential Topology

Independence in Probability

Stochastic Processes

Transition matrices

Logic

Set Theory

1. Algebra

1.1. Group Theory

1.1.1. Axioms

By definition a *group* is a nonempty set G and a map

$$G \times G \ni \{x, y\} \mapsto xy \in G$$

subject to the following axioms:

i. If $x, y, z \in G$ then $x(yz) = (xy)z$ (*associativity*).
ii. There is in G an element denoted e with two properties:
    iia. if $x \in G$ then $ex = x$ (e is a *left identity*);
    iib. if $x \in G$ there is in G a *left inverse* y such that $yx = e$.

Consequences of these axioms are:

iii. There is only one left identity e.
iv. For each x in G there is only one left inverse.
v. The left identity is a right identity: $xe = x$, $x \in G$, and there is only one right identity.
vi. The unique left inverse of an element x is a right inverse of x: $yx = e \Rightarrow xy = e$, $x, y \in G$, and there is only one right inverse of x.

The unique (left and right) *inverse* of x is denoted x^{-1}.

The axiom ii is replaceable by:

ii'. There is in G an element denoted e with two properties:
$ii'a$. if $x \in G$ then $xe = x$ (e is a *right identity*);
$ii'b$. if $x \in G$ there is in G a *right inverse* y such that $xy = e$.

or by

vii. For each pair $\{a, b\}$ in $G \times G$:
$viia$. there is a solution x for the equation $ax = b$;
$viib$. there is a solution y for the equation $ya = b$.

However, assumptions about left identities and right inverses may not be mixed. In other (more formal) terms, if ii is replaced either by:

ii''. There is in G an element denoted e with two properties:
iia. if $x \in G$ then $ex = x$ (e is a *left identity*);
$ii'b$. if $x \in G$ there is in G a *right inverse* y such that $xy = e$;

or by

ii'''. There is in G an element denoted e with two properties:
$ii'a$. if $x \in G$ then $xe = x$ (e is a *right identity*);
iib. if $x \in G$ there is in G a *left inverse* y such that $yx = e$;

then G may fail to be a group.

Example 1.1.1.1. Assume that G is a set consisting of at least two elements and that $x, y \in G \Rightarrow xy = y$. A direct check shows that i (associativity) obtains. Nevertheless in G every element may serve as a left identity (iia is satisfied) but, since there are at least two elements in G, there is no unique left identity (iii is denied).

Furthermore if one element, say e, is singled out to serve as a left identity then $xe = e$ for every x in G and so every element has a right inverse e ($ii'b$ is satisfied) but if $x \neq e$ then x has no left inverse since $yx = x \neq e$ for every y (iib is denied).

Furthermore in G $viia$ obtains but $viib$ does not: b is the solution of $ax = b$ but if $a \neq b$ then $ya = b$ has no solution.

[**Remark 1.1.1.1:** A similar difficulty arises if, in ii, one rephrases iib as:

$ii''b$. If $x \in G$ there is in G a right inverse y such that xy is a left identity.]

1.1.2. Subgroups

Let $\#(S)$ denote the *cardinality* of the set S. If G is a group then $\#(G)$ is the *order* of G. What follows is a classical theorem about a finite group and the orders of it and of its subgroups.

THEOREM **1.1.2.1.** (LAGRANGE) IF G IS A FINITE GROUP AND H IS A *subgroup* THEN $\#(H)$ IS A FACTOR OF $\#(G)$: $\#(H)|\#(G)$.

On the other hand, the converse of the statement above is false.

Exercise 1.1.2.1. Show that in the symmetric group $\mathcal{S}_4$ the subgroup H consisting of the following twelve permutations contains no subgroup of order six.

$$\begin{pmatrix} 1,2,3,4 \\ 1,2,3,4 \end{pmatrix} \quad \begin{pmatrix} 1,2,3,4 \\ 1,3,4,2 \end{pmatrix}$$
$$\begin{pmatrix} 1,2,3,4 \\ 2,1,4,3 \end{pmatrix} \quad \begin{pmatrix} 1,2,3,4 \\ 1,4,2,3 \end{pmatrix}$$
$$\begin{pmatrix} 1,2,3,4 \\ 3,4,1,2 \end{pmatrix} \quad \begin{pmatrix} 1,2,3,4 \\ 3,2,4,1 \end{pmatrix}$$
$$\begin{pmatrix} 1,2,3,4 \\ 4,3,2,1 \end{pmatrix} \quad \begin{pmatrix} 1,2,3,4 \\ 4,2,1,3 \end{pmatrix}$$
$$\begin{pmatrix} 1,2,3,4 \\ 2,3,1,4 \end{pmatrix} \quad \begin{pmatrix} 1,2,3,4 \\ 2,4,3,1 \end{pmatrix}$$
$$\begin{pmatrix} 1,2,3,4 \\ 3,1,2,4 \end{pmatrix} \quad \begin{pmatrix} 1,2,3,4 \\ 4,1,3,2 \end{pmatrix}.$$

Thus if G is a finite group and k is a factor of $\#(G)$, G need not contain a subgroup of order k.

A subgroup H of a group G engenders a decomposition of G into *equivalence classes* according to the equivalence relation R: xRy iff $x \in yH$, i.e., iff x is in the *coset* yH.

Exercise 1.1.2.2. Show that R as described above is an *equivalence relation*, i.e., for all x, y, z in G, a) xRx (R is *reflexive*), b) xRy if yRx (R is *symmetric*), and c) if xRy and yRz then xRz, (R is *transitive*).

Exercise 1.1.2.3. Find the error, via a counterexample, in the argument that symmetry and transitivity of a relation R imply reflexivity.

A subgroup H of a group G is *normal* iff for all x in G, $x^{-1}Hx = H$.

Exercise 1.1.2.4. Show that H is a normal subgroup of a group G iff for all x in G, $xH = Hx$ ("every x-*left coset* is the same as the corresponding x-*right coset*"), iff every left coset is some right coset, iff every right coset is some left coset.

Exercise 1.1.2.5. Show that if the *index* i.e., $G : H \stackrel{\text{def}}{=} \#(G)/\#(H)$ (in $\mathbb{N}$) of H in G is 2 then H is a normal subgroup. Show that the index of normal subgroup H of a group G need not be two.

At some time in the early 1940s Ernst G. Straus, sitting in a group theory class, saw the proof of the first result in **Exercise 1.1.2.5** and immediately conjectured (and proved that night):

THEOREM 1.1.2.2. IF $G : H$ IS THE SMALLEST PRIME DIVISOR p OF $\#(G)$ THEN H IS A NORMAL SUBGROUP.

PROOF. As the next lines show, if $a \notin H$ the p *cosets*

$$H, aH, \ldots, a^{p-1}H$$

are pairwise disjoint. Indeed, otherwise there is a least r and a least s such that

$$0 \leq r < s \leq p - 1 \text{ and } a^r H \cap a^s H \neq \emptyset.$$

Then, since left cosets are *R-equivalence classes* with respect to $cRd \Leftrightarrow c \in dH$, it follows that

$$a^r H = a^s H, \ a^{s-r} H = H.$$

Hence the minimality of r implies

$$r = 0, \ s \leq p - 1, \text{ and } a^s H = H.$$

Let m be the *order* of a, i.e., m is the least natural number such that $a^m = e$. Hence $\{a, a^2, \ldots, a^m\} \stackrel{\text{def}}{=} K$ is a subgroup of G,

$$\#(K) = m, \ m \geq p > s, \ m | \#(G),$$

and there are natural numbers q, t such that

$$m = qs + t, \ 1 \leq q, \ 0 \leq t < s$$
$$e = a^m = a^t (a^s)^q$$
$$H = eH = a^m H = a^t (a^s)^q H = a^t (a^s)^{q-1} a^s H = a^t (a^s)^{q-1} H = \cdots = a^t H.$$

Since s is minimal it follows that $t = 0$, $s | \#(G)$, in contradiction of the definition of p. However $\#(G) = p \cdot \#(H)$, whence $G = \bigcup_{j=0}^{p-1} a^j H$.

If $b \in G \backslash H$, $h \in H$ and $bhb^{-1} \stackrel{\text{def}}{=} a \notin H$ then there is a natural number r and in H a k such that $b = a^r k$. Hence $a = khk^{-1} \in H$, a contradiction, i.e., H is normal. □

1.1.3. Exact versus splitting sequences

Let G, H, K be a set of groups. If the *homomorphisms* $G \stackrel{\phi}{\mapsto} H$ and $H \stackrel{\psi}{\mapsto} K$ are such that the image $\phi(G)$ ($\stackrel{\text{def}}{=} \text{im}(\phi)$) is a subset of the kernel $\psi^{-1}(e)$ ($\stackrel{\text{def}}{=} \text{ker}(\psi)$), i.e., $\text{im}(\phi) \subset \text{ker}(\psi)$, the situation is symbolized by the *sequence*

$$G \stackrel{\phi}{\mapsto} H \stackrel{\psi}{\mapsto} K.$$

If $\mathrm{im}(\phi) = \ker(\psi)$ the sequence is *exact*. In particular the sequence is exact if it *splits*, i.e., if H is the *direct product* $G \times K$, ϕ is the *injection*: $\phi(g) = \{g, e\} \in G \times K$, whence $\phi(G) = G \times \{e\}$, and ψ is the *surjection*: $\psi(\{g, k\}) = k$, whence $\psi(G \times K) = K$.

Example 1.1.3.1. If G and K are arbitrary, $\phi(G) = H$, and $\psi(H) = e$ then

$$ G \overset{\phi}{\mapsto} H \overset{\psi}{\mapsto} K $$

is exact. If $\#(G) + \#(K) > 2$ and $H = \{e\}$ the sequence does not split.

1.1.4. The functional equation: $f(x + y) = f(x) + f(y)$

Let G be a *locally compact topological group* and let μ be a *Haar measure* on the σ-*ring* $\mathcal{S}(\mathsf{K})$ (generated by the set K of *compact* sets of G) [**Halm, Loo**]. Let H be a topological group for which there is a *homomorphism*: $h : G \mapsto H$. Then h is:

▷ *continuous* iff $h^{-1}(U)$ is *open* for every open set U in H;
▷ *open* iff $h(V)$ is open for every open set V in G;
▷ *measurable* iff $h^{-1}(U) \in \mathcal{S}(\mathsf{K})$ for every open set U in H.

THEOREM **1.1.4.1.** IF H CONTAINS A *countable dense* SET $S \overset{\text{def}}{=} \{s_n\}_{n=1}^{\infty}$ AND IF THE HOMOMORPHISM $h : G \mapsto H$ IS *measurable* ON SOME SET P OF POSITIVE MEASURE THEN h IS CONTINUOUS (EVERYWHERE).

PROOF. Let W and U in H be *neighborhoods* of e and such that $UU^{-1} \subset W$. It may be assumed that $\mu(P)$ is finite. Then, since S is dense,

$$ H = \bigcup_{n=1}^{\infty} U s_n. $$

If $P_n \overset{\text{def}}{=} h^{-1}(U s_n)$ then $P_n \cap P$ is measurable and

$$ \bigcup_{n=1}^{\infty} (P_n \cap P) = P. $$

Hence there is an n_0 such that $\mu(P_{n_0} \cap P) > 0$.

If $A \overset{\text{def}}{=} P_{n_0} \cap P$ then there is in G an open set V containing e and contained in AA^{-1}. Indeed, χ_A denoting the *characteristic function* of A,

$$ x \mapsto \mu(A \cap xA) \left(= \int_G \chi_A(y) \chi_A(x^{-1}y)\, dy \right) $$

is:

▷ a *uniformly continuous* function of x;
▷ positive at e and hence in a neighborhood V of e;
▷ zero off AA^{-1}.

Hence $V \subset AA^{-1}$. It follows that $h(V) \subset UU^{-1} \subset W$ whence h is continuous at e. Because h is a homomorphism continuous at e, h is continuous everywhere.

$\square$

The set $\mathbb{R}$ may be regarded as *vector space* over $\mathbb{Q}$. Since $\mathbb{R}$ is *uncountable* there is an infinite set that is *linearly independent* over $\mathbb{Q}$. According to *Zorn's lemma* there is a set B that is linearly independent over $\mathbb{Q}$ and properly contained in no other set that is linearly independent over $\mathbb{Q}$: B is a maximal linearly independent set, i.e., a *Hamel basis* for $\mathbb{R}$ over $\mathbb{Q}$. Then B is *uncountable* and hence there is in $\mathbb{R}$ a limit point b of B. Hence there is in B an infinite sequence $S \overset{\text{def}}{=} \{x_{\lambda_n}\}_{n=1}^{\infty}$ such that $\lim_{n \to \infty} x_{\lambda_n} = b$. Define $f : \mathbb{R} \mapsto \mathbb{R}$ as follows:

$$
f(x) = \begin{cases} n & \text{if } x = x_{\lambda_n} \\ \sum_{\lambda \in \Lambda} a_\lambda f(x_\lambda) & \text{if } x = \sum_{\lambda \in \Lambda} a_\lambda x_\lambda \in \text{span}(S) \\ 0 & \text{if } x \in B \setminus \text{span}(S). \end{cases}
$$

Then $f(x+y) = f(x) + f(y)$, $x, y \in \mathbb{R}$, and f is not continuous (at b). The argument that proved THEOREM **1.1.4.1. 5** shows that if f is *Lebesgue measurable* then f is continuous everywhere. Hence f is not Lebesgue measurable and hence there is an open set U such that $f^{-1}(U)$ is not Lebesgue measurable. (In **Section 2.2** there is an alternative proof of the existence in $\mathbb{R}$ of a subset that is not Lebesgue measurable. Nevertheless, the Axiom of Choice is part of the argument.)

> The *Axiom of Choice*, which implies the existence of a Hamel basis for $\mathbb{R}$ over $\mathbb{Q}$, implies the existence in $\mathbb{R}$ of a set that is not Lebesgue measurable.

Exercise 1.1.4.1. Let f (in $\mathbb{R}^{\mathbb{R}}$) be a *nonmeasurable function* that is a solution of the functional equation $f(x+y) = f(x) + f(y)$.

i. Show that f is *unbounded* both above and below in every nonempty open *interval*.

ii. Let R stand for one of the relations $<, \leq, >, \geq$ and let $E_{R,\alpha}$ be

$$\{x : f(x) \, R \, \alpha\}.$$

Show that for all α in $\mathbb{R}$ and for every open set U, $E_{R,\alpha} \cap U$ is dense in U.

[*Hint:* Show that the *discontinuity* of f at 0 implies there is a positive ϵ and a sequence $\{x_n\}_{n=1}^{\infty}$ such that $\lim_{n\to\infty} x_n = 0$ and $|f(x_n)| \geq \epsilon$. For each m consider the set $\{f(mx_n)\}_{n=1}^{\infty}$.]

If U is an open subset of $\mathbb{R}$, a function f in $\mathbb{R}^U$ is *convex* iff whenever $t \in [0,1]$, $x, y, tx + (1-t)y \in U$ then $f(tx + (1-t)y) \leq tf(x) + (1-t)f(y)$: "the curve lies below the chord." It follows [**Roy, Rud**] that a convex function is continuous everywhere and *differentiable a.e.*

A less restrictive definition of convexity for a function f is the requirement that f be *midpoint-convex*: "at the midpoint of an interval the curve lies below the chord," i.e.,

$$f\left(\frac{x+y}{2}\right) \leq \frac{1}{2}f(x) + \frac{1}{2}f(y).$$

Exercise 1.1.4.2. Show that Axiom of Choice implies that there are nonmeasurable midpoint-convex functions.

THEOREM **1.1.4.2.** THERE IS FOR $\mathbb{R}$ OVER $\mathbb{Q}$ A HAMEL BASIS B SUCH THAT $\lambda(B) = 0$.

The PROOF is a consequence of the conclusions in **Exercises 1.1.4.3** and **1.1.4.4**.

Exercise 1.1.4.3. Let C_0 be the *Cantor set*:

$$C_0 = \left\{ \sum_{k=1}^{\infty} \epsilon_k 3^{-k} \ : \ \epsilon_k = 0 \text{ or } 2, \ k \in \mathbb{N} \right\}.$$

Show that $\frac{1}{2}C_0 + \frac{1}{2}C_0 \overset{\text{def}}{=} \{\, x + y \ : \ x, y \in C_0 \,\} = [0,1]$.

[*Hint:* For t in $[0,1]$ consider a *binary representation* of t.]

[**Note 1.1.4.1:** The PROOF of THEOREM **1.1.4.1. 5** shows that if A is a measurable set of positive (Haar) measure in a locally compact group then AA^{-1} contains a neighborhood of the identity.

When the group is abelian and the binary operation of the group is symbolized by $+$ the set AA^{-1} is written $A - A$. The Haar measure (Lebesgue measure) of C_0 in **Exercise 1.1.4.3** is zero. Hence the measure of the set $A \overset{\text{def}}{=} C_0 \cup -C_0$ is zero and $-A = A$. Since $A - A = A + A = [-2, 2]$ the condition: *measure of A is positive* is a sufficient but not necessary condition for the conclusion that $A - A$ contains a neighborhood of the identity.]

Exercise 1.1.4.4. Let B be a *maximally $\mathbb{Q}$-linearly independent subset* of C_0 (or of $\frac{1}{2}C_0$). Show B is a Hamel basis for $\mathbb{R}$ over $\mathbb{Q}$. $\qquad\square$

For further properties of Hamel bases in $\mathbb{R}$ see **Section 2.2**.

In the *category* $\mathcal{G}$ of groups and homomorphisms the following phenomenon often occurs.

There is a property $P(G)$ of (some) groups G and whenever

$$\{0\} \hookrightarrow B \hookrightarrow A \overset{\phi}{\mapsto} C \overset{\psi}{\mapsto} \{0\}$$

is a *short exact sequence* of groups then

$$P(B) \wedge P(C) \Rightarrow P(A). \tag{1.1.4.1}$$

For simplicity, let a property P for which (1.1.4.1) (or its analog in some other category) holds be called a *Quotient Lifting* (QL) property.

Example 1.1.4.1. In the context just described, e.g.,

i. (1.1.4.1) is valid if $P(G)$ means "G is finite;"
ii. (1.1.4.1) is valid if $P(G)$ means "G is infinite."

In next two **Examples** there are illustrations of both the absence and the presence of the QL property.

Example 1.1.4.2. Let $P(G)$ mean "G is abelian." Then (1.1.4.1) fails for P. Indeed, if $\mathcal{S}_3$ is the *symmetric group* of order 6, i.e., $\mathcal{S}_3$ is the set of all permutations of the sequence $1, 2, 3$, if $\mathcal{A}_3$ is the *alternating subgroup* of $\mathcal{S}_3$, i.e., the set of *even permutations* in $\mathcal{S}_3$, and $C = \mathcal{S}_3/\mathcal{A}_3$, then $\#(\mathcal{A}_3) = 3$, $\#(C) = 2$ and so (THEOREM **1.1.2.2. 4**) $\mathcal{A}_3$ is a normal, cyclic, hence abelian subgroup, C is cyclic, hence also abelian, but $\mathcal{S}_3$ is not abelian, i.e., "abelianity" is not a QL property.

Exercise 1.1.4.5. Show that "solvability" is a QL property.

[*Hint:* Assume H is a normal subgroup of the group G and that both H and $G/H \overset{\text{def}}{=} K$ are solvable. If

$$K \overset{\text{def}}{=} K_0 \supset K_1 \supset \cdots \supset K_{r-1} \supset \{e\}$$
$$H \overset{\text{def}}{=} H_0 \supset H_1 \supset \cdots \supset H_{s-1} \supset \{e\}$$

are finite sequences of subgroups, if each subgroup is normal in its predecessor, and if all the corresponding *quotient groups* are abelian then there are in G subgroups $N_1, \ldots, N_{r-1}$ such that in the sequence

$$K = G/H \supset N_1/H \supset N_2/H \cdots \supset N_{r-1}/H \supset H/H = \{e\}$$

each subgroup is normal in its predecessor and the corresponding quotient group is abelian. It follows that each subgroup in

$$G \supset N_1 \supset \cdots \supset N_{r-1} \supset$$
$$H_0 \supset \cdots \supset H_{s-1} \supset \{e\}$$

is normal in its predecessor and the corresponding quotient groups are all abelian.]

The QL theme ((1.1.4.1), page 8) is repeated in a number of other categories, cf. **Subsection 2.3.3, Section 2.4**.

Example 1.1.4.3. In the category $\mathcal{LCG}$ of locally compact topological groups and continuous open homomorphisms let $P(G)$ mean "G is compact." Then (1.1.4.1), page 8 is valid for P.

[PROOF. Let V be a *compact neighborhood* of the identity in A. Then $A = \bigcup_{a \in A} aV$, $C = A/B = \bigcup_{a \in A} \phi(aV)$. Since C is compact, there are in A elements $a_1, \ldots, a_n$ such that $C = \bigcup_{i=1}^n \phi(a_i V)$ whence $\bigcup_{i=1}^n a_i V B = A$. Since B is compact it follows that A is compact. $\square$]

1.1.5. Free groups; free topological groups

If X is a nonempty set, a *free group on X* is a group $F(X)$ such that:

i. $F(X)$ contains a *bijective* image of X (by abuse of language,

$$X \subset F(X));$$

ii. if G is a group and $\phi : X \mapsto G$ is a map then ϕ may be extended to a homomorphism $\Phi : F(X) \mapsto G$.

Exercise 1.1.5.1. Show that if X is a set there is a free group on X. [*Hint:* Consider the set

$$W(X) \stackrel{\text{def}}{=} \{\, x_1^{\epsilon_1} \cdots x_n^{\epsilon_n} \ : \ x_i \in X, \ \epsilon_i = \pm 1, \ n \in \{0\} \cup \mathbb{N} \,\}$$

of all *words*. (If $n = 0$ the corresponding word is the empty word $\tilde{\emptyset}$.) For $w_1 \stackrel{\text{def}}{=} x_1^{\epsilon_1} \cdots x_n^{\epsilon_n}$ and $w_2 \stackrel{\text{def}}{=} y_1^{\delta_1} \cdots y_m^{\delta_m}$, define their product $w_1 w_2$ to be

$$x_1^{\epsilon_1} \cdots x_n^{\epsilon_n} y_1^{\delta_1} \cdots y_m^{\delta_m}.$$

Each symbol $x_i^{\epsilon_i}$ is a *factor* of the (nonempty) word

$$w_1 \stackrel{\text{def}}{=} x_1^{\epsilon_1} \cdots x_n^{\epsilon_n}.$$

Call two words w_1 and w_2 *adjacent* if there are words u and v and in X an x such that $w_1 = ux^\epsilon x^{-\epsilon} v$ and $w_2 = uv$. (The word w_1 is said to *simplify* or to *reduce* and w_2 is a *simplification* or a *reduction* of w_1.) Call two words u and v *equivalent* $(u \sim v)$ iff there are words $w_1, \ldots, w_n$ such that $u = w_1$, w_i and w_{i+1} are adjacent, $1 \le i \le n-1$, and $w_n = v$. Show that $\sim$ is an equivalence relation. If $w \in W(X)$ let $[w]$ denote the equivalence class of w. Show that the set $F(X) \overset{\text{def}}{=} W(X)/\sim$ of equivalence classes with multiplication of equivalence classes defined by multiplication of their representatives is a group, a free group on X. In particular:

   i. the equivalence class $[\widetilde{\emptyset}]$ of the empty word is the identity;

   ii. the equivalence class $[x_n^{-\epsilon_n} \cdots x_1^{-\epsilon_1}]$ of $x_n^{-\epsilon_n} \cdots x_1^{-\epsilon_1}$ is the inverse of the equivalence class $[x_1^{\epsilon_1} \cdots x_n^{\epsilon_n}]$ of $x_1^{\epsilon_1} \cdots x_n^{\epsilon_n}$;

   iii. if $x \in X$, the equivalence class $[x]$ of x may be identified with x and X is in bijective correspondence with a subset of $F(X)$.

   For details see [**Hal**].]

[**Note 1.1.5.1:** If $X \subsetneqq \widetilde{X}$ then $F(X)$ is *isomorphic* to a *proper subgroup* $H(\widetilde{X})$ of $F(\widetilde{X})$: $F(X) \cong H(\widetilde{X}) \subsetneqq F(\widetilde{X})$. If $\phi : X \mapsto G$ is a map then ϕ may be extended to a map of $\widetilde{X}$ into G and hence to a homomorphism $\Phi : F(\widetilde{X}) \mapsto G$. However if $\mathcal{F}$ is any group containing X then the *bijection* $\phi : X \ni x \mapsto x \in \mathcal{F}$ may be extended to a *monomorphism* (an *injective* homomorphism) $\Phi : F(X) \mapsto \mathcal{F}$. Hence $F(X)$ may be regarded as a minimal free group on X, i.e., $F(X)$ is **the** free group on X.]

Exercise 1.1.5.2. Show that any group G may be regarded as the quotient group of some free group $F(X)$ on some set X.

[*Hint:* Let G be a group and regard G as a set X. Then $F(X)$ is the free group on the set X and, if ϕ is the identity map:

$$\phi : X \ni x \mapsto x \in G,$$

ϕ may be extended to a homomorphism

$$\Phi : F(X) \mapsto G.$$

Consider $F(X)/\Phi^{-1}(e) \ (= F(X)/\ker(\Phi))$.]

   If G is a group and is regarded as the quotient group of a free group $F(X)$ according to the procedure in the *Hint* above then G is called a *free group* iff Φ is an isomorphism. If S is a subset of a group G and $w \in W(S)$ there is the element $\gamma(w)$ calculated by multiplying the factors in w according to the multiplication defined in G. The set S is called *free* iff for each word w in $W(S)$:

$$\gamma(w) = e \Leftrightarrow w \sim \tilde{\emptyset}.$$

[**Note 1.1.5.2:** Although every group G is the quotient group of a free group F, there need not be just one free group of which G is a quotient group, e.g., if $\#(G) = 1$ then G is a quotient group of every free group F: $G = F/F$. Thus there arises the notion of the *presentation* of a group G, namely the definition of a set X of *generators* of a free group $F(X)$ and the definition of an *epimorphism* $\Phi : F(X) \mapsto G$. The normal subgroup $N \overset{\text{def}}{=}$ $\ker(\Phi) \overset{\text{def}}{=} \Phi^{-1}(e)$ then defines (a set of) *relations* among the elements of X. These relations may be regarded as constituting in $W(X)$ a subset R of words corresponding to a minimal set of generators of N or, alternatively as a set of *identities* imposed on those words. The group G is said to be *presented* by the set X of generators and the set R of relations. If both X and R are finite the group is *finitely presented*.

If a group G is *presented* in the manner described above, there arises the *word problem*, i.e., whether there is an algorithm that successfully determines whether a word in $W(X)$ is equivalent to $\tilde{\emptyset}$. Boone and Novikov independently showed that there are groups for which there are presentations that admit no such algorithm. Their work was shortened by Britton [**Boo, Brit, Rot**].

Baumslag, Boone, and Neumann [**BBN**] gave an example of a finitely presented group containing a finitely generated subgroup for which there is no finite presentation.

Yet another related and very old problem is the *Burnside question*:

If $X \overset{\text{def}}{=} \{x_1, \ldots, x_n\}$, if $k_i \in \mathbb{N}$, $1 \leq i \leq n$, and if the identities $x_i^{k_i} \sim \tilde{\emptyset}$, $1 \leq i \leq n$, are imposed, is the group G presented in this way finite?

The question remained open for many years until 1968 when Novikov and Adian answered it negatively by means of a counterexample [**Ad, NovA**].]

In a similar vein Morse and Hedlund [**MoH**] exhibited a *semigroup* Σ containing 0 and such that:

i. Σ is generated by three elements denoted 1, 2, 3;

ii. $0a = a0 = 0$, $a \in \Sigma$; $1^2 = 2^2 = 3^2 = 0$;

iii. for no k in $\mathbb{N}$ is it true that every product of k different elements of Σ is 0 (Σ is not *nilpotent*).

What follows is a sketch of the Morse-Hedlund development. Assume

$$a_0 = 1, \, b_0 = 2$$
$$a_1 = a_0 b_0, \, b_1 = b_0 a_0$$
$$\cdots \cdots$$
$$a_{n+1} = a_n b_n, \, b_{n+1} = b_n a_n;$$
$$c_0 c_1 \cdots c_{2^n - 1} = a_n, \, n = 0, 1, \ldots$$
$$c_i = 1 \text{ or } 2, \, c_{-i} = c_{i-1}, \, i \in \mathbb{N}$$
$$T \overset{\text{def}}{=} \cdots c_{-2} c_{-1} c_0 c_1 c_2 \cdots .$$

Thus, e.g.,

$$c_0 c_1 \cdots = 1221 \, 2112 \, 2112 \, 1221 \, 2112 \, 1221 \, 1221 \, 2112 \, 2112 \, 1221 \cdots,$$

and there are no more than two successive 1's or 2's in T.

In T let B_i, $i \in \mathbb{Z}$, be the block $c_i c_{i+1}$, whence each B_i has one of the four forms: 11, 12, 21, 22. Denote these forms by 1, 2, 3, 4. Then

$$S \overset{\text{def}}{=} \cdots B_{-2} B_{-1} B_0 B_1 B_2 \cdots$$
$$B_0 B_1 \cdots = 2432 \, 3124 \, 3123 \, 2432 \, 3123 \, 2431 \, \cdots$$

and there is in S no block PQ (of any size) for which $P = Q$.

In S replace each 4 by 1 and call the result U. Thus U contains the block

$$2132 \, 3121 \, 3123 \, 2132 \, 3123 \, 2131.$$

Let Σ be the semigroup generated by the three symbols 1, 2, 3 and assume $1^2 = 2^2 = 3^2 = 0$. The set of nonzero elements of Σ is the set of all blocks (of any size) in U. Thus Σ is a semigroup enjoying the properties described at the start of the discussion.

If G is a topological group, it and all its subsets are *completely regular* topological spaces. Hence in the *category* of topological groups and continuous homomorphisms the counterpart of a free topological group on a set X is definable only if X is a completely regular topological space.

If X is a completely regular topological space a *free topological group* on X is a topological group $F_{top}(X)$ such that:

i'. $F_{top}(X)$ contains a topological image of X (by abuse of language,

$$X \subset F_{top}(X));$$

ii'. if G is a topological group and $\phi : X \mapsto G$ is a continuous map then ϕ may be extended to a continuous homomorphism $\Phi : F_{top}(X) \mapsto G$.

The following facts about $\mathbb{H}$, the *noncommutative field (division ring, skew field, sfield) of quaternions* (cf. **Subsection 1.2.1**) will prove useful in the development that follows.

▷ The quaternions constitute a four-dimensional algebra over $\mathbb{R}$.
▷ There is for $\mathbb{H}$ a Hamel basis $\{1, \mathbf{i}, \mathbf{j}, \mathbf{k}\}$ over $\mathbb{R}$ and

$$1 \cdot \mathbf{q} = \mathbf{q}, \ \mathbf{q} \in \mathbb{H}$$
$$\mathbf{i}^2 = \mathbf{j}^2 = \mathbf{k}^2 = -1$$
$$\mathbf{ij} = -\mathbf{ji} = \mathbf{k}, \ \mathbf{jk} = -\mathbf{kj} = \mathbf{i}, \ \mathbf{ki} = -\mathbf{ik} = \mathbf{j}.$$

▷ If

$$\mathbb{H} \ni \mathbf{q} \overset{\text{def}}{=} a\mathbf{1} + b\mathbf{i} + c\mathbf{j} + d\mathbf{k}, \ \{a, b, c, d\} \subset \mathbb{R}$$

the *conjugate* of $\mathbf{q}$ is

$$\overline{\mathbf{q}} \overset{\text{def}}{=} a\mathbf{1} - b\mathbf{i} - c\mathbf{j} - d\mathbf{k}$$

and the *norm* of $\mathbf{q}$ is

$$|\mathbf{q}| \overset{\text{def}}{=} \sqrt{\mathbf{q}\overline{\mathbf{q}}} = \sqrt{a^2 + b^2 + c^2 + d^2} \ (\geq 0).$$

(Hence $|\mathbf{q}| = 0$ iff $\mathbf{q} = 0\mathbf{1} + 0\mathbf{i} + 0\mathbf{j} + 0\mathbf{k} \overset{\text{def}}{=} \mathbf{0}$.) The norm of the product $\mathbf{ab}$ of two quaternions $\mathbf{a}$ and $\mathbf{b}$ is the product of their norms:

$$|\mathbf{ab}| = |\mathbf{a}| \cdot |\mathbf{b}|.$$

▷ If $\mathbf{q} \neq \mathbf{0}$ then the *inverse* $\mathbf{q}^{-1}$ of $\mathbf{q}$ exists,

$$\mathbf{q}^{-1} = \frac{\overline{\mathbf{q}}}{|\mathbf{q}|^2},$$

(and $\mathbf{q}\mathbf{q}^{-1} = \mathbf{1}$).
▷ A quaternion of the form $b\mathbf{i} + c\mathbf{j} + d\mathbf{k}$ is a *pure quaternion*.

Exercise 1.1.5.3. Show that every quaternion $\mathbf{q}$ is a square: there is a quaternion $\mathbf{r}$ such that $\mathbf{q} = \mathbf{r}^2$.

Exercise 1.1.5.4. Let $\mathbf{q}_m \overset{\text{def}}{=} b_m\mathbf{i} + c_m\mathbf{j} + d_m\mathbf{k}$, $m = 1, 2$ be two pure quaternions. Show that they *commute* ($\mathbf{q}_1\mathbf{q}_2 = \mathbf{q}_2\mathbf{q}_1$) iff they are linearly dependent over $\mathbb{R}$.

[*Hint:* Show that they commute iff the *rank* of the matrix

$$\begin{pmatrix} b_1 & c_1 & d_1 \\ b_2 & c_2 & d_2 \end{pmatrix}$$

is not more than 1.]

THEOREM **1.1.5.1.** IF X IS A COMPLETELY REGULAR TOPOLOGICAL
SPACE THERE IS A FREE TOPOLOGICAL GROUP $F_{top}(X)$ ON X.

PROOF outline:

▷ Let $\mathbb{H}_1$ be the set of quaternions of norm 1 and let $\mathcal{F}$ be the set of
  continuous maps $f : X \mapsto \mathbb{H}_1$.

  ▷▷ In $\mathbb{H}_1$ there is an infinite set $S \overset{\text{def}}{=} \{s_n\}_{n \in \mathbb{N}}$ that generates a *free
     subgroup* of $\mathbb{H}_1$, i.e., $F(S)$ is isomorphic to the intersection of all
     subgroups of $\mathbb{H}_1$ that contain S [**Grood, Hau**], cf. also **Remark
     1.1.5.1. 17**. As a subgroup of $\mathbb{H}_1$, $F(S)$ is a topological group on
     S.

  ▷▷ If $p_1, \ldots, p_n$ are n different points of X and if $\epsilon_1 = \pm 1, \ldots, \epsilon_n \pm 1$
     then, because X is completely regular, there is in $\mathcal{F}$ an f such that
     $f(p_k) = s_k^{\epsilon_k}$, $1 \leq k \leq n$.

▷ For each f in $\mathcal{F}$ let $\mathbb{H}_f$ be a copy of $\mathbb{H}_1$ and let $\mathbb{H}_\infty$ be the (compact)
  topological group that is the topological Cartesian product $\prod_{f \in \mathcal{F}} \mathbb{H}_f$.

▷ For x in X let $\theta(x) \overset{\text{def}}{=} \mathbf{x}$ in $\mathbb{H}_\infty$ be the vector for which the fth
  component is $f(x) \overset{\text{def}}{=} x_f$:

$$\theta : X \ni x \mapsto \mathbf{x} \overset{\text{def}}{=} (x_f)_{f \in \mathcal{F}} \in \mathbb{H}_\infty.$$

Then θ is a topological embedding of X in $\mathbb{H}_\infty$.

▷ Correspondingly embed $F(X)$ in $\mathbb{H}_\infty$: if $x_1^{\epsilon_1} \cdots x_n^{\epsilon_n}$ represents an element ξ in $F(X)$ let $\Theta(\xi)$ be the vector

$$\Xi \overset{\text{def}}{=} (f(x_1)^{\epsilon_1} \cdots f(x_n)^{\epsilon_n})_{f \in \mathcal{F}}.$$

So embedded $F(X)$ inherits a topology that makes $F(X)$ a topological
group in which X is topologically embedded.

▷ Let T_{max} be the *supremum* of the (nonempty!) set of topologies T such
  that:

  ▷▷ $F(X)$ is a topological group in the topology T;
  ▷▷ X inherits its original topology from T.

Topologized by T_{max}, $F(X)$ is a topological group $F_{top}(X)$ and con-
forms to the requirements i', ii'. For details see [**Ge4, Ge5**] and for alternative approaches see [**Kak2, Ma**].

The construction described above is a streamlined version of the construction described next. The latter provides added insight into the subject.

Again let X be a completely regular topological space. Let $\mathbb{H}^*$ be the
multiplicative group of nonzero quaternions and this time let $\mathcal{F}$ be the set of

bounded continuous $\mathbb{H}^*$-valued functions. In $\mathcal{F}$ let $\mathcal{Q}$ be the *group* consisting of elements that have reciprocals in $\mathcal{F}$, i.e., $\mathcal{Q}$ is the set of invertible elements in the multiplicative structure of $\mathcal{F}$. (Alternatively, $f \in \mathcal{Q}$ iff $f \in \mathcal{F}$ and $\frac{1}{f}$ is bounded.)

In analogy with the procedure used before, for each f in $\mathcal{Q}$ let $\mathbb{H}_f^*$ be a copy of $\mathbb{H}^*$ and let $\mathbb{H}_\infty^*$ be the topological group that is the topological Cartesian product $\prod_{f \in \mathcal{Q}} \mathbb{H}_f^*$. The embedding $X \ni x \mapsto \mathbf{x} \stackrel{\text{def}}{=} (f(x))_{f \in \mathcal{Q}}$ is a topological embedding and the procedure outlined earlier leads to the free topological group $F(X)$.

If G is a group and if $Q(G)$ is the subgroup generated by all elements of the form $aba^{-1}b^{-1}$ (*commutators*) then $Q(G)$, the *commutator subgroup* of G, is a normal subgroup and the quotient group $G/Q(G)$ is *abelian*, whence $G/Q(G)$ is called an *abelianization* of G. Since G/G is abelian the set of abelianizing subgroups of G is nonempty and $Q(G)$ is the intersection of all normal subgroups $\widetilde{G}$ such that $G/\widetilde{G}$ is abelian. By abuse of language $Q(G)$ is the smallest of all normal subgroups $\widetilde{G}$ such that $G/\widetilde{G}$ is abelian. Thus $G/Q(G)$ is **the** abelianization of G.

In the discussion that follows the next result will be helpful.

THEOREM **1.1.5.2.** A QUATERNION $\mathbf{q}$ IS OF NORM 1: $|\mathbf{q}| = 1$ IFF $\mathbf{q}$ IS A COMMUTATOR.

[**Note 1.1.5.3:** The kernel of the homomorphism

$$t : \mathbb{H}^* \ni \mathbf{q} \mapsto t(\mathbf{q}) \stackrel{\text{def}}{=} |\mathbf{q}| \in \mathbb{R}^+$$

is $S \stackrel{\text{def}}{=} \{ \mathbf{q} \; : \; |\mathbf{q}| = 1 \}$. Since the multiplicative group $\mathbb{R}^+$ of positive real numbers is abelian it follows that $S \supset Q(\mathbb{H}^*)$. Hence a corollary to the THEOREM is the equality: $S = Q(\mathbb{H}^*)$.]

PROOF. If $\mathbf{q}$ is a commutator the equality $|\mathbf{ab}| = |\mathbf{a}| \cdot |\mathbf{b}|$ implies that $|\mathbf{q}| = 1$.

If $|\mathbf{q}| = 1$ and $\mathbf{q} \neq -\mathbf{1}$ then $\mathbf{q} + \mathbf{1} \stackrel{\text{def}}{=} \alpha$ is such that $\alpha \overline{\alpha}^{-1} = \mathbf{q}$. If $\mathbf{q} = -\mathbf{1}$ then $\mathbf{q} = \mathbf{i}\overline{\mathbf{i}}^{-1}$. In short if $|\mathbf{q}| = 1$ there is an α such that $\mathbf{q} = \alpha\overline{\alpha}^{-1}$.

If $\mathbf{q} = \mathbf{1}$ then

$$\mathbf{q} = \mathbf{1}\mathbf{1}^{-1}\mathbf{1}\mathbf{1}^{-1}.$$

Thus it may be assumed that $\mathbf{q} \neq \mathbf{1}$.

Since $\mathbf{q} \neq \mathbf{1}$ it follows that there are real numbers d, e, f, not all 0, and a real number c and such that

$$\alpha = c\mathbf{1} + d\mathbf{i} + e\mathbf{j} + f\mathbf{k} \stackrel{\text{def}}{=} c\mathbf{1} + \beta.$$

The nonzero quaternion β is a pure quaternion.

For any $\mathbf{q}$, both $\mathbf{q}$ and $\overline{\mathbf{q}}$ are zeros of the polynomial

$$p_{\mathbf{q}}(x) \overset{\text{def}}{=} x^2 - (\mathbf{q} + \overline{\mathbf{q}})x + \mathbf{q}\overline{\mathbf{q}}$$

in which the coefficients are multiples of $\mathbf{1}$, i.e., $p_{\mathbf{q}}$ is a polynomial over $\mathbb{R}$. It follows that $p_\beta(\beta) = p_\beta(\overline{\beta}) = 0$. Hence the $\mathbb{R}$-span of $\mathbf{1}$ and β is a two-dimensional commutative *proper subfield* $\mathbb{K}$ of $\mathbb{H}$: $\mathbb{K} \subsetneq \mathbb{H}$. The dimension of the set of pure quaternions is three and thus there is a pure quaternion γ not in the span of the pure quaternion β. However

$$\delta \overset{\text{def}}{=} \beta\gamma - \gamma\beta = \mathbf{0} \Leftrightarrow \text{span}(\beta) = \text{span}(\gamma)$$

(**Exercise 1.1.5.4. 13**) whence $\delta \neq \mathbf{0}$. Furthermore since β is pure, $\overline{\beta} = -\beta$ and so $\beta^2 = -|\beta|^2 \mathbf{1}$. Thus, because $\delta \neq \mathbf{0}$ it follows that δ^{-1} exists and so

$$\overline{\beta}\delta = -\beta\delta$$
$$= -[-|\beta|^2\gamma - \beta\gamma\beta]$$
$$= \delta\beta$$
$$\overline{\beta} = \delta\beta\delta^{-1}$$
$$\overline{\alpha} = c\mathbf{1} + \overline{\beta} = \delta(c\mathbf{1} + \beta)\delta^{-1} = \delta\alpha\delta^{-1}$$
$$\mathbf{q} = \alpha\overline{\alpha}^{-1} = \alpha\delta\alpha^{-1}\delta^{-1}.$$

□

The added interest in the second method of construction of the free group on a set X comes from the notion of a *free abelian group* $A(X)$ on a set X. The equivalence relation $\sim$ is replaced by a new equivalence relation $\sim'$: $w_1 \sim' w_2$ iff $w_1 \sim w_2$ OR there are words u and v such that $w_1 = uv$ and $w_2 = vu$. Then $A(X) = W(X)/\sim'$.

The free abelian group $A(X)$ on X may be viewed as the minimal group, by abuse of language, containing X and such that if $\phi : X \mapsto A$ is a map of X into an abelian group A there is an extension Φ of ϕ that is a homomorphism of $A(X)$ into A.

The second construction of the free topological group on X can be mimicked for the construction of $A_{top}(X)$, the free topological abelian group on the (completely regular) set X: $\mathbb{H}^*$ is replaced by $\mathbb{R}^+$, the abelianization of $\mathbb{H}^*$, $\mathcal{Q}$ is replaced by $\mathcal{R}$, the set of bounded continuous functions $f : X \ni x \mapsto f(x) \in \mathbb{R}^+$ such that $\frac{1}{f}$ is also bounded.

To find an infinite free subgroup of $\mathbb{R}^+$ let $B \overset{\text{def}}{=} \{r_\lambda\}_{\lambda \in \Lambda}$ be a Hamel basis for $\mathbb{R}$ over $\mathbb{Q}$. Then Λ is necessarily infinite. In fact, since $B \subset \mathbb{R}$ it follows that $\#(\Lambda) \leq \#(\mathbb{R})$. On the other hand, the set Φ_Λ of finite subsets of Λ has the same cardinality as that of Λ: $\#(\Phi_\Lambda) = \#(\Lambda)$. If $\phi \overset{\text{def}}{=} \{x_{\lambda_1}, \ldots, x_{\lambda_n}\} \in \Phi_\Lambda$ the cardinality of the set of those real numbers expressible as

$$\sum_{k=1}^{n} a_k x_{\lambda_k}, \ a_k \in \mathbb{Q}$$

is $[\#(\mathbb{Q})]^n$ $(= \#(\mathbb{Q}) = \#(\mathbb{N}))$. Hence

$$\#(\mathbb{R}) = \#(\mathbb{N})\#(\Phi_\Lambda) = \#(\Phi_\Lambda) = \#(\Lambda).$$

The set $R \stackrel{\text{def}}{=} \{2^{r_\lambda}\}_{l\in\Lambda}$ generates a free subgroup of $\mathbb{R}^+$ and is used in place of S in the first construction.

[**Remark 1.1.5.1:**   The abelianizing map

$$\theta : \mathbb{H}^* \ni \mathbf{q} \mapsto |\mathbf{q}| \in \mathbb{R}^+$$

may be used to demonstrate the existence in $\mathbb{H}^*$ of a set T free in $\mathbb{H}^*$ and such that $\#(T) = \#(\mathbb{R})$. Indeed, if $T = \theta^{-1}(R)$ then T is free and $\#(\mathbb{R}) \geq \#(T) \geq \#(\Lambda) = \#(\mathbb{R})$. As noted earlier, in $\mathbb{R}$ there must be an infinite set Φ linearly independent over $\mathbb{Q}$. The existence of such a set is independent of Zorn's lemma and engenders the set $\theta^{-1}(\Phi)$ that is perforce an infinite free subset of $\mathbb{H}^*$.

Let $F(T)$ be the (free) group generated in $\mathbb{H}^*$ by T. Let C be the set of all commutators $xyx^{-1}y^{-1}, x, y \in T$, $x \neq y$. Then since T is free so is C. Hence there is in $Q(\mathbb{H}^*)$ the free set C and $\#(C) = \#(\mathbb{R})$.]

In [**Ma**] it is shown that $A_{top}(X)$ is the abelianization of $F_{top}(X)$. Hence the second construction of $F_{top}(X)$, the topological free group on X, leads to the following parallel:

The underlying structure or *source* $\mathbb{R}^+$ for constructing the abelianization $A_{top}(X)$ of $F(X)$ is the abelianization of the underlying structure or *source* $\mathbb{H}^*$ for constructing $F_{top}(X)$.

The parallel above may be viewed as a kind of *commutative diagram* (1.1.5.1) if α is used as the generic symbol for the *quotient map* arising from abelianization:

$$\begin{array}{ccc}
\{X , & \mathbb{H}^*\} & \to & F_{top}(X) \\
\text{id} \downarrow & \downarrow \alpha & & \downarrow \alpha \\
\{X , & \mathbb{R}^+\} & \to & A_{top}(X)
\end{array} \qquad (1.1.5.1)$$

Let G be a group, Y be a set, and $P \stackrel{\text{def}}{=} \{W_\lambda\}_{\lambda\in\Lambda}$ be a subset of $W(Y)$. The elements y of Y may be viewed as "parameters" the "values" of which may be taken as elements g of G. Thus a word $y_1^{\epsilon_1} \cdots y_n^{\epsilon_n}$ is replaced by $g_1^{\epsilon_1} \cdots g_n^{\epsilon_n}$. (Some of the elements $g_1, \ldots, g_n$ of G may be the same, e.g., $g_1 = g_3$.) Let $N(P, F(Y))$ be the normal subgroup generated in $F(Y)$ by P. Correspondingly let $N(P, G)$ be the normal subgroup generated in G after replacing in all possible ways the parameters y by elements g of G. Of particular interest are $N(P, \mathbb{H}^*)$, and, in the *norm-induced* topology of $\mathbb{H}^*$, the closure $\overline{N(P, \mathbb{H}^*)}$ of $N(P, \mathbb{H}^*)$.

If X is a completely regular topological space the set $\mathcal{N}(P, F_{top}(X))$ is taken as the closed normal subgroup generated in $F_{top}(X)$ after replacing in all possible ways the parameters y by elements g of $F_{top}(X)$. If ω is the generic symbol for the quotient map arising from dividing $\mathbb{H}^*$ resp. $F_{top}(X)$ by $\overline{N(P, \mathbb{H}^*)}$ resp. $\mathcal{N}(P, F_{top}(X))$ the diagram that corresponds to (1.1.5.1) looks like this:

$$
\begin{array}{ccccc}
\{X & , & \mathbb{H}^* & \} & \to & F_{top}(X) \\
\text{id} \downarrow & & \downarrow \omega & & \downarrow \omega \\
\{X & , & \mathbb{H}^*/\overline{N(P, \mathbb{H}^*)} & \} & \to & F_{top}(X)/\mathcal{N}(P, F_{top}(X))
\end{array}
\qquad (1.1.5.2)
$$

Regrettably, as the next few lines show, the diagram (1.1.5.2) is not necessarily commutative.

Example 1.1.5.1. Let X be $\mathbb{T}$, the set of complex numbers of absolute value 1, and let P be $\{yy\}$. If f is $\mathbb{T} \ni (a+bi) \mapsto a\mathbf{1}+b\mathbf{i} \in \mathbb{H}^*$ then there is in $\mathcal{Q}$ no function h such that $(h(a\mathbf{1} + b\mathbf{i}))^2 = f(a + bi)$, cf. **Exercise 1.1.5.5.** below. Thus $f \notin \mathcal{N}(P, F_{top}(X))$ and so $F_{top}(X)/\mathcal{N}(P, F_{top}(X))$ consists of more than one element. Since every quaternion is a square (**Exercise 1.1.5.3. 13**) it follows that $N(P, \mathbb{H}^*) = \mathbb{H}^*$ and so $\mathbb{H}^*/\overline{N(P, \mathbb{H}^*)} = \{\mathbf{1}\}$. The set of $\mathcal{Q}_1$ of continuous bounded functions $f : X \ni x \mapsto f(x) \in \{\mathbf{1}\}$ consists of one element and cannot be the source in the second construction of the quotient $F_{top}(X)/\mathcal{N}(P, F_{top}(X))$.

Exercise 1.1.5.5. Show that there is in $\mathbb{C}^{\mathbb{T}}$ no continuous function h such that for z in $\mathbb{T}$, $(h(z))^2 = z$. ("The square root function is not continuous on $\mathbb{T}$.")

[*Hint:* For each z in $\mathbb{T}$ there are in $[0, 2\pi)$ a unique θ such that $z = e^{i\theta}$ and a unique $\phi(\theta)$ such that $h(z) = e^{i\phi(\theta)}$. For each θ in $[0, 2\pi)$,

$$-2\pi < 2\phi(\theta) - \theta < 4\pi \text{ and } 2\phi(\theta) - \theta \in 2\pi\mathbb{Z},$$

whence, for any θ in $[0, 2\pi)$,

$$\text{a) } 2\phi(\theta) - \theta = 2\pi \text{ or b) } 2\phi(\theta) - \theta = 0.$$

If ϕ is discontinuous, i.e., if the switch a) $\leftrightarrow$ b) occurs, then h switches to $-h$. If only one of a) or b) obtains for all θ in $[0, 2\pi)$ then $\lim_{\theta \uparrow 2\pi} e^{i\phi(\theta)} \neq e^{i\phi(0)}$. Thus h is discontinuous on $\mathbb{T}$.]

1.1.6. Finite simple groups

No discussion of group theory can ignore the achievement in early 1981 of the classification of all finite simple groups. The success culminated more than 30 years of research by tens of mathematicians publishing hundreds of papers amounting to thousands of pages.

One of the great achievements in the early part of the effort was the result of Feit and Thompson to the effect that every group of odd order is solvable or, equivalently, every finite simple nonabelian group is of even order. Their paper [FeT] occupied an entire issue of the Pacific Journal of Mathematics.

[**Note 1.1.6.1:**   In [FeT] there arises the question:

For two (different) primes p and q, are the natural numbers

$$\frac{p^q - 1}{p - 1} \text{ and } \frac{q^p - 1}{q - 1}$$

relatively prime?

Simple illustrations, e.g., with the first 100 primes, suggest that the answer is affirmative. Had the answer been known, [FeT] would have been considerably shorter. To the writers' knowledge, the question remains unresolved.]

In effect, every finite simple group is either a "group of Lie type" (cf. **Subsection 1.2.2**) or, for some n in $\mathbb{N}$, the *alternating group* $\mathcal{A}_n$, or one of precisely 26 "sporadic" groups. The largest of the sporadic groups consists of approximately 10^{54} elements. For a thorough exposition, together with a good deal of motivation and history, the interested reader is urged to consult Gorenstein's books [Gor1, Gor2].

1.2. Algebras

1.2.1. Division algebras ("noncommutative fields")

By definition the binary operation dubbed multiplication in a *field* $\mathbb{K}$ is commutative: for a, b in $\mathbb{K}$, $ab = ba$. A *noncommutative* field S or *skew field* or *sfield* or *division algebra* is a set with two binary operations, addition and multiplication that behave exactly like the binary operations in a field except that multiplication is not necessarily commutative: the possibility $ab \neq ba$ is admitted.

If p is an nth degree polynomial with coefficients in $\mathbb{C}$ then p has at most n different zeros. If $\mathbb{C}$ is replaced by $\mathbb{H}$, the noncommutative field of quaternions (cf. **Subsection 1.1.5**), an nth degree polynomial may have more than n zeros.

Example 1.2.1.1.   The polynomial $p(x) \overset{\text{def}}{=} x^2 + 1$ regarded as a polynomial with coefficients from $\mathbb{H}$ has *infinitely many* zeros. Indeed, if $\mathbf{q}$ is any nonzero quaternion then $\mathbf{r_q} \overset{\text{def}}{=} \mathbf{q} i \mathbf{q}^{-1}$ is a zero of p.

Exercise 1.2.1.1. Show that there are infinitely many different quaternions of the form $\mathbf{r_q}$.

[*Hint:* Assume

$$a, b \in \mathbb{R} \text{ and } a^2 + b^2 = 1.$$

Let $\mathbf{q}$ be $a\mathbf{1} + b\mathbf{j}$.]

THEOREM **1.2.1.1.** LET $\mathbf{r}$ BE A QUATERNION SUCH THAT $\mathbf{r}^2 + \mathbf{1} = \mathbf{0}$. THEN THERE IS A NONZERO QUATERNION $\mathbf{q}$ SUCH THAT $\mathbf{r} = \mathbf{r_q} \stackrel{\text{def}}{=} \mathbf{qiq}^{-1}$.

[**Note 1.2.1.1:** See **Exercise 1.2.1.1** above.]

PROOF. Let $\mathbf{q} \stackrel{\text{def}}{=} a\mathbf{1} + b\mathbf{i} + c\mathbf{j} + d\mathbf{k}$ be such that $|\mathbf{q}|^2 = 1$. Then $\mathbf{q}^{-1} = a\mathbf{1} - b\mathbf{i} - c\mathbf{j} - d\mathbf{k}$. If $\mathbf{r} \stackrel{\text{def}}{=} \alpha\mathbf{1} + \beta\mathbf{i} + \gamma\mathbf{j} + \delta\mathbf{k}$ the equation $\mathbf{r}^2 + \mathbf{1}$ implies

$$(\alpha^2 - \beta^2 - \gamma^2 - \delta^2)\mathbf{1} + 2\alpha\beta\mathbf{i} + 2\gamma\alpha\mathbf{j} + 2\alpha\delta\mathbf{k} = -\mathbf{1}.$$

If $\alpha \neq 0$ then $\beta = \gamma = \delta = 0$ and so $\alpha^2 = -1$, an impossibility since $\alpha \in \mathbb{R}$. Hence $\alpha = 0$, i.e., $\mathbf{r}$ is pure. To find a nonzero $\mathbf{q}$ such that $\mathbf{r} = \mathbf{r_q}$ is to find a nonzero $\mathbf{q}$ such that $\mathbf{rq} = \mathbf{qi}$. Hence $\mathbf{q}$ should be such that

$$\begin{pmatrix} \beta & 0 & -\delta & \gamma \\ 0 & \beta & \gamma & \delta \\ -\delta & \gamma & -\beta & 0 \\ \gamma & \delta & 0 & -\beta \end{pmatrix} \begin{pmatrix} a \\ b \\ c \\ d \end{pmatrix} = \begin{pmatrix} a \\ b \\ c \\ d \end{pmatrix} \stackrel{\text{def}}{=} \mathbf{x}. \tag{1.2.1.1}$$

In matrix-vector form (1.2.1.1) is $U\mathbf{x} = \mathbf{x}$. Viewed as vectors in $\mathbb{R}^4$, the rows of U are pairwise orthogonal. Furthermore, $U \neq I$, $U = U^t$, and $UU^t = U^2 = I$, i.e., U is an *orthogonal* self-adjoint matrix and its *minimal polynomial* is $z^2 - 1$, whence one of its eigenvalues is 1. Hence (1.2.1.1) has a solution $\mathbf{x}$ that is a (nonzero) eigenvector corresponding to the eigenvalue 1, i.e., the quaternion $\mathbf{q}$ exists.

$\square$

1.2.2. General algebras

If one pares away the various restrictive axioms that are used to define an algebra, there emerge interesting classes of structures that behave like algebras in some ways and yet violate the discarded axioms.

A *nonassociative algebra* over a field $\mathbb{K}$ is one in which multiplication is not necessarily associative, i.e., in which the identity $x(yz) = (xy)z$ is not necessarily valid. If A is an *algebra* in which multiplication *is* associative but not necessarily commutative, there is a counterpart algebra $\{A\}$ in which "multiplication" is defined as follows:

$$x \circ y \stackrel{\text{def}}{=} xy - yx.$$

Exercise 1.2.2.1. Let A be the algebra of $n \times n$ matrices over a field $\mathbb{K}$. Show that $\{A\}$ is a nonassociative algebra. Show that if A is any (associative) algebra over a field $\mathbb{K}$ then $\{A\}$ is associative, i.e., $(x \circ y) \circ z = x \circ (y \circ z)$, iff

$$yxz + zxy = xzy + yzx.$$

Exercise 1.2.2.2. Show that if A is an associative algebra over a field $\mathbb{K}$ then the binary operation $\circ$ is such that for x, y, z in A and c in $\mathbb{K}$,

$$(cx) \circ y = c(x \circ y)$$
$$x \circ y + y \circ x = 0$$
$$x \circ (y \circ z) + z \circ (x \circ y) + y \circ (z \circ x) = 0. \qquad (1.2.2.1)$$

The last is a version of the *Jacobi identity*.

[**Remark 1.2.2.1:** The equations (1.2.2.1) are the starting point for the definition and study of *Lie algebras*, which play a fundamental rôle in the concept of finite groups of *Lie type*, which in turn are the building blocks for the classification of all finite simple groups (cf. **Subsection 1.1.6**).

The formalism for passing from a Lie algebra to a group of Lie type is rather complex, depending, as it does, on a profound analysis of the structure of Lie algebras. Nevertheless an outline of the ideas can be given in the following manner.

Let $\mathcal{L}$ be a Lie algebra in which the product of two elements p and q is denoted $[pq]$. For a fixed element a of $\mathcal{L}$, the map

$$T_a : \mathcal{L} \ni x \mapsto [xa]$$

is a *linear* endomorphism of $\mathcal{L}$. For special kinds of Lie algebras there are singled out finitely many special elements a_i, $1 \leq i \leq N$, for which each corresponding map T_{a_i} is *nilpotent*: for some n_i in $\mathbb{N}$, $T_{a_i}^{n_i} = O$. If $\mathcal{L}$ is an algebra over a field $\mathbb{K}$ and if $t \in \mathbb{K}$ then the formal power series for

$$\exp(tT_{a_i}) \stackrel{\text{def}}{=} I + \sum_{k=1}^{\infty} \frac{(tT_{a_i})^k}{k!}$$

has only finitely many nonzero terms, whence $\exp(tT_{a_i})$ is well-defined and is an invertible endomorphism of $\mathcal{L}$, i.e., an *automorphism*. If the field $\mathbb{K}$ is finite then the finite set

$$\{ tT_{a_i} : t \in \mathbb{K}, \ 1 \leq i \leq N \}$$

generates a finite *group of Lie type* of automorphisms of $\mathcal{L}$. Finite simple groups of Lie type constitute one of the three classes of finite simple groups (cf. **Subsection 1.1.6**).]

The set $\mathbb{C}$ of complex numbers is a field that is also a finite-dimensional vector space over $\mathbb{R}$: dim $(\mathbb{C}) = 2$. The set $\mathbb{H}$ of quaternions is an example of a division algebra that is a finite-dimensional vector space over $\mathbb{R}$: dim $(\mathbb{H}) = 4$.

Exercise 1.2.2.3. Let $\mathcal{C}$ be the set $\mathbb{H} \times \mathbb{H}$ regarded as an eight-dimensional vector space over $\mathbb{R}$. Define a binary operation ("multiplication") according to the following formula:

$$\cdot : \mathcal{C} \ni ((\mathbf{a}, \mathbf{b}), (\mathbf{c}, \mathbf{d})) \mapsto (\mathbf{a}, \mathbf{b}) \cdot (\mathbf{c}, \mathbf{d}) \stackrel{\mathrm{def}}{=} (\mathbf{ac} - \mathbf{d}\overline{\mathbf{b}}, \mathbf{cb} + \overline{\mathbf{a}}\mathbf{d}).$$

Show that the *Cayley algebra* $\mathcal{C}$ so structured is an *alternative (division) algebra*, i.e., $\mathcal{C}$ behaves just like a division algebra except that multiplication is neither (universally) commutative nor (universally) associative.

[*Hint:* Show that $\mathbf{I} \stackrel{\mathrm{def}}{=} (\mathbf{1}, \mathbf{0})$ is the multiplicative identity and that if $(\mathbf{a}, \mathbf{b}) \neq (\mathbf{0}, \mathbf{0})$ then there is a $(\mathbf{c}, \mathbf{d})$ such that $(\mathbf{a}, \mathbf{b}) \cdot (\mathbf{c}, \mathbf{d}) = \mathbf{I}$. To prove absence of universal associativity examine products of three elements, each of the form

$$(\mathbf{a}, \mathbf{b}), \ \mathbf{a}, \mathbf{b} \in \{\mathbf{i}, \mathbf{j}, \mathbf{k}\}.]$$

Milnor [**Miln2**] showed that the only vector spaces (over $\mathbb{R}$) that can be structured, via a second binary operation, to become a field, a division algebra, or an alternative division algebra are: $\mathbb{R}, \mathbb{C}, \mathbb{H}$ and $\mathcal{C}$. See also the book by Tarski [**T**], where it is shown that if a vector space V over a *real-closed* field $\mathbb{K}$ is an alternative algebra then dim(V) must be $1, 2, 4$, or 8.

1.2.3. Miscellany

A field $\mathbb{K}$ is *ordered* iff there is in $\mathbb{K}$ a subset P such that:

i.
$$x, \ y \in P \Rightarrow x + y \in P \text{ and } xy \in P;$$

ii. P, $\{0\}$, and $-P$ are pairwise disjoint and $P \cup \{0\} \cup -P = \mathbb{K}$, i.e., $\mathbb{K} = P \cup \{0\} \cup - P$ (whence $P \neq \emptyset$).

By definition $x > y$ iff $x - y \in P$.

Exercise 1.2.3.1. Show that $\mathbb{C}$ cannot be ordered.

[*Hint:* If $i \in P$ then $i^2, i^4 \in P$ and yet $i^2 + i^4 = 0$; mutatis mutandis, the same argument obtains if $i \in -P$.]

Exercise 1.2.3.2. Show that the field

$$\mathbb{K} \overset{\text{def}}{=} \mathbb{Q}(\sqrt{2}) \overset{\text{def}}{=} \left\{ r + s\sqrt{2} \; : \; r, s \in \mathbb{Q} \right\}$$

can be ordered by defining P to be either the set of all positive numbers in $\mathbb{K}$ or by the rule

$$r + s\sqrt{2} \in P \Leftrightarrow r - s\sqrt{2} > 0.$$

Show also that these two orders are different.

An ordered field $\mathbb{K}$ is *complete* iff every nonempty set S that is bounded above and contained in $\mathbb{K}$ has a *least upper bound* or *supremum* (lub or sup) in $\mathbb{K}$, viz.:

If $S \neq \emptyset$ and there is a b such that every s in S does not exceed b then there is in $\mathbb{K}$ an l such that:

  ▷ every s in S does not exceed l;
  ▷ if $l' < l$ there is in S an s' such that $l' < s'$.

The number l is unique and $\text{lub}(S) = \sup(S) \overset{\text{def}}{=} l$.

Exercise 1.2.3.3. Show that $\mathbb{Q}$ in its usual order is not complete, e.g., that $\left\{ x \; : \; x \in \mathbb{Q}, \; x^2 \leq 2 \right\}$ is bounded above and yet has no lub.

An ordered field $\mathbb{K}$ is *Archimedean* iff $\mathbb{N}$ (necessarily a subset of an ordered field) is not bounded above.

Exercise 1.2.3.4. Show that the field $\mathbb{K}$ consisting of all rational functions of a single indeterminate x and with coefficients in $\mathbb{R}$:

$$\mathbb{K} = \left\{ \frac{f}{g} \; : \; f, \, g \in \mathbb{R}(x), \; \text{degree}[\text{GCD}(f, g)] = 0 \right\}$$

is ordered but not Archimedeanly ordered when P is the set of elements $\frac{f}{g}$ in which the leading coefficients of f and g have the same sign.

[**Note 1.2.3.1:** If $\mathbb{K}$ and $\mathbb{K}'$ are complete Archimedeanly ordered fields then they are *order-isomorphic*. Customarily the equivalence class of order-isomorphic, complete, Archimedeanly ordered fields is denoted $\mathbb{R}$ [**O1**].]

Exercise 1.2.3.5. Show that $\mathbb{K}$ as in **Exercise 1.2.3.4** cannot be embedded in $\mathbb{R}$ so that the orders in $\mathbb{R}$ and in $\mathbb{K}$ are consistent.

[*Hint:* The set $\mathbb{N}$ is naturally a subset of both $\mathbb{R}$ and $\mathbb{K}$ but is unbounded in $\mathbb{R}$ and not in $\mathbb{K}$:

$$\frac{x - n}{1} = x - n \in P, \; n \in \mathbb{N}, \text{ i.e.,}$$

for all n in $\mathbb{N}$, $x > n$.]

A *net* in a set S is a map $\Lambda \ni \lambda \mapsto a_\lambda \in S$ of a *directed* set $\{\Lambda, \succ\}$ (a *diset*). When S is endowed with a topology derived from a *uniform structure* U, e.g., that provided by a metric, a net $\{a_\lambda\}$ is a *Cauchy net* iff for each element (*vicinity*) U of the uniform structure U there is in Λ a λ_0 such that $(a_\lambda, a_\mu) \in U$ if $\lambda, \mu \succ \lambda_0$. A net $\{a_\lambda\}$ is *convergent* iff there is in S an a such that for each *neighborhood* V of a there is a λ_0 such that $a_\lambda \in V$ if $\lambda \succ \lambda_0$. If every Cauchy net is convergent (*"converges"*) S is *Cauchy complete* (cf. [**Du, Ke, Tu1**]).

[**Remark 1.2.3.1:**   Let Λ be the set of finite subsets of $\mathbb{N}$. If $\lambda, \mu \in \Lambda$ let $\lambda \succ \mu$ mean $\lambda \supset \mu$. Then $\{\Lambda, \succ\}$ is a diset. If $\lambda \in \Lambda$ let n_λ be the largest member in λ. For each sequence $\{x_n\}_{n \in \mathbb{N}}$ there is a net $\{x_\lambda\}_{\lambda \in \Lambda}$ defined by the equation $x_\lambda = x_{n_\lambda}$. The sequence $\{x_{n_\lambda}\}_{n \in \mathbb{N}}$ is a Cauchy resp. convergent sequence iff the net $\{x_\lambda\}_{\lambda \in \Lambda}$ is a Cauchy resp. convergent net.]

Two Cauchy nets $\{a_\lambda\}_{\lambda \in \Lambda}$ and $\{b_\gamma\}_{\gamma \in \Gamma}$ are *equivalent* ($\{a_\lambda\} \sim \{b_\gamma\}$) iff for each vicinity U there is a pair $\{\lambda_0, \gamma_0\}$ such that $(a_\lambda, b_\gamma) \in U$ if $\lambda \succ \lambda_0$ and $\gamma \succ \gamma_0$. The *Cauchy completion* S_{Cauchy} is the set of $\sim$-equivalence classes of Cauchy nets. The set S_{Cauchy} is Cauchy complete.

An ordered field $\mathbb{K}$ has a uniform structure provided by P: a vicinity is determined by an ϵ in P and is the set of all pairs (a, b) such that $-\epsilon < a - b < \epsilon$.

Exercise 1.2.3.6. Show that a complete Archimedeanly ordered field, i.e., essentially $\mathbb{R}$, is Cauchy complete.

[*Hint:* Let $\{a_\lambda\}_{\lambda \in \Lambda}$ be a Cauchy net in $\mathbb{R}$. For each n in $\mathbb{N}$ choose λ_n so that $-\frac{1}{n} < a_\lambda - a_\mu < \frac{1}{n}$ if $\lambda,\ \mu \succ \lambda_n$. Then

$$-\infty < a_{\lambda_n} - \frac{1}{n} \le \inf_{\lambda \succ \lambda_n} a_\lambda \overset{\text{def}}{=} l_n \le L_n \overset{\text{def}}{=} \sup_{\lambda \succ \lambda_n} a_\lambda \le a_{\lambda_n} + \frac{1}{n} < \infty$$

$l_n \le l_{n+1} \le L_{n+1} \le L_n$.

In other words, the sequences

$$\{l_n\}_{n \in \mathbb{N}} \text{ resp. } \{L_n\}_{n \in \mathbb{N}}$$

are *monotone increasing* resp. *decreasing* and $\{a_\lambda\}_{\lambda \in \Lambda}$ converges to

$$a \overset{\text{def}}{=} \lim_{n \to \infty} l_n \ (= \lim_{n \to \infty} L_n).]$$

If $\{a_\lambda\}_{\lambda \in \Lambda}$ is a net in $\mathbb{R}$ one may define

$$L_\mu \overset{\text{def}}{=} \sup\{ a_\lambda \ : \ \lambda \succ \mu \}$$

$$l_\mu \overset{\text{def}}{=} \inf\{ a_\lambda \ : \ \lambda \succ \mu \}.$$

Then $\mu \succ \nu \Rightarrow l_\nu \le l_\mu \le L_\mu \le L_\nu$. Hence there are defined

$$\limsup_{\lambda \in \Lambda} a_\lambda \overset{\text{def}}{=} \inf_{\mu \in \Lambda} L_\mu$$

$$\liminf_{\lambda \in \Lambda} a_\lambda \overset{\text{def}}{=} \sup_{\mu \in \Lambda} l_\mu.$$

Exercise 1.2.3.7. Show that a net $\{a_\lambda\}_{\lambda \in \Lambda}$ in $\mathbb{R}$ is a Cauchy net iff

$$\limsup_{\lambda \in \Lambda} a_\lambda = \liminf_{\lambda \in \Lambda} a_\lambda \ (\overset{\text{def}}{=} \lim_{\lambda \in \Lambda} a_\lambda).$$

Example 1.2.3.1. The ordered field $\mathbb{K}$ in **Exercise 1.2.3.4. 23** has a Cauchy completion. Nevertheless that Cauchy completion is an ordered field that is perforce Cauchy complete and yet, owing to **Exercise 1.2.3.5. 23**, is not embeddable in $\mathbb{R}$.

Ordered fields are special instances of algebraic objects endowed with (usually Hausdorff) topologies with respect to which the algebraic operations are continuous. For example, a topological division algebra $\mathcal{A}$ is a division algebra endowed with a Hausdorff topology such that the maps

$$\mathcal{A} \times \mathcal{A} \ni (a, b) \mapsto a - b \in \mathcal{A}$$
$$\mathcal{A} \times \mathcal{A} \ni (a, b) \mapsto ab \in \mathcal{A}$$
$$\mathcal{A} \setminus \{O\} \ni a \mapsto a^{-1} \in \mathcal{A}$$

are continuous.

1.3. Linear Algebra

1.3.1. Finite-dimensional vector spaces

If V is a finite-dimensional vector space and $T : V \mapsto V$ is a linear transformation of V into itself, i.e., T is an *endomorphism* of V, the *eigenvalues* of T are the numbers λ such that $T - \lambda I$ is *singular*. The *eigenvalue problem* — the problem of finding the eigenvalues, if they exist, of an endomorphism T — is central in the study of *endomorphisms* of finite-dimensional vector spaces. If a *vector space* V is *n-dimensional* over $\mathbb{C}$ then the set $[V]$ of its endomorphisms may, via the choice of a Hamel basis, be regarded as the set of all $n \times n$ *matrices (over $\mathbb{C}$)*: $[V] = \mathbb{C}^{n^2}$.

If $A \overset{\text{def}}{=} (a_{ij})_{i,j=1}^{m,n}$ is an $m \times n$ matrix its *transpose* $A^t \overset{\text{def}}{=} (b_{ji})_{j,i=1}^{n,m}$ is the $n \times m$ matrix in which the jth row is the jth column of A: $b_{ji} = a_{ij}$. The *adjoint* $A^* \overset{\text{def}}{=} (c_{ji})_{j,i=1}^{n,m}$ is the matrix $\overline{A^t}$, i.e., the matrix A^t in which each entry is replaced by its complex conjugate: $c_{ji} = \overline{a_{ij}}$. If $\mathbb{K}$ is a field then $\mathbb{K}^n$ resp. $\mathbb{K}_n$ is the set of all $n \times 1$ matrices (*column vectors*) resp. the set of all $1 \times n$ matrices (*row vectors*) with entries in $\mathbb{K}$.

THEOREM **1.3.1.1.** THE SET $[V]_{sing}$ OF SINGULAR ENDOMORPHISMS OF AN n-DIMENSIONAL VECTOR SPACE V IS CLOSED IN $\mathbb{C}^{n^2}$ AND THE LEBESGUE MEASURE OF $[V]_{sing}$ IS 0: $\lambda_{n^2}([V]_{sing}) = 0$.

PROOF. If $T \in [V]$ and T^{-1} exists let M be $\|T^{-1}\|$, the Euclidean norm of T^{-1} $\left(\text{in } \mathbb{C}^{n^2}\right)$. If $A \in [V]$ and $\|A\| < \frac{1}{M}$ then $\|T^{-1}A\| \le \|T^{-1}\| \|A\| < 1$, $I + \sum_{n=1}^{\infty} (T^{-1}A)^n$ converges in $\mathbb{C}^{n^2}$ to say, B, and $B(I - T^{-1}A) = I$. Hence $I - T^{-1}A$ and $T - A$ $\left(= T(I - T^{-1}A)\right)$ are invertible. In sum, all elements of the open ball $\left\{ T - A \ : \ \|A\| < \|T^{-1}\|^{-1} \right\}$ are invertible. Hence $[V]_{inv} \overset{\text{def}}{=} [V] \setminus [V]_{sing}$, the set of invertible elements of $[V]$, is open, i.e., $[V]_{sing}$ is closed, in $\mathbb{C}^{n^2}$.

The *Identity Theorem* for analytic functions of a complex variable implies that if a function f is analytic on a nonempty open subset U of $\mathbb{R}$ then either f is constant in U or for every constant a

$$\lambda[f^{-1}(a) \cap U] = 0.$$

It follows by induction [Ge5] that if f is a real- or complex-valued function on $\mathbb{R}^k$ and if for some constant a the Lebesgue measure $\lambda(f^{-1}(a))$ is positive then $f \equiv a$ in any region R where f is analytic and such that $R \supset f^{-1}(a)$.

If $A \overset{\text{def}}{=} (a_{ij})_{i,j=1}^{n,n} \in [V]$ there are on $\mathbb{R}^{n^2}$ polynomial functions p, q such that $\det(A) = p(a_{11}, \dots, a_{nn}) + iq(a_{11}, \dots, a_{nn})$. The result cited above and applied in the present instance shows that $\lambda([V]_{sing}) = 0$. $\square$

COROLLARY **1.3.1.1.** THE SET $[V] \setminus [V]_{sing} \overset{\text{def}}{=} [V]_{inv}$ IS A DENSE (OPEN) SUBSET OF $\mathbb{C}^{n^2}$.

PROOF. Since $[V]_{sing}$ is a closed null set it follows that $[V]_{sing}$ is nowhere dense and hence that $[V]_{inv}$ is (open and) dense in $\mathbb{C}^{n^2}$. $\square$

A SQUARE matrix A is *diagonable* iff there is an invertible matrix P such that $P^{-1}AP$ is a *diagonal* matrix. There is a unique *minimal polynomial* m_A such that a) $m_A(A) = O$, b) the leading coefficient of m_A is 1, and c) the degree of m_A is least among the degrees of all polynomials satisfying a) and b). The matrix A is diagonable iff the zeros of its minimal polynomial are *simple* [Ge9].

Exercise **1.3.1.1.** Show that in $\mathbb{C}^{n^2}$ the set $\mathcal{D}$ of diagonable $n \times n$

matrices is nowhere dense, that its complement is open and dense and that $\lambda_{n^2}(\mathcal{D}) = 0$. (Note how the conclusions here are parallel to those in **Theorem 1.3.1.1** and **Corollary 1.3.1.1**. All these results are in essence reflections of elaborations, cited above, of the Identity Theorem.)

[*Hint:* A polynomial p has simple zeros iff p and p' have no nonconstant common factor, i.e., iff their *resultant* vanishes. (The resultant of two polynomials: $f(x) \overset{\text{def}}{=} a_0 x^m + \cdots + a_m$ and $g(x) \overset{\text{def}}{=} b_0 x^n + \cdots + b_n$ is if, e.g., $m < n$, the determinant of the matrix

$$
\begin{matrix}
 & & \overbrace{\hspace{5cm}}^{m+n+2} & \\
\left.\vphantom{\begin{matrix}a\\a\\a\\b\\b\\b\end{matrix}}\right\}{\scriptstyle m+1} & \begin{pmatrix} a_0 & \cdots & a_m & & & & \\ & a_0 & \cdots & a_m & & & \\ & & \cdots & \cdots & \cdots & & \\ & & & a_0 & \cdots & a_m & \\ b_0 & \cdots & \cdots & b_n & & & \\ & b_0 & \cdots & \cdots & b_n & & \\ & & \cdots & \cdots & \cdots & & \\ & & b_0 & \cdots & \cdots & b_n \end{pmatrix} & & \\
\end{matrix}
$$

and thus is a polynomial function of the coefficients of f and g.)]

If $\mathcal{M}$ is a finite set of $n \times n$ diagonable matrices then they are *simultaneously diagonable* iff they *commute in pairs*, i.e., there is an invertible matrix P such that for every A in $\mathcal{M}$ the matrix $P^{-1}AP$ is a diagonal matrix iff each pair A, B of matrices in $\mathcal{M}$ is such that $AB = BA$ **[Ge9]**.

Exercise 1.3.1.2. Show that the matrices

$$A \overset{\text{def}}{=} \begin{pmatrix} 0 & 1 \\ 0 & 0 \end{pmatrix} \text{ and } B \overset{\text{def}}{=} \begin{pmatrix} 0 & 2 \\ 0 & 0 \end{pmatrix}$$

commute and that neither is diagonable. Show also that there is no invertible matrix P such that both $P^{-1}AP$ and $P^{-1}BP$ are in *Jordan normal form:*

$$\begin{pmatrix} 0 & 1 \\ 0 & 0 \end{pmatrix}.$$

THEOREM **1.3.1.2.** IF V IS A FINITE-DIMENSIONAL VECTOR SPACE OVER $\mathbb{R}$ OR OVER $\mathbb{C}$, IF $\mathcal{W} \overset{\text{def}}{=} \{W_k\}_{1 \le k \le K}$ IS A SET OF SUBSPACES OF V, AND IF

$$V = \bigcup_{k=1}^{K} W_k$$

THEN THERE IS A k_0 SUCH THAT $V = W_{k_0}$.

PROOF. If no W_k is V it may be assumed that $\mathcal{W}$ is minimal:

$$1 \le k' \le K \Rightarrow \bigcup_{k \ne k'} W_k \subsetneqq V.$$

Thus in each W_k there is a vector $\mathbf{x}_k$ not in the union of the other $W_{k'}$.

In $S \overset{\text{def}}{=} \{\, t\mathbf{x}_1 + (1-t)\mathbf{x}_2 \; : \; 0 \le t \le 1 \,\}$ there are infinitely many vectors and so two different ones among them must belong to some subspace, say $W_{k'}$. But then

$$S \subset W_{k'}$$

whence

$$\mathbf{x}_1, \mathbf{x}_2 \in W_{k'}$$

and so

$$W_{k'} = W_1 = W_2,$$

a contradiction.   □

[**Remark 1.3.1.1:**   The space V need not be finite-dimensional. The argument can be generalized somewhat. If the underlying field is merely infinite or if it is finite and its cardinality exceeds K the argument remains valid.]

Exercise 1.3.1.3. Show that if $\mathbb{K}$ is the finite field $\{0, 1\}$, i.e.,

$$\mathbb{K} = \mathbb{Z}/2 \overset{\text{def}}{=} \mathbb{Z}_2,$$

if $\mathbf{x}$ and $\mathbf{y}$ are *indeterminates*,

$$V \overset{\text{def}}{=} \{\, a\mathbf{x} + b\mathbf{y} \; : \; a, b \in \mathbb{K} \,\},$$

and

$$W_1 \overset{\text{def}}{=} \{\, a\mathbf{x} \; : \; a \in \mathbb{K} \}$$
$$W_2 \overset{\text{def}}{=} \{\, a\mathbf{y} \; : \; a \in \mathbb{K} \}$$
$$W_3 \overset{\text{def}}{=} \{\, a(\mathbf{x} + \mathbf{y}) \; : \; a \in \mathbb{K} \}$$

then $V = W_1 \cup W_2 \cup W_3$ and yet V is none of W_1, W_2, W_3, i.e., THEOREM **1.3.1.2** does not apply to V.

Exercise 1.3.1.4. Let $A \overset{\text{def}}{=} (a_{ij})_{i,j=1}^{m,n}$ be an $m \times n$ matrix. Show that there is an $n \times m$ matrix $T \overset{\text{def}}{=} (t_{pq})_{p,q=1}^{n,m}$ such that $ATA = A$. The matrix T is the *Moore-Penrose* or *pseudo-inverse* A^+ of A, cf. [**Ge9**].

[*Hint:* If V resp. W is an m-dimensional resp. n-dimensional vector space then for every choice of bases for V and W there is a natural correspondence

$$[V, W] \ni T \leftrightarrow A \in \mathrm{Mat}_{mn}$$

between the set $[V, W]$ of linear maps of V into W and the set Mat_{mn} of $m \times n$ matrices. Fix bases in $\mathbb{C}^m$ and $\mathbb{C}^n$ let T^A in $[\mathbb{C}^m, \mathbb{C}^n]$ correspond to A given above. Choose a Hamel basis Y' for $\mathrm{im}\left(T^A\right)$ and let X' be a set such that $T^A(X') = Y'$ and $\#(X') = \#(Y')$. Fill Y' out to a Hamel basis Y for $\mathbb{C}^n$ and fill X' out to a Hamel basis X for $\mathbb{C}^m$. Define the linear transformation $S \in [\mathbb{C}^n, \mathbb{C}^m]$ by the rule:

$$S(Y') = X', \ T^A S = I, \ S(Y \setminus Y') = \{\mathbf{O}\}.$$

Then $T^A S T^A = T^A$. Let S correspond to the matrix A^+.]

The *Gauß-Seidel algorithm* is one of the accepted recursive techniques for approximating the solution(s) of a system $A\mathbf{x} = \mathbf{b}$ of linear equations. Like *Newton's algorithm* (cf. **Example 2.1.3.9. 95**) for finding the real root(s) of an equation $f(x) = 0$, the Gauß-Seidel algorithm can fail by producing a divergent sequence of "approximants."

Example 1.3.1.1. Let the system $A\mathbf{x} = \mathbf{b}$ be

$$\begin{pmatrix} 2 & -1 \\ 1 & -2 \end{pmatrix} \begin{pmatrix} x_1 \\ x_2 \end{pmatrix} = \begin{pmatrix} b_1 \\ b_2 \end{pmatrix}.$$

Then

$$A = \begin{pmatrix} 2 & -1 \\ 1 & -2 \end{pmatrix} = \begin{pmatrix} 2 & 0 \\ 1 & -2 \end{pmatrix} - \begin{pmatrix} 0 & 1 \\ 0 & 0 \end{pmatrix} \overset{\mathrm{def}}{=} P - Q$$

and a direct calculation shows

$$P^{-1} = \begin{pmatrix} \frac{1}{2} & 0 \\ \frac{1}{4} & -\frac{1}{2} \end{pmatrix} \text{ and } P^{-1}Q = \begin{pmatrix} 0 & \frac{1}{2} \\ 0 & \frac{1}{4} \end{pmatrix}.$$

The eigenvalues of $P^{-1}Q$ are 0 and $\frac{1}{4}$ and, ρ_M denoting the *spectral radius* of the matrix M, $\rho_{P^{-1}Q} = \frac{1}{4} < 1$. If

$$\mathbf{x}_0 = \begin{pmatrix} a \\ b \end{pmatrix}$$

then, via the Gauß-Seidel algorithm, there arises the recursion

$$\mathbf{x}_{n+1} \overset{\mathrm{def}}{=} (P^{-1}Q)^{n+1}\mathbf{x}_0 + \sum_{k=0}^{n} (P^{-1}Q)^k P^{-1}\mathbf{b}, \ n \in \mathbb{N}.$$

The identity $(I - B^{n+1}) = \sum_{k=0}^{n} B^k (I - B)$, stemming from the algebraic identity $1 - z^{n+1} = \sum_{k=0}^{n} z^k(1 - z)$, is valid for any SQUARE matrix B. Since $\rho_{P^{-1}Q} = \frac{1}{4}$ it follows that

$$I = \lim_{n \to \infty} \sum_{k=0}^{n} (P^{-1}Q)^k (I - P^{-1}Q)$$

$$\lim_{n \to \infty} \mathbf{x}_{n+1} = (I - P^{-1}Q)^{-1} P^{-1} \mathbf{b}$$

$$= \begin{pmatrix} 1 & \frac{2}{3} \\ 0 & \frac{4}{3} \end{pmatrix} \begin{pmatrix} \frac{1}{2} & 0 \\ \frac{1}{4} & -\frac{1}{2} \end{pmatrix} \begin{pmatrix} b_1 \\ b_2 \end{pmatrix}$$

$$= \begin{pmatrix} \frac{2}{3} & -\frac{1}{3} \\ \frac{1}{3} & -\frac{2}{3} \end{pmatrix} \begin{pmatrix} b_1 \\ b_2 \end{pmatrix}$$

$$= \begin{pmatrix} \frac{1}{3}(2b_1 - b_2) \\ \frac{1}{3}(b_1 - 2b_2) \end{pmatrix}. \tag{1.3.1.1}$$

A direct check shows that the (column) vector in the right member of (1.3.1.1) is indeed the solution of $A\mathbf{x} = \mathbf{b}$.

On the other hand, E_{ij} denoting the identity matrix I with rows i and j interchanged, the system may be rewritten $A E_{12} E_{12} \mathbf{x} = \mathbf{b}$, i.e., as follows:

$$\begin{pmatrix} -1 & 2 \\ -2 & 1 \end{pmatrix} \begin{pmatrix} x_2 \\ x_1 \end{pmatrix} = \begin{pmatrix} b_1 \\ b_2 \end{pmatrix}.$$

The matrix B of the system is $A E_{12}$, the unknown $\mathbf{y}$ of the system is $E_{12} \mathbf{x}$ and the right member of the system is unchanged: $B\mathbf{y} = \mathbf{b}$.

This time write

$$B = \begin{pmatrix} -1 & 0 \\ -2 & 1 \end{pmatrix} - \begin{pmatrix} 0 & -2 \\ 0 & 0 \end{pmatrix} \overset{\text{def}}{=} S - T.$$

Then

$$S^{-1} = \begin{pmatrix} -1 & 0 \\ -2 & 1 \end{pmatrix} \text{ and } S^{-1}T = \begin{pmatrix} 0 & 2 \\ 0 & 4 \end{pmatrix}.$$

This time the eigenvalues of $S^{-1}T$ are 0 and 4 whence $\rho_{S^{-1}T} = 4 > 1$. Furthermore, if

$$\mathbf{y}_0 \overset{\text{def}}{=} \begin{pmatrix} c \\ d \end{pmatrix}$$

then

$$(S^{-1}T)^n = \begin{pmatrix} 0 & 2 \cdot 4^{n-1} \\ 0 & 4^n \end{pmatrix}$$

$$\sum_{k=0}^{n} (S^{-1}T)^k = \begin{pmatrix} 1 & \frac{2}{3}(4^n - 1) \\ 0 & \frac{1}{3}(4^{n+1} - 1) \end{pmatrix}$$

$$\mathbf{y}_{n+1} = \begin{pmatrix} 2 \cdot 4^n d \\ 4^{n+1} d \end{pmatrix} + \begin{pmatrix} -b_1 + \frac{2}{3}(4^n - 1)(-2b_1 + b_2) \\ \frac{1}{3}(4^{n+1} - 1)(-2b_1 + b_2) \end{pmatrix}$$

$$= \begin{pmatrix} 4^n(2d + \frac{2}{3}(-2b_1 + b_2)) + \frac{1}{3}(b_1 - 2b_2) \\ 4^{n+1}(d + \frac{1}{3}(-2b_1 + b_2)) + \frac{1}{3}(2b_1 - b_2) \end{pmatrix}. \tag{1.3.1.2}$$

The sequence $\{\mathbf{y}_n\}$ converges iff the coefficients of 4^n and 4^{n+1} in (1.3.1.2) are 0, i.e., iff

$$d = \tfrac{1}{3}(2b_1 - b_2).$$

In that case for all n

$$\mathbf{y}_n = \begin{pmatrix} \tfrac{1}{3}(b_1 - 2b_2) \\ \tfrac{1}{3}(2b_1 - b_2) \end{pmatrix}$$

and the $\mathbf{y}_n$ converge (trivially) to the solution found before. Hence iff one uses for $\mathbf{y}_0$ a vector in which the second component d is the very special number $\tfrac{1}{3}(2b_1 - b_2)$ does the sequence $\{\mathbf{y}_n\}$ converge at all.

1.3.2. General vector spaces

If V is a vector space and $T \in [V]$ then T is *invertible* iff there is in $[V]$ an S, the *inverse* of T, such that $ST = TS = I$. If V is finite-dimensional then **[Ge9]** there is an S such that $ST = I$ iff there is an R such that $TR = I$. If such an S (and hence an R) exists then $R = S$, whence inverses are unique. The last statement is not necessarily valid if V is infinite-dimensional.

Example 1.3.2.1. Let V be the vector space $\mathbb{C}[z]$ of polynomials of a single (complex) variable z. If $f \in V$ let $[0, z]$ be the line segment connecting 0 and z in $\mathbb{C}$ and let $T(f)(z)$ be

$$\int_{[0,z]} f(w)\, dw.$$

Then T is a *monomorphism*: T is *linear* and $T(f) = T(g) \Rightarrow f = g$. If $S(f) = f'$ then $ST = I$. However if $0 \neq a \in \mathbb{C}$ and $f(z) \equiv a$ then $S(f) = 0$ and $TS(f) = 0 \neq I(f) = f$.

[**Remark 1.3.2.1:**   The *range* of T is the vector space W of polynomials with constant term 0, whence $TS(f) = f$ iff $f(0) = 0$. Restricted to W, T does have an inverse: $ST = TS = I$.]

If V is a vector space and $T \in [V]$ the *spectrum* $\sigma(T)$ is the set of numbers λ such that $T - \lambda I$ is not invertible. If V is finite-dimensional then $\sigma(T)$ is the (nonempty!) finite set of *eigenvalues* of T. If V is a Banach space and T is continuous then $\sigma(T)$ is compact and nonempty although the set of eigenvalues of T may well be empty.

By contrast, if V is infinite-dimensional without further restriction then the continuity of T may be meaningless and, as the **Examples** below reveal, the neat results cited above are absent in rather striking ways:

i. T may fail to have even one eigenvalue;
ii. T may have a nonempty open spectrum;
iii. T may have an empty spectrum;
iv. T may have as its spectrum the noncompact, open, and closed set $\mathbb{C}$.

Example 1.3.2.2. Let V be the set of all *two-sided sequences*

$$\mathbf{a} \overset{\text{def}}{=} \{a_n\}_{-\infty < n < \infty}$$

of complex numbers of which only finitely many are nonzero. Let T (the *negative shift operator*) be defined according to the rule:

if $T\mathbf{a} \overset{\text{def}}{=} \mathbf{b} \overset{\text{def}}{=} \{b_n\}_{-\infty < n < \infty}$ then $b_n = a_{n+1}$, $-\infty < n < \infty$.

If λ is an eigenvalue of T and $\mathbf{x} \overset{\text{def}}{=} \{x_n\}_{-\infty < n < \infty}$ is a corresponding eigenvector then

$$x_{n+1} = \lambda x_n, \quad -\infty < n < \infty.$$

There are integers K and L such that

$$x_n \begin{cases} = 0 & \text{if } n < K \text{ or } n > L \\ \neq 0 & \text{if } n = K \text{ or } n = L. \end{cases}$$

Hence $x_{L+1} = 0 = \lambda x_L$ whence $\lambda = 0$. Hence $x_L = \lambda x_{L-1} = 0$ a contradiction. Thus T has *no* eigenvalues.

Example 1.3.2.3. Let T be as in **Example 1.3.2.2**. Then T^{-1} exists and $T - \lambda I$ is not invertible iff $\lambda \neq 0$, i.e., $\sigma(T) = \mathbb{C} \setminus \{0\}$.

Indeed if $T - \lambda I$ is invertible let $\mathbf{x}$ be the vector such that

$$x_n = \begin{cases} 1 & \text{if } n = 0 \\ 0 & \text{otherwise.} \end{cases}$$

If $(T - \lambda I)^{-1}\mathbf{x} \overset{\text{def}}{=} \mathbf{y}$ then $\mathbf{y} \neq \mathbf{O}$ since otherwise $\mathbf{x} = (T - \lambda I)\mathbf{y} = \mathbf{O}$. Let $T\mathbf{y}$ be $\mathbf{w} \overset{\text{def}}{=} (w_n)_{n \in \mathbb{Z}}$. There is in $\mathbb{N}$ an L such that

$$y_n \begin{cases} = 0 & \text{if } n > L \\ \neq 0 & \text{if } n = L. \end{cases}$$

If $\lambda \neq 0$ then $x_L = y_{L+1} - \lambda y_L = -\lambda y_L \neq 0$, whence $L = 0$. Thus

$$y_0 \neq 0, \quad w_n = y_{n+1}, \quad \lambda y_0 = w_0 = y_1 = 0.$$

whence $\lambda = 0$, a contradiction. Hence $\lambda = 0$ and $\sigma(T) = \mathbb{C} \setminus \{0\}$.

Example 1.3.2.4. Let V be the set $\mathbb{C}(z)$ of all *rational functions*

$$f : \mathbb{C} \ni z \mapsto f(z) \in \mathbb{C}.$$

Each f in V is the quotient of two polynomials p and q: $f = p/q$. Let k be a fixed polynomial of positive degree and let T be defined by the equation

$$T : V \ni f \mapsto T(f) \overset{\text{def}}{=} kf \ (= \frac{kp}{q}) \in V.$$

With respect to the natural definitions of addition and scalar multiplication V is a vector space over $\mathbb{C}$ and $T \in [V]$. If $\lambda \in \sigma(T)$ then

$$g_\lambda \overset{\text{def}}{=} \frac{1}{k - \lambda} \in V$$

and if S_λ is the map

$$V \ni f \mapsto g_\lambda f$$

then

$$S_\lambda(T - \lambda I) = (T - \lambda I)S_\lambda = I,$$

a contradiction. It follows that $\sigma(T) = \emptyset$.

Example 1.3.2.5. If V and T are the objects in **Example 1.3.2.1. 31** then $\sigma(T) = \mathbb{C}$. Indeed, if $\lambda \in \mathbb{C}$ and $(T - \lambda I)^{-1} \overset{\text{def}}{=} R_\lambda$ exists then, because T is not invertible, $\lambda \neq 0$. Let $R_\lambda(1)$ be the polynomial p_λ. It follows that $(T - \lambda I)p_\lambda = 1$. The endomorphism $T - \lambda I$ raises the degree of any nonzero polynomial and since $p_\lambda \neq 0$ a contradiction emerges.

For a vector space V an *inner product*

$$(\, , \,) : V \times V \ni \{\mathbf{x}, \mathbf{y}\} \mapsto (\mathbf{x}, \mathbf{y}) \in \mathbb{C}$$

is assumed to be a) *linear* in the first argument, b) *conjugate symmetric* (whence *conjugate linear* in the second argument), and c) *positive definite*:

    a) $a, b \in \mathbb{C} \Rightarrow (a\mathbf{x} + b\mathbf{y}, \mathbf{z}) = a(\mathbf{x}, \mathbf{z}) + b(\mathbf{y}, \mathbf{z})$

    b) $(\mathbf{x}, \mathbf{y}) = \overline{(\mathbf{y}, \mathbf{x})}$

    c) $(\mathbf{x}, \mathbf{x}) \geq 0$, $(\mathbf{x}, \mathbf{x}) = 0 \Leftrightarrow \mathbf{x} = \mathbf{O}$.

If there is an inner product for the vector space V then V is a *Euclidean* vector space. Two vectors $\mathbf{x}$ and $\mathbf{y}$ are *orthogonal* $(\mathbf{x} \perp \mathbf{y})$ iff $(\mathbf{x}, \mathbf{y}) = 0$.

The *norm* associated with the inner product is

$$\| \ \| : V \ni \mathbf{x} \mapsto \|\mathbf{x}\| \overset{\text{def}}{=} \sqrt{(\mathbf{x}, \mathbf{x})}.$$

Owing to c) and the *Schwarz* and *Minkowski* inequalities:

$$\text{Schwarz} : |(\mathbf{x}, \mathbf{y})| \leq \|\mathbf{x}\| \cdot \|\mathbf{y}\|$$

$$\text{Minkowski} : \|\mathbf{x} + \mathbf{y}\| \leq \|\mathbf{x}\| + \|\mathbf{y}\|$$

(equality obtains in each iff $\mathbf{x}$ and $\mathbf{y}$ are linearly dependent), the norm permits the definition of a *metric* d:

$$d : V \times V \ni \{\mathbf{x}, \mathbf{y}\} \mapsto d(\mathbf{x}, \mathbf{y}) \overset{\text{def}}{=} \|\mathbf{x} - \mathbf{y}\|.$$

A nonempty set $X \overset{\text{def}}{=} \{\mathbf{x}_\lambda\}_{\lambda \in \Lambda}$ in a Euclidean vector space is an *orthogonal* set iff each two vectors in X are orthogonal; X is *orthonormal* iff

$$(\mathbf{x}_\lambda, \mathbf{x}_\mu) = \delta_{\lambda\mu} \overset{\text{def}}{=} \begin{cases} 1 & \text{if } \lambda = \mu \\ 0 & \text{otherwise.} \end{cases}$$

In a finite-dimensional Euclidean vector space V (endowed with an inner product), associated to an endomorphism T is another endomorphism T^*, its *adjoint* T^*, such that $(T\mathbf{x}, \mathbf{y}) = (\mathbf{x}, T^*\mathbf{y})$. In fact if X is an orthonormal Hamel basis for V and if A_T is the matrix associated via X to T then T^* is the endomorphism associated via X to A_T^*.

If V is an n-dimensional vector space over $\mathbb{C}$ then V is essentially $\mathbb{C}^n$. Every dense subspace of V is V itself and every endomorphism of V is automatically continuous. A basis yields the isomorphism $[V] \leftrightarrow \mathrm{Mat}_{nn}$, which permits a complete analysis of all endomorphisms of V.

If V is an infinite-dimensional Banach space then: a) In V there can be dense proper subspaces, e.g., the set of polynomials in $C([0,1], \mathbb{C})$. b) Schauder bases for V may fail to exist (**Subsection 2.3.1**). c) Matrices are of little value. d) Hamel bases used as in **Exercise 1.1.4.1. 6** lead to discontinuous algebraic endomorphisms of V. e) If T is an algebraic endomorphism of V, T is continuous iff there is for the *dual space* V^* an algebraic endomorphism T^* such that for $\mathbf{x}^*$ in V^* and $\mathbf{x}$ in V,

$$T^* (\mathbf{x}^*) (\mathbf{x}) = \mathbf{x}^* (T(\mathbf{x})). \qquad (*)$$

[PROOF. If T is continuous then $(*)$ defines T^*. If T^* exists, $\mathbf{x}_n \to \mathbf{x}$, $T(\mathbf{x}_n) \to \mathbf{y}$, and $\mathbf{x}^* \in V^*$, then $\mathbf{x}^*(T(\mathbf{x}_n)) = T^*(\mathbf{x}^*)(\mathbf{x}_n) \to \mathbf{x}^*(\mathbf{y})$. On the other hand, $T^*(\mathbf{x}^*)(\mathbf{x}_n) \to T^*(\mathbf{x}^*)(\mathbf{x}) = \mathbf{x}^*(T(\mathbf{x}))$. Since $\mathbf{x}^*$ is arbitrary, $T(\mathbf{x}) = \mathbf{y}$, i.e., the *graph* of T is closed. The *closed graph theorem* [**Rud**] implies T is continuous. $\square$]

Example 1.3.2.6. Let $\mathcal{P}$ be the set of all polynomials

$$p : [0,1] \ni x \mapsto p(x) \overset{\text{def}}{=} \sum_{k=1}^{n} a_k x^k, \ a_k \in \mathbb{C}, \ n \in \mathbb{N}.$$

Then $\mathcal{P}$ is infinite-dimensional. Introduce into $\mathcal{P}$ the inner product

$$(\ ,\ ) : \mathcal{P} \times \mathcal{P} \ni \{p, q\} \mapsto \int_0^1 p(x)\overline{q(x)}\, dx \overset{\text{def}}{=} (p, q).$$

Then $\mathcal{P}$ is a Euclidean vector space and

$$T : V \ni p \mapsto \frac{dp}{dx}$$

is a linear transformation. However T has no adjoint. Indeed if T^* is the adjoint of T, if $p_n(x) \overset{\text{def}}{=} x^n$, $0 \le n < \infty$, then

$$\begin{aligned}
(p_n, T^* p_0) &= \int_0^1 x^n \overline{T^* p_0(x)}\, dx \\
&= \int_0^1 T(x^n)\overline{p_0}\, dx \\
&= \int_0^1 n x^{n-1}\, dx = 1.
\end{aligned}$$

Let p be the polynomial T^*p_0. From the Schwarz inequality

$$|(\mathbf{x}, \mathbf{y})| \le \|\mathbf{x}\| \cdot \|\mathbf{y}\|$$

it follows that

$$|(p_n, p)| = 1 \le \|p_n\| \cdot \|p\| = \sqrt{\frac{1}{2n+1}} \cdot \|p\|.$$

If n is such that $\|p\| < \sqrt{2n+1}$ there emerges the contradiction: $1 < 1$.

Furthermore, if V is finite-dimensional, a norm-preserving endomorphism U, i.e., an *isometry*, is automatically *unitary*: U^{-1} exists and

$$(U\mathbf{x}, U\mathbf{y}) = (\mathbf{x}, \mathbf{y}).$$

When V is infinite-dimensional the statements above need not hold.

Example 1.3.2.7. If

$$V \overset{\text{def}}{=} l^2 \overset{\text{def}}{=} \left\{ \{a_n\}_{1 \le n < \infty} : a_n \in \mathbb{C}, \sum_{n=1}^{\infty} |a_n|^2 < \infty \right\}$$

there is definable the inner product

$$(,) : V \times V \ni \{\{a_n\}, \{b_n\}\} \mapsto \sum_{n=1}^{\infty} a_n \overline{b_n} \overset{\text{def}}{=} (\{a_n\}, \{b_n\}).$$

If U is defined as follows:

$$U : l^2 \ni \mathbf{a} \overset{\text{def}}{=} (a_1, a_2, \ldots) \mapsto U\mathbf{a} \overset{\text{def}}{=} (0, a_1, a_2, \ldots) \in l^2$$

then U is an isometry and $(U\mathbf{a}, U\mathbf{b}) = (\mathbf{a}, \mathbf{b})$. If S is defined by:

$$S : l^2 \ni \mathbf{a} \overset{\text{def}}{=} (a_1, a_2, \ldots) \mapsto S\mathbf{a} \overset{\text{def}}{=} (a_2, a_3, \ldots) \in l^2$$

then $S = U^*$ and $U^*U = I$ but $UU^* \ne I$.

In the study of *quadratic forms* an important result may be described as follows.

Let A be a *self-adjoint* matrix $(a_{ij})_{i,j=1}^{n,n}$. For each nonsingular matrix T there is the *congruent matrix* T^*AT that is also self-adjoint. For A fixed, the necessarily real eigenvalues of T^*AT fall into three sets: the set P_T of positive eigenvalues, the set N_T of negative eigenvalues, and the set Z_T of zero eigenvalues.

THEOREM 1.3.2.1. (SYLVESTER'S LAW OF INERTIA). FOR A FIXED, THE CARDINALITIES $\#(P_T) \overset{\text{def}}{=} p_T$, $\#(N_T) \overset{\text{def}}{=} n_T$, AND $\#(Z_T) \overset{\text{def}}{=} z_T$ ARE INDEPENDENT OF T.

PROOF. The argument uses *homotopy*. The set $[\mathbb{C}^n]$ of endomorphisms of $\mathbb{C}^n$ may be viewed as $\mathbb{C}^{n^2}$. There is a unitary matrix U such that $U^*AU \ (= U^{-1}AU) \stackrel{\text{def}}{=} \Delta$ is a diagonal matrix in which the diagonal entries are the (necessarily real) eigenvalues of A.

Assume first that A is invertible. Because A is invertible each of its eigenvalues is not zero. For a given T, because the set $\mathcal{I}$ of invertible endomorphisms is both open and connected, there is a continuous map $[0,1] \ni t \mapsto T_t \in \mathcal{I}$ such that $T_0 = T$, $T_1 = U$. The *Weyl minmax theorem* about the eigenvalues of a self-adjoint matrix implies that each eigenvalue is a continuous function of the matrix entries. It follows that the eigenvalues of $T_t^* A T_t$ vary continuously as t varies between 0 and 1. For every t, $T_t^* A T_t$ is invertible and so each eigenvalue is not zero, whence the positive resp. negative eigenvalues of $T_t^* A T_t$ remain positive resp. negative as t varies, i.e., the numbers p_{T_t} and n_{T_t} remain constant as t varies. However, $p_{T_0} = p_T$ and $n_{T_0} = n_T$ while p_{T_1} resp. n_{T_1} are the numbers of positive resp. negative eigenvalues of A, numbers that are independent of T.

If A is singular then consideration of $\Delta \pm aI$ shows that there is a positive ϵ such that if $0 < a < \epsilon$ then the numbers of positive eigenvalues of $A - aI$ resp. A are the same and the numbers of negative eigenvalues of $A + aI$ resp. A are the same. Hence, for each t, the numbers of positive resp. negative eigenvalues of A are the same as the numbers of positive resp. negative eigenvalues of $T_t^* A T_t$.

$\square$

Crucial in the discussion above is the connectedness of the set of invertible elements in the *Banach algebra* $[\mathbb{C}^n]$ of endomorphisms of $\mathbb{C}^n$.

THEOREM 1.3.2.2. THE SET $\mathcal{I}$ OF CONTINUOUS INVERTIBLE ENDOMORPHISMS OF *Hilbert space* $\mathcal{H}$ IS CONNECTED.

PROOF. If T is an arbitrary invertible continuous endomorphism of $\mathcal{H}$, then there is a *polar decomposition*: $T = PU$, (P positive definite and U unitary).

The *spectral theorem* implies that there are positive numbers ϵ and M and *spectral measures* E_λ and F_θ such that

$$P = \int_\epsilon^M \lambda \, dE_\lambda \text{ and } U = \int_0^1 e^{2\pi i \theta} \, dF_\theta$$

(cf. [**Mu, Sto1**]).

Hence if, for t in $[0, 1]$,

$$P_t \stackrel{\text{def}}{=} \int_\epsilon^M e^{(1-t)\ln\lambda} \, dE_\lambda \text{ and } U_t \stackrel{\text{def}}{=} \int_0^1 e^{2\pi i(1-t)\theta} \, dF_\theta,$$

then $P_0 = P$, $P_1 = I$ and $U_0 = U$ and $U_1 = I$. The dependence of P_t and U_t on t is continuous, each P_t is invertible, each U_t is unitary, and

$t \mapsto P_t$ resp. $t \mapsto U_t$ are continuous curves linking P resp. U to I in the set of continuous invertible endomorphisms of $\mathcal{H}$. Hence $t \mapsto T_t \overset{\text{def}}{=} P_t U_t$ is a continuous curve connecting T to I in the set of continuous invertible endomorphisms of $\mathcal{H}$.

$\square$

However there are *Banach algebras* — even commutative Banach algebras — in which the set of invertible elements is not connected.

Example 1.3.2.8. Let B be the Banach algebra $C\,(\mathbb{T}, \mathbb{C})$ of continuous $\mathbb{C}$-valued functions defined on $\mathbb{T} \overset{\text{def}}{=} \{\, z \ : \ z \in \mathbb{C}, \ |z| = 1 \,\}$. Thus B may be viewed as the set of continuous $\mathbb{C}$-valued functions h defined on $[0,1]$ and such that $h(0) = h(1)$. The set S of invertible elements of B is precisely the set of functions that never vanish. In the norm-induced topology of B the set S is open. Inside each norm-induced neighborhood of a point there is a convex, hence connected, norm-induced neighborhood of the same point.

Hence if S is connected then it is *arcwise connected*. Thus if f and g are two elements of S, then as functions they are *homotopic* in S. Two particular elements of S are $f : [0,1] \ni \theta \mapsto e^{2\pi i \theta}$ and $g : [0,1] \ni \theta \mapsto e^{4\pi i \theta}$. The numbers

$$\frac{1}{2\pi i} \int_f \frac{dz}{z} \ (= 1) \quad \text{and} \quad \frac{1}{2\pi i} \int_g \frac{dz}{z} \ (= 2)$$

are the *indices* with respect to 0 of f and g regarded as curves in $\mathbb{C}$. If there is a homotopy f_t, $t \in [0,1]$ such that $f_0 = f$, $f_1 = g$, and for each t in $[0,1]$, $f_t \in S$ then the continuous map

$$[0,1] \ni t \mapsto \frac{1}{2\pi i} \int_{f_t} \frac{dz}{z}$$

takes on only integral values and yet is not constant, a contradiction, whence S is not connected.

1.3.3. Linear programming

If $\mathbf{x} \overset{\text{def}}{=} (x_1, \ldots, x_n)^t \in \mathbb{R}^n$ and $\mathbf{x}_i \geq 0$, $1 \leq i \leq n$, one writes $\mathbf{x} \succeq \mathbf{O}$. Similar interpretations are attached to $\mathbf{x} \succeq \mathbf{y}$, $\mathbf{x} \succ \mathbf{y}$, etc. Let $\mathbb{R}^{(n,+)}$ denote the nonnegative *orthant* of $\mathbb{R}^n$:

$$\mathbb{R}^{(n,+)} \overset{\text{def}}{=} \{\, \mathbf{x} \ : \ \mathbf{x} \in \mathbb{R}^n, \ \mathbf{x} \succeq \mathbf{O} \,\}.$$

If A is an $m \times n$ matrix over $\mathbb{R}$, if $\mathbf{p}$ is a real $m \times 1$ (column) vector, and if $\mathbf{c}$ is a real $1 \times n$ (row) vector the *primal linear programming problem* (**PLPP**) is that of finding in $\mathbb{R}^{(n,+)}$ an *optimal* vector $\mathbf{x}_{Opt}$ such that

$$A\mathbf{x}_{Opt} \succeq \mathbf{p}$$

$$(A\mathbf{x} \succeq \mathbf{p}) \wedge \left(\mathbf{x} \in \mathbb{R}^{(n,+)}\right) \Rightarrow \mathbf{c}\mathbf{x}_{Opt} \leq \mathbf{c}\mathbf{x} \overset{\text{def}}{=} Cost.$$

Dantzig's *simplex algorithm* [**Dan**] for dealing with the **PLPP** proceeds by:

  i. finding a vertex $\mathbf{x}_0$, i.e., an *extreme point*, in the *convex polyhedron* $\mathcal{Q}$ defined by the inequalities $A\mathbf{x} \succeq \mathbf{p}$ and $\mathbf{x} \succeq \mathbf{O}$;

  ii. passing from $\mathbf{x}_0$ through a succession $\mathbf{x}_1, \mathbf{x}_2, \ldots,$ of $\mathcal{Q}$-vertices to an *optimal vertex* (if one exists) [**Ge9**]; each segment $[\mathbf{x}_k, \mathbf{x}_{k+1}]$ is an *edge* of $\mathcal{Q}$ on which the *gradient* of *Cost* is most negative at $\mathbf{x}_k$.

Exercise 1.3.3.1. Show that for m, n in $\mathbb{N}$, there is a number $f(m, n)$ such that the number of $\mathcal{Q}$-vertices does not exceed $f(m, n)$.

For a given solvable **PLPP**, if $\mathbf{x}_0 = \mathbf{x}_{Opt}$, testing all *neighboring vertices* confirms the optimality and the simplex algorithm is quickly concluded. If $\mathbf{x}_0 \neq \mathbf{x}_{Opt}$ then the simplex algorithm, cleverly applied, leads to an optimal vertex and if $\mathbf{x}_0$ is a neighbor of an $\mathbf{x}_{Opt}$ the simplex algorithm, cleverly applied, reaches a conclusion *quickly*.

For a given $A, \mathbf{p}$, and $\mathbf{c}$, $\mathcal{Q}$ depends only on A and $\mathbf{p}$, but not on $\mathbf{c}$. For a given $\mathbf{c}$ let $l_{\mathbf{c}}(\mathcal{Q})$ denote the largest number of $\mathcal{Q}$-vertices that can be visited in the course of applying the simplex algorithm. In the worst case $l_{\mathbf{c}}(\mathcal{Q}) \leq f(m, n)$. For m, n in $\mathbb{N}$, the maximum $M(m, n)$ of $l_{\mathbf{c}}(\mathcal{Q})$, as $\mathcal{Q}$ varies over the set all convex polyhedra corresponding to the pair m, n and $\mathbf{c}$ varies over $\mathbb{R}^n$, is of great practical interest.

Klee and Minty [**KlM**] showed that there is on $\mathbb{N}^2$ no polynomial function p such that $M(m, n) \leq p(m, n)$.

Example 1.3.3.1. By contrast, if A is an $m \times n$ matrix, the number of steps performed in *Gaußian elimination* applied to the system $A\mathbf{x} = \mathbf{b}$ does not exceed

$$\mu(m, n) \stackrel{\text{def}}{=} \begin{cases} \frac{m(m-1)(3n-m-2)}{6} & \text{if } m < n \\ \frac{n(3mn+3(m-n)-n^2-2)}{6} & \text{otherwise} \end{cases}$$

[**Ge9**].

With respect to a reasonable method for averaging over the totality of **PLPP**s for a fixed pair (m, n), Smale [**Sm2, Sm3**] showed that the corresponding number $\overline{M}(m, n)$ is dominated by a polynomial function $\overline{q}(m, n)$. Finally, Karmarkar [**Karm, Ge9**] developed a replacement for the simplex algorithm. In Karmarkar's approach, the optimum is successively approximated by a recursive algorithm that, for any degree of accuracy given a priori, yields a solution in $K(m, n)$ steps and $K(m, n)$ is dominated by a polynomial function $k(m, n)$: $K(m, n)$ is *polynomially dominated*.

Another troublesome phenomenon that occurs occasionally in the execution of the simplex algorithm is *cycling*:

The algorithm defines a sequence $\{\mathbf{x}_n\}_{0 \leq n < \infty}$ of vertices such that:

    i. the algorithm identifies none as optimal;

    ii. for some k_0 and n_0 in $\mathbb{N}$

$$k \geq k_0 \Rightarrow \mathbf{x}_{k+n_0} = \mathbf{x}_k.$$

Example 1.3.3.2. Assume

$$A \stackrel{\text{def}}{=} \begin{pmatrix} 1 & 0 & 0 & 1 & 1 & 1 & 1 \\ 0 & 1 & 0 & 0.5 & -5.5 & -2.5 & 9 \\ 0 & 0 & 1 & 0.5 & -1.5 & -0.5 & 1 \end{pmatrix}$$

$$\mathbf{p} \stackrel{\text{def}}{=} \begin{pmatrix} 1 \\ 0 \\ 0 \end{pmatrix} \ (= \mathbf{e}_1)$$

$$\mathbf{c} \stackrel{\text{def}}{=} (\,0 \quad 0 \quad 0 \quad -1 \quad 7 \quad 1 \quad 2\,).$$

Let the associated **PLPP** be to find in $\mathbb{R}^{(7,+)}$ an $\mathbf{x}_{Opt}$ such that

$$A\mathbf{x}_{Opt} = \mathbf{p}$$
$$\mathbf{c}\mathbf{x}_{Opt} = minimum\,\{\,\mathbf{cx} \ : \ \mathbf{x} \succeq \mathbf{O}, \ A\mathbf{x} = \mathbf{p}\,\}.$$

At the start one may, via Gaußian elimination, choose as *basic variables*

$$x_1, x_2, x_3,$$

expressed as linear combinations of the *free variables*

$$x_4, x_5, x_6, x_7.$$

Thereupon

$$C \stackrel{\text{def}}{=} \mathbf{cx} = -x_4 + 7x_5 + x_6 + 2x_7.$$

For each choice of a pair (basic variables, free variables), setting the free variables at 0 and calculating the values of the basic variables gives rise to a vector (in the current instance $(x_1, x_2, x_3, 0, 0, 0, 0)$) that is a vertex of $\mathcal{Q}$.

    Since $\mathbf{x} \in \mathbb{R}^{(n,+)}$ the only way C *might* be minimized is by allowing x_4 to increase while $x_5 = x_6 = x_7 = 0$. However if $x_4 > 0 = x_5 = x_6 = x_7$ then $x_2, x_3 < 0$, whence x_4 may not be increased, but the *SWITCH: basic variable $x_2 \leftrightarrow$ free variable x_4* leads to the formula:

$$C = 2x_2 - 4x_5 - 4x_6 + 20x_7.$$

(Each SWITCH changes the pair (basic variables, free variables) and determines a new vertex.)

Again increases of x_5 or of x_6 *might* decrease C but again if either of x_5 or x_6 is positive then $x_4, x_3 < 0$.

Each of the SWITCHes in following sequence

$$x_2 \leftrightarrow x_4$$
$$x_3 \leftrightarrow x_5$$
$$x_4 \leftrightarrow x_6$$
$$x_5 \leftrightarrow x_7$$
$$x_6 \leftrightarrow x_2$$
$$x_7 \leftrightarrow x_3$$

is consonant with the simplex algorithm and *might* lead to a decrease in the value of C. However if that execution of the simplex algorithm is performed the sequence x_3, x_5, x_7 of recurs endlessly ($k_0 = n_0 = 3$) without leading to the conclusion that the minimal value of C is zero.

The difficulty lies in the failure of the simplex algorithm to offer guidance when several permissible SWITCHes present themselves, as they do above in the second round of the algorithm: either x_5 or x_6 may be chosen as the right half of a SWITCH. The choice of x_5 in the SWITCH $x_3 \leftrightarrow x_5$ opens the way for cycling.

On the other hand, the sequence of SWITCHes

$$x_2 \leftrightarrow x_4$$
$$x_4 \leftrightarrow x_6$$
$$x_5 \leftrightarrow x_2$$

leads to

$$C = 2x_3 + 4x_5 + 4x_7,$$

which shows that the minimal achievable value of C is indeed zero: any increase from zero of any of the free variables x_3, x_4, x_5, x_7 cannot reduce C. Hence no further SWITCHes are indicated and the simplex algorithm leads to the conclusion that the minimal value of C is zero.

Bland **[Bl]** and Charnes **[Char]** provided different but effective modifications of the simplex algorithm.

Bland's algorithm uses the equivalence of $Ax \succeq p$ and $Ax - w = p$ for some w satisfying $w \succeq O$. Thus if

$$A \stackrel{\text{def}}{=} m \begin{pmatrix} \overset{n}{A} & \overset{m}{-I} \end{pmatrix}, \text{ and } \mathbf{X} \stackrel{\text{def}}{=} \begin{matrix} n \\ m \end{matrix} \begin{pmatrix} \overset{1}{\mathbf{x}} \\ \mathbf{w} \end{pmatrix}$$

then $(Ax \succeq p) \wedge \left(x \in \mathbb{R}^{(n,+)} \right)$ is equivalent to $\left(A\mathbf{X} = p \right) \wedge \left(\mathbf{X} \in \mathbb{R}^{(m+n,+)} \right)$. By simple elimination (*without* row exchanges) $A\mathbf{X} = p$ permits the expression of (new) basic variables X_{b_i} in terms of (new) free variables X_{f_j}:

$$X_{b_i} = q_i - \sum_{j=1}^{n} \alpha_{ij} X_{f_j}, \ 1 \le i \le m. \tag{$*$}$$

An *eligible* free variable is one that occurs with a negative coefficient in the expression for the (new) *Cost* function and occurs as well with some positive coefficient in $(*)$. Bland's rule is to choose the eligible free variable with the smallest index, say $X_{f_{j'}}$. An eligible basic variable X_{b_i} is one for which $\alpha_{ij'} > 0$ and

$$\frac{q_i}{\alpha_{ij'}} = \inf \left\{ \frac{q_k}{\alpha_{kj'}} \ : \ \alpha_{kj'} > 0 \right\}.$$

The basic variable for the SWITCH is the eligible basic variable with the smallest index, say $X_{b_{i'}}$ and the Bland SWITCH is $X_{b_{i'}} \leftrightarrow X_{f_{j'}}$. The three noncycling SWITCHes in **Example 1.3.3.2. 39** are Bland SWITCHes.

The Charnes technique employs a perturbation A, $\mathbf{p}(\epsilon)$, $\mathbf{c}$ of the set A, $\mathbf{p}$, $\mathbf{c}$ of original data of the **PLPP**. It can be shown for the associated **PLPP**(ϵ) that there is a positive ϵ_0 such that if $0 < \epsilon < \epsilon_0$ then a) no cycling can occur and b) any optimal solutions $\mathbf{x}_{Opt}(\epsilon)$ converge as $\epsilon \to 0$ to optimal solutions $\mathbf{x}_{Opt}$ for **PLPP**.

Both the Bland and Charnes techniques are explained and illustrated in [**Ge9**].

2. Analysis

2.1. Classical Real Analysis

2.1.1. $\mathbb{R}^X$

Let $f, g, \ldots$ be $\mathbb{R}$-valued functions defined on a set X endowed with a *topology*. (In many instances X is, for some n, a subset of $\mathbb{R}^n$.) The following facts about $\mathbb{R}$ and functions $f, g, \ldots$ are the basis of much of the subsequent discussion. The set of points of continuity resp. discontinuity of a function f is denoted $\mathrm{Cont}(f)$ resp. $\mathrm{Discont}(f)$. If $f \in \mathbb{R}^\mathbb{R}$ the set of points where f' exists (does not exist) is denoted $\mathrm{Diff}(f)$ ($\mathrm{Nondiff}(f)$).

Typical of the results that are used without proof or further comment are the following [**HeSt, O1, Rud**].

  i. If f is continuous on a *compact* set K then $f(K)$ is compact (hence f is bounded on K) and there are in K points x_m and x_M such that for x in K
$$f(x_m) \leq f(x) \leq f(x_M).$$

  iii. If f is continuous on A and g is continuous on $f(A)$ then their *composition* $g \circ f$ is continuous on A.

  iv. If f is continuous and bijective on a compact set K then f^{-1} is continuous on $f(K)$ (f is *bicontinuous*).

  v. If $X = \mathbb{R}$ and f is *monotone* then $\mathbb{R} \setminus \mathrm{Cont}(f)$ is empty, finite, or countable; f is differentiable a.e. (cf. **Exercise 2.1.1.14. 49**).

vi. The only *connected* sets in $\mathbb{R}$ are intervals:

$$[a,b] \overset{\text{def}}{=} \{\, x \,:\, -\infty < a \le x \le b < \infty \,\} \text{ (closed)}$$

$$(a,b] \overset{\text{def}}{=} \{\, x \,:\, -\infty \le a < x \le b < \infty \,\} \text{ (half-open)}$$

$$[a,b) \overset{\text{def}}{=} \{\, x \,:\, -\infty < a \le x < b \le \infty \,\} \text{ (half-open)}$$

$$(a,b) \overset{\text{def}}{=} \{\, x \,:\, -\infty \le a < x < b \le \infty \,\} \text{ (open)}$$

$$\emptyset \overset{\text{def}}{=} \{\, x \,:\, x \ne x \,\} \text{ (open and closed)}.$$

vii. If f is continuous on an interval I containing $[a,b]$ and if $f(a) < v < f(b)$ or $f(a) > v > f(b)$, i.e., if v is between $f(a)$ and $f(b)$, there is between a and b a c such that $f(c) = v$: f enjoys the *intermediate value property* on I.

THEOREM **2.1.1.1.** THE SET $\text{CONT}(f)$ OF POINTS OF CONTINUITY OF f IS A COUNTABLE INTERSECTION OF OPEN SETS, i.e., $\text{CONT}(f)$ IS A G_δ [**HeSt**].

Exercise 2.1.1.1. Show that $\text{Discont}(f)$ is a countable union of *closed sets*, i.e., an F_σ.

Example 2.1.1.1. Every closed set is an F_σ. However $\mathbb{Q}$ is an F_σ but $\mathbb{Q}$ is not closed.

A set S is of the *first category* if it is the union of countably many nowhere dense sets. A set that is not of the first category is of the *second category*.

The next result is frequently cited as *Baire's (category) theorem* although the term *category* is used first in COROLLARY **2.1.1.1.** The collection of these results has wide application, e.g., in the proofs of the *open mapping* and *closed graph* theorems, which play a vital role in the study of Banach spaces [**Ban**].

THEOREM **2.1.1.2.** IF (X,d) IS A COMPLETE METRIC SPACE AND IF $\{U_n\}_{n\in\mathbb{N}}$ IS A SEQUENCE OF DENSE OPEN SUBSETS OF X, THEN $G \overset{\text{def}}{=} \bigcap_{n\in\mathbb{N}} U_n$ IS DENSE IN X [**HeSt, Rud**].

The complement of a dense open set is a *nowhere dense* (closed) set.

COROLLARY **2.1.1.1.** A NONEMPTY OPEN SUBSET OF A COMPLETE METRIC SPACE IS OF THE SECOND CATEGORY, i.e., IS NOT THE UNION OF COUNTABLY MANY NOWHERE DENSE SETS.

COROLLARY **2.1.1.2.** IF X IS A COMPLETE METRIC SPACE AND $\{F_n\}_{n\in\mathbb{N}}$ IS A SEQUENCE OF CLOSED SETS SUCH THAT

$$\bigcup_{n\in\mathbb{N}} F_n$$

CONTAINS A NONEMPTY OPEN SUBSET THEN AT LEAST ONE OF THE F_n CONTAINS A NONEMPTY OPEN SUBSET.

Exercise 2.1.1.2. Show that the conclusion of Baire's theorem obtains if each U_n is not necessarily open but does contain a dense open subset.

However, in Baire's theorem the dense open sets U_n may not be replaced by arbitrary dense sets D_n with merely nonempty *interiors* D_n°.

Example 2.1.1.2. Let $\mathbb{Q} \overset{\text{def}}{=} \{t_n\}_{n\in\mathbb{N}}$ be the set of rational numbers and let V_n be $(-n, n)\cup(\mathbb{Q}\setminus\{t_1, \ldots, t_n\})$. Then each V_n is dense in the complete metric space $\mathbb{R}$ and has a nonempty interior but $\bigcap_{n\in\mathbb{N}} V_n = (-1, 1)\setminus\mathbb{Q}$, which is not dense in $\mathbb{R}$.

There are yet other aspects of Baire's theorem.

i. The completeness of X plays an important role. For example, $\mathbb{Q}$ in its topology inherited from $\mathbb{R}$ is not complete. If $\mathbb{Q} \overset{\text{def}}{=} \{r_n\}_{n\in\mathbb{N}}$ and $U_n \overset{\text{def}}{=} \mathbb{Q}\setminus\{r_n\}$, $n\in\mathbb{N}$, then each U_n is a dense open subset of $\mathbb{Q}$ and yet $\bigcap_{n\in\mathbb{N}} U_n = \emptyset$. On the other hand, Baire's theorem remains valid if X is replaced by a *perfect* subset S of X or by the intersection $S\cap U$ of a perfect subset S and an open subset U of X.

ii. Although a complete metric space was originally and is now most frequently the context for applying Baire's theorem, it is nevertheless true that a locally compact space X (even if X is not a metric space) is also not of the first category, cf. **Corollary 2.1.1.1.** [PROOF (sketch). If $\{A_n\}_{n\in\mathbb{N}}$ is a sequence of nowhere dense subsets of X and if $X = \bigcup_{n\in\mathbb{N}} A_n$ then the closures $\overline{A_n}$, $n\in\mathbb{N}$ are also nowhere dense and so it may be assumed a priori that each A_n is closed. In $V_1 \overset{\text{def}}{=} X\setminus A_1$ there is a nonempty open set containing an nonempty open subset U_1 for which the closure $K_1 \overset{\text{def}}{=} \overline{U_1}$ is a compact subset of V_1. Then $V_2 \overset{\text{def}}{=} U_1\setminus A_2$ is a nonempty open set containing a nonempty open subset U_2 for which the closure $K_2 \overset{\text{def}}{=} \overline{U_2}$ is a compact subset of $K_1, \ldots$. There is an inductively definable sequence $\{K_n\}_{n\in\mathbb{N}}$ consisting of compact closures of open sets and such that $K_{n+1} \subset K_n$, $n\in\mathbb{N}$. The intersection $\bigcap_{n\in\mathbb{N}} K_n \overset{\text{def}}{=} K$ is a nonempty compact set by virtue of the *finite intersection property* of the sequence $\{K_n\}_{n\in\mathbb{N}}$ of closed subsets of the compact set K_1. On the other hand K meets none of the sets in $\{A_n\}_{n\in\mathbb{N}}$, i.e., K is not in X, a contradiction.]

iii. In its discrete topology $\mathbb{N}$ may be regarded as both a complete metric space and as a locally compact space. Thus $\mathbb{N}$ is a countable topological space that, on two scores, is a space of the second category.

THEOREM **2.1.1.3.** IF EACH f_n IS CONTINUOUS ON $\mathbb{R}$ AND

$$\lim_{n\to\infty} f_n = f$$

ON $\mathbb{R}$ THEN CONT(f) IS DENSE IN $\mathbb{R}$:

$$\overline{\mathrm{CONT}(f)} = \mathbb{R}$$

[HeSt].

[**Remark 2.1.1.1:**   If $\mathbb{R}$ is replaced by a (Cauchy) complete metric space X the conclusion remains valid.]

Let S° denote the *interior* of a set S: S° is the union of all the open subsets of S.

Exercise 2.1.1.3. Show that if F is a closed set and its *interior* F° is empty then F is nowhere dense.

Exercise 2.1.1.4. Prove THEOREM **2.1.1.3** with $\mathbb{R}$ replaced by a complete metric space X.

[*Hint:* The sets

$$F_{km} \overset{\text{def}}{=} \left[\bigcap_{m\le n} \left\{ x \; : \; |f_m(x) - f_n(x)| \le \frac{1}{k} \right\} \right]$$

$$F_k \overset{\text{def}}{=} \bigcup_{m\in\mathbb{N}} F_{km}$$

$$G_{km} \overset{\text{def}}{=} F_{km}^\circ$$

$$G_k \overset{\text{def}}{=} \bigcup_{m\in\mathbb{N}} G_{km}$$

$$G \overset{\text{def}}{=} \bigcap_{k\in\mathbb{N}} G_k$$

have a number of important properties listed below.

▷ Each F_{km} is closed because the f_n are continuous.
▷ Each F_k is X because the f_n converge everywhere.
▷ If $F_{km}^{\circ} = \emptyset$ then F_{km} is nowhere dense because F_{km} is closed.
▷ Not all F_{km}° are empty because X is Cauchy complete and hence not of the first category.
▷ The set $R_{km} \stackrel{\text{def}}{=} F_{km} \setminus F_{km}^{\circ}$ is closed and its interior is empty, whence R_{km} is nowhere dense and $R_k \stackrel{\text{def}}{=} \bigcup_{m \in \mathbb{N}} R_{km}$ is of the first category.
▷ Since
$$G_k = X \setminus R_k$$
it follows that G_k, as the complement of a nowhere dense set R_k in a complete metric space X, is dense. The set G_k, as a union of open sets, is open and so G_k is a dense open set.
▷ Baire's Theorem implies that G is dense.
▷ At each point x of G the limit function f is continuous. [PROOF. For each k in $\mathbb{N}$ there is in $\mathbb{N}$ an m_k such that $x \in F_{km_k}^{\circ}$, i.e.,

$$|f_{m_k}(x) - f_n(x)| \le \frac{1}{k}, \ m_k \le n.$$

Since $F_{km_k}^{\circ}$ is open it contains a neighborhood $U(x)$ and for every z in $U(x)$

$$|f_{m_k}(z) - f_n(z)| \le \frac{1}{k}, \ m_k \le n,$$

whence
$$|f_{m_k}(z) - f(z)| \le \frac{1}{k}$$

and so
$$|f(z) - f(x)| \le |f(z) - f_{m_k}(z)| + |f_{m_k}(z) - f_{m_k}(x)| \\ + |f_{m_k}(x) - f(x)|.$$

The first and third terms in the right member of the last display do not exceed $\frac{1}{k}$. Owing to the continuity of f_{m_k}, contained in $U(x)$ is a neighborhood $W(x)$ such that if $z \in W(x)$ then the second term is less than $\frac{1}{k}$. Hence f is continuous at x, as required. □]

Thus $\text{Cont}(f)$ is dense in X.]

Let χ_S denote the *characteristic function* of the set S:

$$\chi_S(x) = \begin{cases} 1 & \text{if } x \in S \\ 0 & \text{otherwise.} \end{cases}$$

Exercise 2.1.1.5. Show that there is in $\mathbb{R}^{\mathbb{R}}$ a function:

i. f_1 that is continuous nowhere and yet $|f_1|$ is constant (hence continuous everywhere);

ii. f_2 that is nonmeasurable and yet $|f_2|$ is constant (hence measurable).

[*Hint:* For f_p choose a set E_p and the function χ_{E_p}, $p = 1, 2$.]

Exercise 2.1.1.6. Show that there is in $\mathbb{R}^{\mathbb{R}}$ a function:

i. g_1 that is continuous somewhere and yet is not the limit of a sequence of continuous functions;

ii. g_2 that is not measurable but continuous somewhere;

iii. g_3 that is continuous a.e. but is not continuous everywhere;

iv. g_4 that is equal to a continuous function a.e. but is not itself continuous;

v. h_k that is not measurable but somewhere differentiable of order k.

[*Hint:* For v choose a nonmeasurable set E and consider $x \mapsto x^k \left(\chi_E - \chi_{\mathbb{R} \setminus E} \right).$]

Exercise 2.1.1.7. Let S be a noncompact subset of $\mathbb{R}$. Show that:

i. if S is unbounded and $f(x) = x$ on S then f is continuous and unbounded on S;

ii. if S is bounded there is in $\overline{S} \setminus S$ a point a and then if on S

$$f(x) \overset{\text{def}}{=} \frac{1}{x - a}$$

f is continuous and unbounded on S.

[**Remark 2.1.1.2:** In i and ii above the function f is *locally bounded*: if $x \in S$ there is an open set $N(x)$ containing x and such that f is bounded on $S \cap N(x)$.]

Exercise 2.1.1.8. Assume S is a noncompact subset of $\mathbb{R}$.

i. Show that if S is unbounded above there in S a sequence $\{a_n\}_{n \in \mathbb{N}}$ such that $n < a_n < a_{n+1}$. Show that if, for each x in S,

$$f(x) = \begin{cases} (-1)^n n & \text{if } x = a_n, \; n = 2, 3, \ldots \\ t f(a_n) + (1 - t) f(a_{n+1}) & \text{if } x = t a_n + (1 - t) a_{n+1}, \; 0 < t < 1 \\ -1 & \text{if } x \in (-\infty, a_1) \end{cases}$$

then f is continuous on S and achieves neither a maximum nor a minimum on S.

ii. Carry out a similar construction if S is unbounded below.

iii. Show that if S is bounded and $a \in \overline{S} \setminus S$ then f as in **Exercise 2.1.1.7**ii is a bijective bicontinuous map of S onto an unbounded set $f(S) \overset{\text{def}}{=} S_1$. For S_1 show how to use i or ii to define on S_1 a continuous function f_1 that achieves no maximum or minimum on S_1. Show that $f \overset{\text{def}}{=} f_1 \circ f$ is continuous on S and achieves neither a maximum nor a minimum on S.

Exercise 2.1.1.9. Let K be the compact set $[-1, 1]$. Show that if

$$f(x) = \begin{cases} x & \text{if } x \in (-1, 1) \\ 0 & \text{if } x = \pm 1 \end{cases}$$

then f is bounded on K and achieves neither a maximum nor a minimum on K.

Exercise 2.1.1.10. Show that if $f \in \mathbb{R}^{\mathbb{R}}$ and

$$f(x) = \begin{cases} \frac{(-1)^n m}{n+1} & \text{if } x \in \mathbb{Q} \cap [0, 1] \text{ and } x = \frac{m}{n}, \, m, n \in \mathbb{N} \text{ and } (m, n) = 1 \\ 0 & \text{otherwise} \end{cases}$$

then on $[0, 1]$, $\liminf f(x) \equiv -1 < f(x) < 1 = \limsup f(x) \equiv 1$. Hence f is bounded, is nowhere *semicontinuous*, and achieves neither a maximum nor a minimum on the compact set $[0, 1]$.

THEOREM **2.1.1.4.** IF f IS A *periodic* NONCONSTANT FUNCTION ON $\mathbb{R}$ AND IF f IS CONTINUOUS AT ONE POINT, SAY a, THEN THERE IS A *period* p SUCH THAT $p > 0$ AND IF $0 < x < p$ THEN x IS NOT A PERIOD OF f.

PROOF. Otherwise there is a sequence $\{a_n\}_{n \in \mathbb{N}}$ such that $a_n \downarrow 0$ and each a_n is a period of f. Since the set of periods of any function is an additive group, the group G_f of periods of f is dense in $\mathbb{R}$. Since f is not constant, let b be such that $f(b) \neq f(a)$. Then there is a sequence $\{c_n\}_{n \in \mathbb{N}}$ of periods such that $b + c_n \to a$ whence $f(b) = f(b + c_n) \to f(a)$, a contradiction.

$\square$

Exercise 2.1.1.11. Show that $\chi_{\mathbb{Q}}$ is a nonconstant periodic function without a smallest positive period.

Not only is the set $\text{Discont}(f)$ of points of discontinuity of a function f an F_σ (cf. **Exercise 2.1.1.1.** 43) but, as the next **Exercise** reveals, every F_σ is, for some f, $\text{Discont}(f)$.

Exercise 2.1.1.12. Show that if A is an F_σ then:

i. there is a sequence $\{F_n\}_{n\in\mathbb{N}}$ of closed sets such that $F_n \subset F_{n+1}$, $n \in \mathbb{N}$, and $A = \bigcup_{n\in\mathbb{N}} F_n$;

ii. if $F_0 \overset{\text{def}}{=} \emptyset$, if

$$B_n \overset{\text{def}}{=} (F_n \setminus F_{n-1}) \setminus (F_n \setminus F_{n-1})^\circ,$$

and if

$$f(x) = \begin{cases} 2^{-n} & \text{if } x \in B_n \\ 0 & \text{if } x \notin \bigcup_{n\in\mathbb{N}} B_n \end{cases}$$

then $\text{Discont}(f) = A$.

[*Hint:* For c in A there is an n such that such that $c \in F_{n+k}$, $k \in \mathbb{N}$, in which case c is a limit point of $D_n \overset{\text{def}}{=} B_{n-1} \cup B_{n+1}$. If $x \in D_n$ then $|f(c) - f(x)| \geq 2^{-n-1}$. If $c \notin A$ then $f(c) = 0$. For a positive ϵ choose N in $\mathbb{N}$ so that $2^{-N} < \epsilon$ and then a neighborhood $N(c)$ of c so that $N(c) \cap F_N = \emptyset$. Then $|f(x) - f(c)| < 2^{-N} < \epsilon$ if $x \in N(c)$.]

Exercise 2.1.1.13. Show that if $f \in \mathbb{R}^{\mathbb{R}}$ and

$$f(x) = \begin{cases} \frac{1}{n} & \text{if } x = \frac{m}{n},\ m \in \mathbb{Z} \setminus \{0\},\ n \in \mathbb{N}, \text{ and } (m,n) = 1 \\ 1 & \text{if } x = 0 \\ 0 & \text{otherwise} \end{cases}$$

then $\text{Discont}(f) = \mathbb{Q}$.

Exercise 2.1.1.14. If $f \in \mathbb{R}^{\mathbb{R}}$, f is a monotone increasing function, and $a \in \text{Discont}(f)$ then

$$\lim_{x\uparrow a} f(x) \overset{\text{def}}{=} f(a-0) < f(a+0) \overset{\text{def}}{=} \lim_{x\downarrow a} f(x).$$

Hence corresponding to a there is in $(f(a-0), f(a+0))$ a number in $\mathbb{Q}$. Hence show that $\text{Discont}(f)$ is at most countable.

Conversely, let $S \overset{\text{def}}{=} \{a_n\}_{n\in\mathbb{N}}$ be a subset of $\mathbb{R}$ and assume that d_n is positive and $\sum_{n\in\mathbb{N}} d_n < \infty$. Show that if

$$j_a(x) = \begin{cases} 0 & \text{if } x < a \\ 1 & \text{if } x \geq a \end{cases}$$

then

$$f \overset{\text{def}}{=} \sum_{n=1}^{\infty} d_n j_{a_n}$$

is a monotone increasing function, $\text{Discont}(f) = S$, and

$$f(a_n + 0) - f(a_n - 0) = d_n,\ n \in \mathbb{N}.$$

[*Hint:* If $x \in \mathbb{R}$ then

$$f(x) = \sum_{n=1}^{N} d_n j_{a_n}(x) + \sum_{n=N+1}^{\infty} d_n j_{a_n}(x) \overset{\text{def}}{=} \sum_{n=1}^{N} d_n j_{a_n}(x) + R_N(x)$$

and if $\epsilon > 0$ there is an N such that $|R_N(x)| < \epsilon$.]

[**Note 2.1.1.1:** The set S might well be *dense* in $\mathbb{R}$, e.g., $S = \mathbb{Q}$.]

Exercise 2.1.1.15. Show that if $f \in \mathbb{R}^{\mathbb{R}}$, if $f(x + 1) \equiv f(x)$ (f is periodic with period 1), $f(x) = |x|$, $|x| \le \frac{1}{2}$, and

$$g(x) \overset{\text{def}}{=} \sum_{n=1}^{\infty} \frac{f(4^{n-1}x)}{4^{n-1}} \overset{\text{def}}{=} \sum_{n=1}^{\infty} g_n(x)$$

then g is a) continuous everywhere, b) monotone on no *nondegenerate* interval, and c) for k in $\mathbb{N}$ there is in $\mathbb{N}$ an N_k such that for all nonzero h, all x, and $G_k(x) \overset{\text{def}}{=} g(N_k x)$,

$$|G_k(x + h) - G_k(x)| > k|h|$$

("all difference quotients are large everywhere").

Show also that if $H(x) = \sum_{m=0}^{M} a_m \cos 2m\pi x + b_m \sin 2m\pi x$ and if $\mathbb{R} \ni \alpha \neq 0$ then for some K_k in $\mathbb{N}$, a) – c) obtain for $H + \alpha G_{K_k}$.

[*Hint:* For $a \in \mathbb{R}$ and for h_n at least one of $\pm 4^{-n-1}$,

$$|g_m(a + h_n) - g_m(a)| = \begin{cases} |h_n| & \text{if } m \le n \\ 0 & \text{if } m > n. \end{cases}$$

Furthermore

$$\frac{g(a + h_m) - g(a)}{h_m}$$

is an even resp. odd integer according as m is even resp. odd, whence $g'(a)$ exists nowhere.

If g is monotone on some *nondegenerate* interval $[a, b]$ then g is differentiable a.e. on (a, b), a contradiction.]

The function g is based on a construction of Knopp. Weierstraß constructed the nowhere differentiable function

$$W : \mathbb{R} \ni x \mapsto W(x)$$

$$\overset{\text{def}}{=} \sum_{k=0}^{\infty} b^k \cos[a^k \pi x], \ 0 < b < 1, \ ab > 1 + \frac{3\pi}{2},$$

a uniform limit of *analytic*, hence *infinitely differentiable*, functions, i.e., functions in C^∞. Hardy [**Har**] showed that the condition $ab > 1 + \frac{3\pi}{2}$ may be replaced by $ab \geq 1$ without disturbing the conclusion. In [**Har**] Hardy showed also the validity of Riemann's conjecture: the continuous function

$$R(x) \overset{\text{def}}{=} \sum_{n=1}^{\infty} \frac{\sin[n^2 \pi x]}{n^2}$$

is nowhere differentiable.

There is a veritable plethora of continuous nowhere differentiable functions. In fact, in $C\left([0,1], \mathbb{R}\right)$, the *Banach space* of continuous $\mathbb{R}$-valued functions on $[0,1]$, the nowhere differentiable functions constitute a dense set of the *second category* (cf. THEOREM **2.3.3.1. 166**).]

Example 2.1.1.3. On C_0 (cf. **Exercise 1.1.4.3. 7**) the map

$$\phi : C_0 \ni \sum_{n=1}^{\infty} 2\epsilon_n 3^{-n} \mapsto -\frac{1}{2} + \sum_{n=1}^{\infty} \epsilon_n 2^{-n}$$

(cf. **Example 2.1.2.1. 55**) is continuous and $\phi(C_0) = [-\frac{1}{2}, \frac{1}{2}]$. The map $\gamma : \left(-\frac{1}{2}, \frac{1}{2}\right) \ni x \mapsto \tan \pi x$ is continuous, $\operatorname{im}(\gamma) = \mathbb{R}$, $h_0 \overset{\text{def}}{=} \gamma \circ \phi$ is defined on $D_0 \overset{\text{def}}{=} C_0 \setminus \{0, 1\}$, and $\operatorname{im}(h_0) = \mathbb{R}$. Since $[0,1] \setminus C_0$ is the union of countably many pairwise disjoint open intervals, for each of these intervals a similar construction may be carried out. There emerges in each of the intervals treated a new sequence of disjoint open intervals in which the process may be repeated. By endless repetition, i.e., by *inbreeding*, there emerges a sequence of sequences of sequences of ..., re-indexed $\{D_n\}_{n=1}^{\infty}$, of pairwise disjoint sets. Each D_n is, on some open interval, the analog of D_0 on $(0,1)$. Inside each nonempty open interval is an interval containing some D_n. Furthermore, the process of inbreeding ("intervals within intervals within ..."), assures that $D \overset{\text{def}}{=} D_0 \uplus D_1 \uplus D_2 \uplus \cdots$ is dense in $[0,1]$. For each D_n define h_n in analogy with h_0 for D_0. Since D_0 as well as each D_n has Lebesgue measure zero so does D. If

$$H(x) \overset{\text{def}}{=} \begin{cases} 0 & \text{if } x \notin D \\ h_n(x) & \text{if } x \in D_n, \ 0 \leq n < \infty, \end{cases}$$

then H is zero almost everywhere. Since inside every nonempty open subinterval (a,b) of $[0,1]$ there is an interval on which some D_n lies, and since $h_n(D_n) = \mathbb{R}$, it follows that $H[(a,b)] = \mathbb{R}$.

There is in $\mathbb{R}^{[0,1]}$ a function H that is zero a.e. and H maps every nondegenerate interval (a, b) of $[0, 1]$ onto $\mathbb{R}$.

If $t \in [0, 1]$ and $1 < k \in \mathbb{N}$, t has one or two k-ary representations

$$\sum_{n=1}^{\infty} \epsilon_n^{(i)} k^{-n}, \quad \epsilon_n^{(i)} \in \{0, 1, \ldots, k - 1\}, \quad i = 1 \text{ or } i = 1, 2.$$

If there are two, the one for which $\sum_{n=1}^{\infty} \epsilon_n^{(i)} = \infty$ is **the** k-ary representation of t. The number $\epsilon_n^{(i)}$ is *an* nth k-ary marker for t. If $k = 2, 3$, $\epsilon_n^{(i)}$ is an nth *binary, ternary* marker for t; if $k = 10$, $\epsilon_n^{(i)}$ is an nth *decimal* marker.

Exercise 2.1.1.16. Assume $t \in [0, 1]$. Show:

i. for each n in $\mathbb{N}$ there is at most one odd nth decimal marker for t;

ii. if $f_n(t)$ is the sum of the first n binary markers in **the** binary representation of t, $0 \le k \le n$, and $S_{k,n} \overset{\text{def}}{=} f_n^{-1}(k)$, then the binomial theorem and *Stirling's formula* imply

$$\lambda\left(S_{k,n}\right) = 2^{-n} \binom{n}{k}, \quad \lim_{n \to \infty} \lambda\left(S_{n,2n}\right) = 1,$$

$$\int_{[0,1]} \frac{f_n(t)}{n} \, dt = \frac{1}{2}, \quad \lim_{n \to \infty} \int_{[0,1]} \left(\frac{f_n(t)}{n} - \frac{1}{2}\right)^2 dt = 0.$$

(See Problem 479 in **[Ge7]** for more results and interpretations.)

Exercise 2.1.1.17. If $t \in [0, 1]$ let $\sum_{n=1}^{\infty} \epsilon_n 2^{-n}$, $\epsilon_n = 0$ or 1, be **the** binary representation of t and let $\delta(t)$ be $\sum_{n=1}^{\infty} 2\epsilon_n 3^{-n}$ (cf. **Exercise 1.1.4.3. 7**).

i. Show that δ is bijective but not continuous. Give an example of a point t in $\text{Discont}(\delta)$.

ii. Show that ϕ in **Example 2.1.1.3. 51** is continuous but not bijective.

iii. Show that $\phi \circ \delta = \text{id}$ (the identity map) but that $\delta \circ \phi$ is not the identity map.

iv. Give an example of two different points x and y in C_0 and such that $\phi(x) = \phi(y)$.

v. Show that if two maps $S : A \mapsto B$ and $T : B \mapsto A$ are such that $S \circ T = \text{id}$ then T must be bijective whereas S need not be.

In a topological space X a point x is a *point of condensation* of a set S iff for every neighborhood U of x, the set $U \cap S$ is *uncountable*.

Exercise 2.1.1.18. Show that every point in the Cantor set C_0 is a point of condensation of C_0.

2.1.2. Derivatives and extrema

If $f \in \mathbb{R}^{\mathbb{R}}$ and is differentiable then f': a) is measurable and b) enjoys the *intermediate value property*. [PROOF. ad a) Since

$$f'(x) = \lim_{n \to \infty} n[f(x + \frac{1}{n}) - f(x)],$$

as the limit of a sequence of continuous, hence measurable, functions, f' is measurable. ad b) If, e.g., $a < b$, $f'(a) < v < f'(b)$, and if $g(x) = f(x) - vx$ then g is continuous, $g'(a) < 0 < g'(b)$, and the minimum of g in $[a, b]$ is achieved in (a, b) at some c where $g'(c)$ $(= f'(c) - v)$ $= 0.$ $\Box$]

Exercise 2.1.2.1.

i. Show that if $k \in \mathbb{N}$ and if

$$f(x) = \begin{cases} x^{2k} \sin\left(x^{-2k+1}\right) & \text{if } x \neq 0 \\ 0 & \text{if } x = 0 \end{cases}$$

then f is differentiable (on $\mathbb{R}$) but that f' is discontinuous at 0.

ii. Show that if

$$f(x) = \begin{cases} x^2 \sin\left(x^{-2}\right) & \text{if } x \neq 0 \\ 0 & \text{if } x = 0 \end{cases}$$

then f is differentiable on $\mathbb{R}$ but f' is unbounded on $[-1, 1]$.

iii. Show that if

$$f = \begin{cases} x^4 e^{-\frac{x^2}{4}} \sin \frac{8}{x^3} & \text{if } x \neq 0 \\ 0 & \text{if } x = 0 \end{cases}$$

then f is differentiable, f' is bounded on $\mathbb{R}$,

$$\sup_{x \in [-1,1]} f'(x) = 24 = -\inf_{x \in [-1,1]} f'(x),$$

and yet $|f'(x)| \neq 24$ everywhere on $[-1, 1]$.

[**Remark 2.1.2.1:** If f is differentiable then $f'(x)$, as the limit of the sequence $\{n[f(x + \frac{1}{n}) - f(x)]\}_{n \in \mathbb{N}}$ of continuous functions, is such that $\text{Cont}(f')$ is a dense G_δ, whence f' cannot be discontinuous everywhere.]

THEOREM **2.1.2.1.** LET $\{f_n\}_{n \in \mathbb{N}}$ BE A SEQUENCE OF DIFFEREN-TIABLE FUNCTIONS DEFINED ON A COMPACT INTERVAL I. ASSUME:

i. THERE IS IN $\mathbb{R}^I$ A FUNCTION g SUCH THAT $f_n' \overset{u}{\to} g$ ON I;
ii. THERE ARE IN $\mathbb{R}$ AN A AND IN I AN a SUCH THAT $f_n(a) \to A$.

THEN THERE IS A DIFFERENTIABLE FUNCTION f SUCH THAT $f_n \overset{u}{\to} f$ ON I AND $f' = g$ [**Gr, O1**].

[PROOF outline: For m, n in $\mathbb{N}^2$, x, ξ_{mn} in I,

$$f_m(x) - f_n(x) - (f_m(a) - f_n(a)) = [f'_m(\xi_{mn}) - f'_n(\xi_{mn})](x - a).$$

Hence $\{f_n\}_{n \in \mathbb{N}}$ is a uniform Cauchy sequence and THEOREM **2.1.4.1. 97** may be applied. $\square$]

[**Remark 2.1.2.2:** If the hypotheses above are strengthened by adding the assumption that each f'_n is *Riemann integrable* and that

$$f'_n(x) = \frac{d\left[\int_a^x f'_n(t)\, dt\right]}{dx},$$

e.g., by assuming that each f'_n is continuous, then a direct proof by integration is available [**O1, O3**].]

[**Note 2.1.2.1:** If $f_n \equiv n, n \in \mathbb{N}$, then $\{f_n\}_{n \in \mathbb{N}}$ is a sequence for which hypothesis *i* above obtains but hypothesis *ii* does not and the conclusion fails.]

Each function f' in **Exercise 2.1.2.1. 53** is a discontinuous function enjoying the intermediate value property. On the other hand there is the following result.

THEOREM **2.1.2.2.** IF A FUNCTION h DEFINED ON A COMPACT IN-TERVAL I IS OF *bounded variation* ON I ($h \in BV(I)$) AND ALSO ENJOYS THE INTERMEDIATE VALUE PROPERTY THEN h IS CONTINUOUS.

PROOF. As a function of bounded variation, h is the difference of two monotone functions. Hence if $a \in \text{Discont}(h)$ then $h(a \pm 0)$ exist and, for some positive d, $|h(a + 0) - h(a - 0)| \stackrel{\text{def}}{=} d$. It may be assumed that $h(a + 0) - h(a - 0) = d$, whence

$$h(a + 0) - \frac{d}{3} = h(a - 0) + \frac{2d}{3} > h(a - 0) + \frac{d}{3}.$$

There is a positive δ such that if $a - \delta < x < a$ resp. $a < x < a + \delta$ then $h(x) < h(a - 0) + \frac{d}{4}$ resp. $h(x) > h(a + 0) - \frac{d}{4}$. If

$$v \in [h(a - 0) + \frac{d}{4}, h(a + 0) - \frac{d}{4}] \setminus \{h(a)\}$$

then nowhere in $(a - \delta, a + \delta)$ does h assume the value v, a contradiction.
$\square$

COROLLARY **2.1.2.1.** IF A DERIVATIVE f' IS OF BOUNDED VARIATION ON A COMPACT INTERVAL I THEN f' IS CONTINUOUS.

[**Remark 2.1.2.3:** To assume that $f \in BV(I)$ is to assume *less* than that $f \in BV(\mathbb{R})$: for every compact interval I, the function $x \mapsto x$ is in $BV(I)$ but is not in $BV(\mathbb{R})$, i.e., $BV(\mathbb{R}) \subsetneq \bigcap_I BV(I)$.

On the other hand, for the sets $AC(I)$ of functions *absolutely continuous* on every compact interval I, $AC(I) \subset BV(I)$ [**Roy**] and, to boot, $AC(\mathbb{R}) = \bigcap_I AC(I)$.]

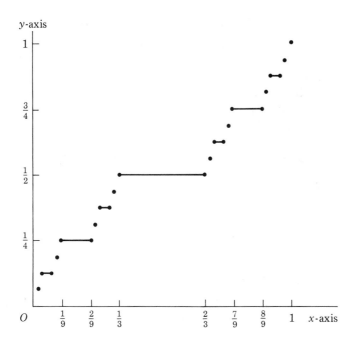

Figure 2.1.2.1. The graph of $y = \mathcal{C}_0(x)$.

Example 2.1.2.1. The *Cantor function* $\mathcal{C}_0$ is defined according to the following prescription: if the nth ternary marker for x in $\mathcal{C}_0$ is (by definition) $2\epsilon_n$ then an nth binary marker for $\mathcal{C}_0(x)$ is ϵ_n. Thus $\mathcal{C}_0$ on $\mathcal{C}_0$ is

$$\sum_{n=1}^{\infty} 2\epsilon_n 3^{-n} \mapsto \overset{\text{def}}{=} \sum_{n=1}^{\infty} \epsilon_n 2^{-n}.$$

The function $\mathcal{C}_0$ is further defined on any one of the countably many intervals that constitute $[0,1] \setminus \mathcal{C}_0$ by (continuous) linear interpolation (whence $\mathcal{C}_0$ is constant on each such interval). Then $\mathcal{C}_0$ is continuous, monotone increasing on $[0,1]$; $\mathcal{C}_0'$ exists a.e. and is 0 on $[0,1] \setminus \mathcal{C}_0$. In **Figure 2.1.2.1** there is an indication of the graph of C_0 (cf. **Exercise 1.1.4.3. 7, Exercise 2.1.1.17. 52**). As in the construction of **Example 2.1.1.3. 51** there is a sequence $\{C_n\}_{n=1}^{\infty}$ of Cantor-like functions, one defined for each of the deleted intervals. Each function is appropriately scaled so that for n in $\mathbb{N}$, $0 \le C_n \le 2^{-n}$. Then on $[0,1]$: a)

$$C_0 + \sum_{n=1}^{\infty} C_n$$

converges uniformly to a function C, continuous and *strictly increasing*, i.e., $0 \le x < y \le 1 \Rightarrow C(x) < C(y)$; b) C' exists and is 0 a.e. Furthermore C may be extended to a function $\widetilde{C}$ continuous and strictly increasing on $\mathbb{R}$ where $\widetilde{C}'$ exists and is 0 a.e.

If $f \in \mathbb{R}^{\mathbb{R}}$ then the results in **Subsection 2.1.1** imply that: a) If $f \in \mathbb{R}^{\mathbb{R}}$ then Discont(f) is an F_σ. b) If $E \subset \mathbb{R}$ and E is an F_σ then for some f in $\mathbb{R}^{\mathbb{R}}$, $E = \text{Discont}(f)$. The result below is an almost flawless parallel.

THEOREM **2.1.2.3.** a') IF $f \in BV$ THEN $\lambda(\text{Nondiff}(f)) = 0$. b') IF $E \subset \mathbb{R}$ AND $\lambda(E) = 0$ THEN THERE IS IN BV A CONTINUOUS f SUCH THAT $E \subset \text{Nondiff}(f)$.

PROOF. The proof of a') is standard [**Gr, HeSt, Roy, Rud, Sz-N**]. The proof of b') follows from the results in **Exercise 2.1.2.2** below.

Exercise 2.1.2.2. 1) Show that if $E \subset \mathbb{R}$ then $\lambda(E) = 0$ (E is a *null* set) iff for each positive ϵ, there is a sequence $\mathcal{J} \overset{\text{def}}{=} \{(a_n, b_n)\}_{n \in \mathbb{N}}$ of intervals such that every point x of E belongs to *infinitely many* of the intervals in $\mathcal{J}$ and $\sum_{n=1}^{\infty}(b_n - a_n) < \epsilon$. 2) Let E and $\mathcal{J}$ be as in 1) and let f_{ab} be $x \mapsto \chi_{[a,b]}(x)(x - a) + (b - a)\chi_{(b,\infty)}(x)$. Show that $f \overset{\text{def}}{=} \sum_{n=1}^{\infty} f_{a_n b_n}$ is monotone, continuous, and $E \subset \text{Nondiff}(f)$.

[*Hint:* ad 1). If $\lambda(E) = 0$ and $n \in \mathbb{N}$ there is a sequence $\{(\alpha_{nk}, \beta_{nk})\}_{k \in \mathbb{N}}$ such that $E \subset \bigcup_{k \in \mathbb{N}}(\alpha_{nk}, \beta_{nk})$ and the *length-sum* $\sum_{k=1}^{\infty}(\beta_{nk} - \alpha_{nk}) < \epsilon 2^{-n-1}$. Consider $\{(\alpha_{nk}, \beta_{nk})\}_{n,k \in \mathbb{N}}$.

ad 2). If $c \in E$, $k \in \mathbb{N}$, and (a, b) is the intersection of k $\mathcal{J}$-intervals $\{(a_{n_1}, b_{n_1}), \ldots, (a_{n_k}, b_{n_k})\}$ containing c then for x in (a, b),

$$\frac{f(x) - f(c)}{x - c} \ge \sum_{j=1}^{k} \frac{f_{n_j}(x) - f_{n_j}(c)}{x - c} \ge k.]$$ $\square$

[**Remark 2.1.2.4:** The parallel drawn above is defective: "$\subset$" in b') is not the same as "$=$" in b). To the writers' knowledge, the true analog of b) has not been established.]

Exercise 2.1.2.3. Show that for f in **Exercise 2.1.1.14. 49**,

$$\text{Nondiff}(f) = \text{Discont}(f) = \{a_n\}_{n\in\mathbb{N}}.$$

If f is a differentiable function defined on an open subset of $\mathbb{R}$ then $f'(a) = 0$ at the site a of an extremum (maximum or minimum) of f. Furthermore if $f''(a) < 0$ resp. $f''(a) > 0$ then $f(a)$ is a local maximum resp. local minimum. It is quite possible that $f'(a) = 0$ and that a is not the site of an extremum, e.g, $f(x) = x^3$, $a = 0$, and that a is an extremum and $f''(a) = 0$, e.g., $f(x) = x^4$, $a = 0$. Of greater interest are **Exercises 2.1.2.4, 2.1.2.5** and **Examples 2.1.2.2, 2.1.2.3** that follow.

Exercise 2.1.2.4. Show that if $f \in \mathbb{R}^{\mathbb{R}}$ and

$$f(x) = \begin{cases} x^4 \left[2 + \sin\left(x^{-1}\right)\right] & \text{if } x \neq 0 \\ 0 & \text{if } x = 0 \end{cases}$$

then at 0 f is at an absolute minimum, $f'(0) = 0$, but that in no interval $(a,0)$ or $(0,b)$ is f monotone.

Cantor-like sets permit the construction of a continuous function f such that in every nonempty open subinterval J of $[0, 1]$ there are two points x_J^+ resp. x_J^- such that

$$x \in J \setminus \{x_J^{\pm}\} \Rightarrow f(x_J^+) > f(x) > f(x_J^-).$$

In other words:

The set S_{max} of sites of proper local maxima of f is dense in $[0, 1]$ and the set S_{min} of sites of proper local minima of f is dense in $[0, 1]$.

Example 2.1.2.2. The Cantor set C_0 may be viewed as the interval $[0, 1]$ from which "middle-third" open intervals have been deleted. Let I be $[0, 1]$. Let $\{I_{mn}^1\}_{m\in\mathbb{N}, \ 1\leq n\leq 2^{m-1}}$ be the set of open intervals deleted from I in the construction of C_0. The intervals I_{mn}^1, $m \in \mathbb{N}$, $1 \leq n \leq 2^{m-1}$, are numbered and grouped so that the length of the first is 3^{-1}, the length of each of the next two is $3^{-2},\ldots$, the length of each of the next 2^n is $3^{-(n+1)}$, etc. For each of the intervals I_{mn}^1 define a function g_{mn}^1 for which g_{11}^1 is the paradigm. The graph of g_{11}^1 is given in **Figure 2.1.2.2. 57** below.

Outside the interval of definition $g_{11}^1 = 0$. The *area* of each triangular lobe formed by the graph of $y = g_{11}^1(x)$ and the horizontal axis is $\frac{1}{8}$. Each g_{mn}^1 is situated with respect to I_{mn}^1 as g_{11}^1 is situated with respect to I_{11} and the graph of g_{mn}^1 is similar to the graph of g_{11}^1. Finally,

$$G_1 \stackrel{\text{def}}{=} \sum_{mn} g_{mn}^1.$$

The series converges since if $m \in \mathbb{N}$ and $1 \le n \le 2^{m-1}$ then the *significant domains* $[0,1] \setminus \left(g_{mn}^1\right)^{-1}(0)$ are pairwise disjoint.

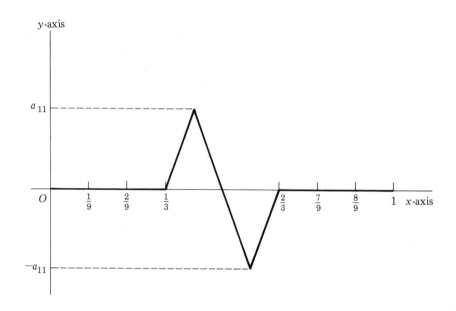

Figure 2.1.2.2. The graph of $y = g_{11}^1(x)$.

The midpoints of the intervals I_{mn}^1 together with the sites of the local maxima and minima of G_1 partition each I_{mn}^1 into four consecutive subintervals: $I_{mn}^{11}, I_{mn}^{12}, I_{mn}^{13}, I_{mn}^{14}$, all of the same length.

From this point on the description of the function f to be constructed will be given verbally rather than by unavoidably impenetrable formulas. On each of the intervals

$$I_{mn}^{11}, \ldots, I_{1n}^{14}$$

construct a Cantor-like set

$$C_{mn}^{11}, \ldots, C_{mn}^{14}$$

and for C_{mn}^{11} and C_{mn}^{14} construct the analogs G_{mn}^{11} and G_{mn}^{14} of G_1. For C_{mn}^{12} and C_{mn}^{13} construct the analogs G_{mn}^{12} and G_{mn}^{13} of $-G_1$. There emerges

$$G_2 \stackrel{\text{def}}{=} \sum_{m,n} \left(\sum_{k=1}^{4} G_{mn}^{1k} \right).$$

Mathematical induction and *inbreeding* lead to a sequence $G_1, G_2, \ldots$, and, owing to the manner of construction,

$$\frac{\max_x |G_{k+1}(x)|}{\max_x |G_k(x)|} = \frac{1}{3}. \tag{2.1.2.1}$$

Hence $f \overset{\text{def}}{=} \sum_{n=1}^{\infty} G_n$ exists and is a continuous function on $[0, 1]$.

In each interval I_{mn}^1 the function G_1 achieves two proper local extrema: $\max_{1n}$ and $\min_{1n}$. Owing to the construction of $G_1, G_2, \ldots$ (on the sites of their significant domains) and (2.1.2.1), $\max_{1n}$ and $\min_{1n}$ persist as proper local extrema of f.

A careful check reveals that a typical segment of the graph of $G_1 + G_2$ has the form depicted in **Figure 2.1.2.3** (over an interval I_{mn}^{11} or I_{1n}^{14}) or in **Figure 2.1.2.4. 60** (over an interval I_{1n}^{12} or I_{1n}^{13}).

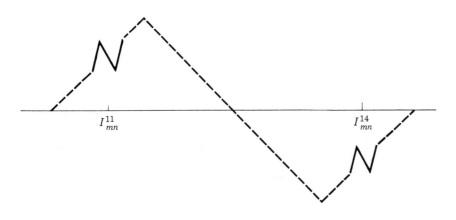

Figure 2.1.2.3. The graph of $y = G_1 + G_2$ over $I_{mn}^{11} \cup I_{mn}^{14}$.

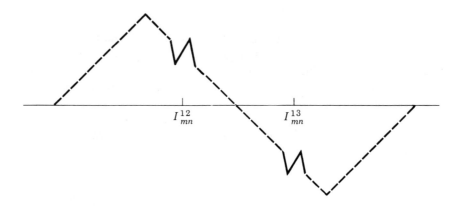

Figure 2.1.2.4. The graph of $y = G_1 + G_2$ over $I_{mn}^{12} \cup I_{mn}^{13}$.

Similarly, the two indicated local extrema of $G_1 + G_2$, one a proper local maximum, the other a proper local minimum, persist as proper local extrema of f, etc. If J is a nonempty open subinterval of $[0, 1]$, infinitely many of the intervals used in the construction of f are subintervals of J. It follows that f has in J infinitely many sites of proper local maxima and infinitely many sites of proper local minima. Hence each of the sets S_{max} and S_{min} is dense in $[0, 1]$. Other constructions can be found in [**Goe**] and [**PV**].

Exercise 2.1.2.5. Show that if h is a continuous function in $\mathbb{R}^{\mathbb{R}}$ then:

    i. if h achieves a local maximum at only one point and h is unbounded above then h achieves a local minimum somewhere;

    ii. if h achieves a local minimum at only one point and h is unbounded below then h achieves a local maximum somewhere.

By contrast there are the functions described next.

Example 2.1.2.3. Each of the continuous functions

$$f : \mathbb{R}^2 \ni (x, y) \mapsto 3xe^y - e^{3y} - x^3$$
$$g : \mathbb{R}^2 \ni (x, y) \mapsto x^2 + y^2(1 + x)^3$$

in $\mathbb{R}^{\mathbb{R}^2}$ achieves only one local extremum (a local maximum at $(1,0)$ for f and a local minimum at $(0,0)$ for the *polynomial* g) and each of f and g is unbounded both above and below.

The function

$$g : \mathbb{R} \ni x \mapsto \begin{cases} \exp\left(-x^{-2}\right) & \text{if } x \neq 0 \\ 0 & \text{if } x = 0 \end{cases} \tag{2.1.2.2}$$

is in C^∞ and, if $a > 0$, g is represented in $(0, 2a)$ by the Taylor series

$$\sum_{n=0}^{\infty} \frac{g^{(n)}(a)}{n!}(x-a)^n.$$

However $g^{(n)}(0) = 0$, $0 \leq n < \infty$, and so the Taylor series at 0, i.e., the Maclaurin series, for g does not represent g in any open interval centered at 0.

[**Remark 2.1.2.5:** The function g in (2.1.2.2) can be used to define a nonmeasurable function g_∞ such that somewhere g_∞ is *infinitely differentiable*, i.e., somewhere each of $g_\infty^{(k)}$, $k \in \mathbb{N}$ exists (cf. **Exercise 2.1.1.6. 47**).]

Computations aside, g has no Maclaurin series representation because 0 is an *essential singularity* of the function

$$(\mathbb{C} \setminus \{0\}) \ni z \mapsto \exp\left(-z^{-2}\right).$$

In this context the next result is derivable.

Example 2.1.2.4. If $f \in \mathbb{C}^{\mathbb{C}}$ and

$$f_c(z) = \begin{cases} \exp\left(-(z-c)^{-2}\right) & \text{if } z \in \mathbb{C} \setminus \{c\} \\ 0 & \text{if } z = c, \end{cases}$$

and $a \neq c$ then f_c may be represented by its Taylor series in any open disk centered at a and not containing c. If x and c are real then $0 \leq f_c(x) \leq 1$. If $\{r_n\}_{n \in \mathbb{N}}$ is an enumeration of $\mathbb{Q}$ then

$$\sum_{n=1}^{\infty} 2^{-n} f_{r_n}(z)$$

a) converges on $\mathbb{C}$, b) converges uniformly on every compact subset of $\mathbb{C} \setminus \mathbb{R}$, and c) defines a function F *holomorphic* in $\mathbb{C} \setminus \mathbb{R}$. Furthermore, d) F is infinitely differentiable on $\mathbb{R}$, e) nevertheless each a in $\mathbb{R}$ is an essential singularity of F, whence F admits no Taylor series representation in any disk centered at any point a of $\mathbb{R}$.

The function g in (2.1.2.2) is related to a class of *bridging functions*. For example, if h is defined on two disjoint closed intervals I and J and is differentiable on each interval, a bridging function H is function such that

i. H is in C^∞ on $\mathbb{R} \setminus (I \cup J)$;
ii. $H = h$ on $I \cup J$;
iii. H is differentiable on $\mathbb{R}$.

The general approach to the construction of such an H is based upon the following function:

$$\beta(x) \overset{\text{def}}{=} \begin{cases} 0 & \text{if } x \le 0 \\ \exp\left[-x^{-2} \exp\left(-(1-x^2)^{-2}\right)\right] & \text{if } 0 < x < 1 \\ 1 & \text{if } x \ge 1. \end{cases}$$

Exercise 2.1.2.6. Show that:

i. β above is in C^∞ and is strictly monotone on $(0,1)$;
ii. if a, b, c, d are real and $a < b$ there are real constants p, q, r, s such that

$$\gamma_{abcd}(x) \overset{\text{def}}{=} p + q\beta(rx + s) = \begin{cases} c & \text{if } x \le a \\ d & \text{if } x \ge b \end{cases}$$

is strictly monotone on (a, b).

Exercise 2.1.2.7. Assume $w < x < y < z$. Find numbers a, b, c, d and a', b', c', d' so that for given numbers A, B, C

$$\delta_{A,B,C}(t) \overset{\text{def}}{=} \gamma_{abcd}(t)\gamma_{a'b'c'd'}(t) = \begin{cases} A & \text{if } t \le w \\ B & \text{if } x \le t \le y \\ C & \text{if } t \ge z \end{cases}$$

and $\delta_{A,B,C}$ is strictly monotone on (w, x) and (y, z).

For k in $\mathbb{N}$ the function F is an *antiderivative of order k* of a function f if $F^{(k)} = f$. By abuse of language F is an antiderivative.

Exercise 2.1.2.8. Show that if $x < y$ and if

$$\{X, Y, X_1, \ldots, X_k, Y_1, \ldots, Y_k\} \subset \mathbb{R}$$

there is an antiderivative η of an appropriate γ_{abcd} so that

$$\eta(x) = X$$
$$\eta(y) = Y$$
$$\eta^{(j)}(x) = X_j, \ 1 \le j \le k$$
$$\eta^{(j)}(y) = Y_j, \ 1 \le j \le k.$$

[*Hint:* Repeatedly integrate some γ_{abcd}.]

Exercise 2.1.2.9. Show how, for a given g defined and k times differentiable on two disjoint closed intervals $I \overset{\text{def}}{=} [a, b]$ and $J \overset{\text{def}}{=} [c, d]$ such that $a < b < c < d$, to construct a bridging function H such that on $I \cup J$, if $1 \leq j \leq k$ then $H^{(j)} = g^{(j)}$.

[*Hint:* Let H have the form $g\left(\delta_{A,B,C} + \delta_{A',B',C'}\right)$.]

A function f is *smooth* on an open set U if $f \in C^{\infty}$ on U, i.e., if, for each k in $\mathbb{N}$, f is k times differentiable on U.

Example 2.1.2.5. If $[a, b] \subset \mathbb{R}$ bridging functions permit the construction of a function f_{ab}, nonnegative, differentiable on $[a, b]$, and such that for given positive numbers ϵ and M_{ab},

$$f_{ab}(x) = \begin{cases} 0 & \text{if } a \leq x \leq a + \frac{b-a}{8} \\ \epsilon + M_{ab}\left(x - \left(a + \frac{1}{4}(b - a)\right)\right) & \text{if } a + \frac{1}{4}(b - a) \leq x \leq a + \frac{3}{8}(b - a) \\ 2\epsilon + \frac{M_{ab}}{8} & \text{if } x = \frac{a+b}{2} \\ f_{ab}(a + b - x) & \text{if } \frac{a+b}{2} < x \leq b. \end{cases}$$

Thus f' is necessarily measurable and its set of points of continuity is (cf. THEOREM **2.1.1.3. 45**) a G_{δ} dense in $[0, 1]$. However

$$\int_{\{x \,:\, f'(x) > 0\}} f'(t)\, dt \geq M_{ab} \frac{b - a}{8}.$$

If $\{[a_n, b_n]\}_{n \in \mathbb{N}}$ is a sequence of pairwise disjoint intervals in $[0, 1]$, and if

$$F(x) \overset{\text{def}}{=} \begin{cases} \sum_{n=1}^{\infty} f_{a_n b_n}(x) & \text{if } x \in \bigcup_{n \in \mathbb{N}} [a_n, b_n] \\ 0 & \text{otherwise} \end{cases}$$

then F is differentiable on $[0, 1]$, F' is measurable, its points of continuity form a G_{δ} dense in $[0, 1]$, and

$$\int_{\{x \,:\, F'(x) > 0\}} F'(t)\, dt \geq \sum_{n=1}^{\infty} M_{a_n b_n} \frac{b_n - a_n}{8}.$$

If

$$M_{a_n b_n}(b_n - a_n) = 1, \ n \in \mathbb{N},$$

then the integral above is infinite: F' is not Lebesgue integrable (hence also not Riemann integrable) although $0 \leq F \leq 1 + 2\epsilon$ on $[0, 1]$.

There are differentiable functions F for which F' is not Lebesgue integrable.

The function F' fails to be Riemann integrable because it is unbounded. It fails to be Lebesgue integrable because it is badly unbounded. In fact,

$\|F'\|_\infty \geq \frac{1}{b_n - a_n}$ on $[a_n, b_n]$. A function h can fail to be Riemann integrable on $[0, 1]$ either because h is unbounded or because the set $\text{Discont}(h)$ of its discontinuities is not a null set. A function k fails to be Lebesgue integrable on $[0, 1]$ because k is not measurable or because the sets where $|k|$ is large do not have sufficiently small measures.

For example, if

$$g(x) = \begin{cases} x^{-\frac{1}{2}} & \text{if } x \neq 0 \\ 0 & \text{otherwise} \end{cases}$$

then g is not Riemann integrable on $[0, 1]$ because g is unbounded on $[0, 1]$. Although g is unbounded on $[0, 1]$, g is Lebesgue integrable on $[0, 1]$ and its Lebesgue integral

$$\int_{[0,1]} g(x)\, dx = 2.$$

In light of the *Fundamental Theorem of Calculus* (FTC), which relates differentiation and integration, it is of interest to note that mild relaxations of the hypotheses lead to invalidation of the conclusion. Among the numerous versions of the FTC is the following [**O3**].

If f is Riemann integrable on every compact subinterval of an interval I, if F is continuous on I and $F' = f$ on the interior I° of I, and if a and b are in I then

$$\int_a^b f(x)\, dx = F(b) - F(a).$$

In particular if f is continuous and

$$F(x) = \int_a^x f(t)\, dt, \ a \in I, \ x \in I^\circ,$$

then F' exists on I° and $F' = f$ on I°.

If $S \subset \mathbb{R}$, $f \in \mathbb{R}^S$, and if $F' = f$ then F is (on S) a *primitive* of f. It should be noted that f is *Riemann integrable* on $[a, b]$ iff f is bounded on $[a, b]$ and $\text{Discont}(f)$ is a null set [**Gof**].

Exercise 2.1.2.10. Prove: If $\{r_n\}_{n \in \mathbb{N}} = \mathbb{Q}$, $a < b$, and $n \in \mathbb{N}$, then $\chi_{\{r_1, \dots, r_n\}}$ is, but $\lim_{n \to \infty} \chi_{\{r_1 \dots, r_n\}}$ $(= \chi_\mathbb{Q})$ is not, Riemann integrable on $[a, b]$.

Exercise 2.1.2.11. Show that if $f \in \mathbb{R}^\mathbb{R}$ and

$$f(x) \overset{\text{def}}{=} \text{sgn}(x) = \begin{cases} x|x|^{-1} & \text{if } x \neq 0 \\ 0 & \text{if } x = 0 \end{cases}$$

(the *signum function*) then f is Riemann integrable on $I \overset{\text{def}}{=} [-1, 1]$ but has no primitive on I.

[*Hint:* Show that sgn does not enjoy the intermediate value property on I.]

Exercise 2.1.2.12. Show that if f is monotone on $[0,1]$ and

$$\text{Discont}(f) = \mathbb{Q} \cap [0,1]$$

(cf. **Exercise 2.1.1.14. 49**) then f is Riemann integrable on $[0,1]$ but has no primitive on any subinterval of $[0,1]$.

Exercise 2.1.2.13. Show that for f in **Exercise 2.1.2.1** *ii*. **53** there is no constant K and no function g, Riemann integrable on $[-1,1]$ and such that $f(x) = K + \int_{-1}^{x} g(t)\,dt$ on $[-1,1]$.
   [*Hint:* Note that f' is unbounded on $[-1,1]$.]

Exercise 2.1.2.14. Show that for f in **Exercise 2.1.1.13. 49**, if g is given by

$$g(x) \stackrel{\text{def}}{=} \int_0^x f(t)\,dt$$

then:

   i. g' exists everywhere and $g' = f$ on $\mathbb{R} \setminus \mathbb{Q}$;
   ii. g is not a primitive of f.

Exercise 2.1.2.15. Show that f in $\mathbb{C}^{[0,1]}$ is *absolutely continuous*: $f \in AC\left([0,1],\mathbb{C}\right)$, if *i–iii* below obtain:

   i. f is continuous: $f \in C\left([0,1],\mathbb{R}\right)$;
   ii. f is of bounded variation: $f \in BV\left([0,1],\mathbb{R}\right)$;
   iii. $f(E)$ is a null set in $\mathbb{R}$ whenever E is a null set in $[0,1]$.

Example 2.1.2.6. If, in **Exercise 2.1.2.15**, any one of i - iii fails to obtain then f is not absolutely continuous, i.e., the satisfaction of i -iii is necessary and sufficient for the absolute continuity of f [**HeSt, Rud**]. Since an absolutely continuous function is continuous and of bounded variation, it suffices to remark that if f is the Cantor function $\mathcal{C}_0$ then f is continuous, monotone, and $\lambda\left[f\left(\mathcal{C}_0\right)\right] = 1$, whence it satisfies i and ii but not iii. Since

$$0 < x \le 1 \Rightarrow \int_0^x f'(t)\,dt = 0 < f(x) - f(0)$$

the Cantor function is not absolutely continuous.

Exercise 2.1.2.16. Show that if $f \in C\left([0,1],\mathbb{C}\right) \cap BV\left([0,1],\mathbb{C}\right)$, if T_f is the *total variation* of f on $[0,1]$, and if $W < T_f$ there is a positive number a such that if P is a *partition* of $[0,1]$ and $|P| < a$ then $T_{fP} > W$.

Example 2.1.2.7. If f is of bounded variation and is not continuous the preceding conclusion can fail to obtain. Indeed if $f = \chi_{[\frac{1}{2}, 1]}$, then f is not continuous, $T_f = 1$, but if

$$P_n = \{[\frac{k}{2^n}, \frac{(k+1)}{2^n}), \ 0 \le k \le 2^n - 1\} \ n \in \mathbb{N}$$

then $|P_n| = 2^{-n}$, and $T_{fP_n} = 0$.

Exercise 2.1.2.17. Show that $f : [0,1] \ni x \mapsto \sqrt{x}$ is *absolutely continuous* on $[0, 1]$.

[*Hint:* If

$$0 < a_i < b_i \le a_{i+1} < b_{i+1} \le 1, \ 1 \le i \le n$$

$$\sum_{i=1}^{n+1} b_i - a_i < \delta$$

then, since the derivative of f is bounded on $[a_1, 1]$, f is absolutely continuous on $[a_1, 1]$.]

Example 2.1.2.8. Divide $[0,1]$ by the partition points 2^{-k}, $k \in \mathbb{N}$, and thereby create open intervals $I_k \stackrel{\text{def}}{=} (2^{-k}, 2^{-k+1})$, $k \in \mathbb{N}$. Divide the interval I_k into 2^k equal subintervals I_{km}, $1 \le m \le 2^k$. On I_k let g be piecewise linear, nonnegative, and, starting at one end of I_k, let g take on alternately the values 0 and 2^{-2k} at the endpoints of the intervals I_{km}.

If $g(0) \stackrel{\text{def}}{=} 0$ then $g \in AC$ on $[0,1]$ since g' exists a.e. and if $x \in [0,1]$ then $g(x) = \int_0^x g'(t) \, dt$. However, for example, if f is $x \mapsto \sqrt{x}$ (cf. **Exercise 2.1.2.17**) then $h \stackrel{\text{def}}{=} f \circ g$ is not absolutely continuous because, for k in $\mathbb{N}$, the total variation of h is $2^k \cdot 2^{-k} = 1$ on the (small) interval I_k.

The composition of two absolutely continuous functions can fail to be absolutely continuous.

2.1.3. Convergence of sequences and series

If A is a nonempty subset of a *separable metric space* (X, d), e.g., $A \subset \mathbb{R}$, there is in A a sequence $\{a_1, a_2, \ldots\}$ such that each point of A is limit of some subsequence of a. (In particular every subset of a separable metric space is a separable metric space.) [PROOF: If A is finite, say $A = \{x_1, \ldots, x_n\}$, then

$$a = \{x_1, \ldots, x_n, x_1, \ldots, x_n, \ldots\}.$$

If A is infinite, for each n in $\mathbb{N}$ let $\mathcal{B}_n$ be the set

$$\left\{ U\left(a, \frac{1}{n}\right) \ : \ U\left(a, \frac{1}{n}\right) \stackrel{\text{def}}{=} \left\{ x \ : \ x \in X, \ d(a, x) < \frac{1}{n}, \ a \in A \right\} \right\}$$

of all *open balls*, each centered at some point a of A and of radius $\frac{1}{n}$. Since X is a separable metric space there is in X a countable dense subset $C \overset{\text{def}}{=} \{ c_m : m \in \mathbb{N} \}$. If $B \in \mathcal{B}_n$ let p be a point in $B \cap C$. The set P of all such p for arbitrary choices of B and n is a subset of C and hence P is countable. If $B \in \mathcal{B}_n$ and $p \in P \cap B$ there is in A a point q_p such that $d(p, q_p) < \frac{2}{n}$. The set $Q \overset{\text{def}}{=} \{ q_p : p \in P \}$ is countable, and, owing to its construction, $Q' \supset A$. $\square$]

Exercise 2.1.3.1. Expand the PROOF above to show that if A is a nonempty closed subset of X there is in A a (nonempty!) countable set $\{a_n\}_{n \in \mathbb{N}}$ such that A is precisely the set of limit points of $\{a_n\}_{n \in \mathbb{N}}$.

Exercise 2.1.3.2. Show that if

$$s_n \overset{\text{def}}{=} \sum_{k=1}^{n} \frac{1}{k}, n \in \mathbb{N}$$

then $\lim_{n \to \infty} |s_{n+p} - s_n| = 0$, $p \in \mathbb{N}$.

[**Remark 2.1.3.1:** Nevertheless the sequence $\{s_n\}_{n \in \mathbb{N}}$ is *not* a Cauchy sequence.]

Exercise 2.1.3.3. Assume $\{\nu(n)\}_{n \in \mathbb{N}}$ is a given strictly increasing subsequence of $\mathbb{N}$. Define a sequence $\{a_n\}_{n \in \mathbb{N}}$ such that $\lim_{n \to \infty} a_n$ does not exist and $\lim_{n \to \infty} |a_{\nu(n)} - a_n| = 0$.

[*Hint:* If $S \overset{\text{def}}{=} \{\nu(n) - n\}_{n \in \mathbb{N}}$ is bounded use the idea in **Exercise 2.1.3.2.** If S is unbounded let k be the least natural number such that $\nu(k) > k$. Show that the goal is achieved if

$$a_n = \begin{cases} 1 & \text{if } n \in \{k, \nu(k), \nu[\nu(k)], \dots, \} \\ 0 & \text{otherwise.} \end{cases}]$$

Exercise 2.1.3.4. In **Exercise 2.1.3.3** replace the assumption that ν is strictly increasing by the weaker assumption $\nu(n) \to \infty$ as $n \to \infty$ and impose the requirement that $\{a_n\}$ be unbounded. Define a sequence $\{a_n\}_{n \in \mathbb{N}}$ such that $\lim_{n \to \infty} a_n$ does not exist and $\lim_{n \to \infty} |a_{\nu(n)} - a_n| = 0$.

Exercise 2.1.3.5. Show that the inequalities

$$\liminf_{n \to \infty} a_n + \liminf_{n \to \infty} b_n \le \liminf_{n \to \infty} (a_n + b_n)$$

$$\liminf_{n \to \infty} (a_n + b_n) \le \liminf_{n \to \infty} a_n + \limsup_{n \to \infty} b_n$$

$$\liminf_{n \to \infty} a_n + \limsup_{n \to \infty} b_n \le \limsup_{n \to \infty} (a_n + b_n)$$

$$\limsup_{n \to \infty} (a_n + b_n) \le \limsup_{n \to \infty} a_n + \limsup_{n \to \infty} b_n$$

obtain in sharper form with each "$\leq$" replaced by "$<$" if

$$\{a_n\} = \{0, 1, 2, 1, 0, 1, 2, 1, 0, \ldots\}$$
$$\{b_n\} = \{2, 1, 1, 0, 2, 1, 1, 0, \ldots\}.$$

Show also that if

$$a_{mn} = \begin{cases} 1 & \text{if } m = n \\ 0 & \text{otherwise} \end{cases}$$

then $\limsup_{n \to \infty}(a_{1n} + \cdots + a_{mn} + \cdots) > \sum_{m=1}^{\infty} \limsup_{n \to \infty} a_{mn}$.
    Show that if

$$a_n = \begin{cases} 1 & \text{if } n \text{ is odd} \\ 0 & \text{if } n \text{ is even} \end{cases}$$

$$b_n = \begin{cases} \frac{n}{n+1} & \text{if } n \text{ is odd} \\ -\frac{1}{n+1} & \text{if } n \text{ is even} \end{cases}$$

then, although $a_n \geq b_n, n \in \mathbb{N}$,

$$\liminf_{n \to \infty} a_n < \limsup_{n \to \infty} b_n.$$

    Show that if $\{A_n\}_{n \in \mathbb{N}}$ and $\{B_n\}_{n \in \mathbb{N}}$ are two sequences of subsets of a set S then

$$\liminf_{n \to \infty}(A_n \cap B_n) = \liminf_{n \to \infty} A_n \cap \liminf_{n \to \infty} B_n$$
$$\limsup_{n \to \infty}(A_n \cup B_n) = \limsup_{n \to \infty} A_n \cup \limsup_{n \to \infty} B_n.$$

    Exercise 2.1.3.6. Let A_n be the set of irrational numbers in $\left(-\frac{1}{n}, \frac{1}{n}\right)$: $A_n = \mathbb{I} \cap \left(-\frac{1}{n}, \frac{1}{n}\right)$. Show that $A_{n+1} \subset A_n$ and

$$\bigcap_{n=1}^{\infty} A_n = \emptyset$$

even though $\#(A_n) = \#(\mathbb{R}), \ n \in \mathbb{N}$.
    Let $A \stackrel{\text{def}}{=} \{a_n\}_{n \in \mathbb{N}}$ be a sequence of *vectors* in a vector space V endowed with an *inner product* (denoted $(\ , \ )$) or merely with a *norm* $\| \ \|$. For example, if V is the set $C([0,1])$ of continuous $\mathbb{C}$-valued functions defined on $[0, 1]$ and if $f \in C([0,1])$ then $\|f\| \stackrel{\text{def}}{=} \sup\{ |f(x)| \ : \ x \in [0, 1] \}$. On the other hand, V may be $\mathbb{R}$ or $\mathbb{C}$ in which case a vector is simply a real or complex number. The symbol $\sum_{k=1}^{\infty} a_n$ is a priori the formal designation of an object to be studied. If the *sequence* $\{s_m \stackrel{\text{def}}{=} \sum_{n=1}^{m} a_n\}_{m \in \mathbb{N}}$ converges to a vector s, i.e., if $\lim_{m \to \infty} \|s_m - s\| = 0$, then $\sum_{n=1}^{\infty} a_n$ denotes s as well. If

$V = \mathbb{R}$ the vectorial notation (boldface) $\mathbf{a}_n$, $\mathbf{s}$, etc., is dropped in favor of a_n, s, etc., and the norm $\| \ \|$ is dropped in favor of absolute value $| \ |$. Let:

> α denote some strictly increasing subsequence $\{n_k\}_{k \in \mathbb{N}}$ of $\mathbb{N}$;
>
> ϕ denote some finite subset of $\mathbb{N}$;
>
> Φ denote the set of all ϕ;
>
> $\mathbf{S}_\beta(A)$ denote $\sum_{n \in \beta} \mathbf{a}_n$;
>
> $\Gamma(A)$ denote $\{ \mathbf{S}_\phi(A) \ : \ \phi \in \Phi \}$
>
> Π denote the set of all permutations of $\mathbb{N}$.   (2.1.3.1)

Exercise 2.1.3.7. Show that if $\alpha = \mathbb{N}$ and $V = \mathbb{R}$ then the series $S_\alpha(A) \overset{\text{def}}{=} S(A)$, i.e., $\sum_{n=1}^\infty a_n$, is *absolutely convergent*, i.e.,

$$S_{|\ |}(A) \overset{\text{def}}{=} \sum_{n=1}^\infty |a_n| < \infty,$$

iff any one (and hence each) of the following obtains:

i. for each *permutation (bijection)* $\pi : \mathbb{N} \ni n \mapsto \pi(n) \in \mathbb{N}$ of $\mathbb{N}$ the series

$$S_\pi(A) \overset{\text{def}}{=} \sum_{n=1}^\infty a_{\pi(n)}$$

converges to a (real) number independent of π;

ii. $\Gamma(A)$ is a bounded subset of $\mathbb{R}$;

iii. for each $\mathsf{E} \overset{\text{def}}{=} \{\epsilon_n\}_{n \in \mathbb{N}}, \ \epsilon_n = \pm 1$, the series $S_\mathsf{E}(A) \overset{\text{def}}{=} \sum_{n=1}^\infty \epsilon_n a_n$ converges.

If $S_{|\ |}(A) = \infty$ and $S(A)$ converges ($S(A)$ is *conditionally convergent*), e.g., if $a_n = (-1)^n \frac{1}{n}, \ n \in \mathbb{N}$, then $\Gamma(A)$ is unbounded *both* above and below.

Exercise 2.1.3.8. Show that if $S(A)$ is conditionally convergent then:

i. for each closed interval $J \overset{\text{def}}{=} [p, q]$ there is in Π a permutation π such that the set $\Sigma_\pi(A)$ of partial sums $s_n(\pi)(A) \overset{\text{def}}{=} \sum_{k=1}^n a_{\pi(k)}, n \in \mathbb{N}$, is such that the set $\Sigma_\pi(A)'$ of *limit points* of $\Sigma_\pi(A)$ is J;

ii. for each closed interval J there is a sequence $\mathsf{E} \overset{\text{def}}{=} \{\epsilon_n\}_{n \in \mathbb{N}}, \ \epsilon_n = \pm 1$, such that $S_\mathsf{E}(A)' = J$.

Note the special cases: $p = q$ and $p = -\infty, \ q = \infty$. The latter corresponds for i to the *Riemann derangement theorem*:

IF $S(A)$ IS CONDITIONALLY CONVERGENT, AND IF $x \in \mathbb{R}$ THERE
IS IN Π A π SUCH THAT $x = S_\pi(A)$.

[*Hint:* The series formed of the positive terms and the series
formed of the negative terms are *divergent* and $\lim_{n\to\infty} a_n = 0$.
If $-\infty < p < q < \infty$ choose π so that the terms in a *sub*sequence
of the partial sums in S_π are alternately below p and above q. If
$\epsilon > 0$ there is in $\mathbb{N}$ an $n(\epsilon)$ such that if $n > n(\epsilon)$ and $s_n(\pi)(A) < p$
and $s_{n+k}(\pi)(A) > q$ then $|s_{n+j+1}(\pi)(A) - s_{n+j}(\pi)(A)| < \epsilon$, $j =$
$0, 1, \ldots, k - 1$. If $-\infty = p < q < \infty$ choose π so that for n in
$\mathbb{N}$, the partial sums are alternately below $-n$ and above q. Similar constructions serve for the circumstances $-\infty < p < q = \infty$
and $-\infty = p < q = \infty$. The same kinds of arguments apply for
$S_\mathsf{E}(A)$.]

Let $Z \overset{\text{def}}{=} \{\mathbf{z}_n \overset{\text{def}}{=} (z_{n1}, \ldots, z_{nr})\}_{n\in\mathbb{N}}$ be a sequence of vectors in $V \overset{\text{def}}{=}$
$\mathbb{R}^r$.

In [**St**] Steinitz proved the following generalization of the Riemann
derangement theorem.

THEOREM **2.1.3.1.** IF $\mathbf{S}(Z)$ IS A SERIES OF VECTORS IN A FINITE-
DIMENSIONAL VECTOR SPACE V THEN THE SET

$$\mathbf{S}_\Pi(Z) \overset{\text{def}}{=} \{\, \mathbf{S}_\pi(Z) \; : \; \pi \in \Pi \,\}$$

OF ALL CONVERGENT SUMS OF REARRANGEMENTS OF $\mathbf{S}(Z)$ IS EMPTY OR
THERE IS A SUBSPACE M AND A VECTOR $\mathbf{x}$ SUCH THAT

$$\mathbf{S}_\Pi(Z) = \mathbf{x} + M.$$

[**Note 2.1.3.1:**   If $M = \{\mathbf{O}\}$ then $\mathbf{S}_\Pi = \mathbf{x}$.]

[**Remark 2.1.3.2:**   The Steinitz paper, published in 1913 but
written in 1906, provides a beautiful but somewhat old-fashioned
introduction to the subject of vector spaces. In [**Rosn**] there is a
simplified proof of the Steinitz theorem.]

Sierpinski [**Si1**] proved the following version of the Riemann derangement theorem.

THEOREM **2.1.3.2.**   IF $\sum_{n=1}^{\infty} a_n$ IS A CONDITIONALLY CONVERGENT
SERIES IN WHICH THE TERMS ARE REAL AND IF $s' \le s \overset{\text{def}}{=} \sum_{n=1}^{\infty} a_n$ THEN
THERE IS IN Π A π SUCH THAT $a_n < 0 \Rightarrow a_{\pi(n)} = a_n$ AND $\sum_{n=1}^{\infty} a_{\pi(n)} = s'$.

[**Remark 2.1.3.3:**   In [**Si1**] the author refers to an earlier paper
he wrote in his native language, Polish. In that paper he showed

that if a series of real terms is conditionally convergent and if s is an arbitrary real number there is a rearrangement of the series in which the positions of the $+$ and $-$ signs are those of the original series and yet the sum of the series is s.]

Exercise 2.1.3.9. Let $\{\mathbf{z}_n \overset{\text{def}}{=} (c_n, d_n)\}_{n \in \mathbb{N}}$ be a sequence of vectors in $\mathbb{R}^2$. Use THEOREM **2.1.3.1. 70** as needed to show:

i. if $\mathbf{S}_{\|\ \|} \overset{\text{def}}{=} \sum_{n=1}^{\infty} \|\mathbf{z}_n\| = \infty$ then $\sum_{n=1}^{\infty} |c_n| = \infty$ or $\sum_{n=1}^{\infty} |d_n| = \infty$;

ii. if $\mathbf{S}_{\|\ \|} = \infty$, if $\mathbf{S} \overset{\text{def}}{=} \sum_{n=1}^{\infty} \mathbf{z}_n$ converges (in $\mathbb{R}^2$) ($\mathbf{S}$ is *conditionally convergent*), and if $\sum_{n=1}^{\infty} |c_n| < \infty$ then $\{\, \mathbf{S}_\pi \ : \ \pi \in \Pi, \ \mathbf{S}_\pi \text{ converges} \,\}$ is the set $\sum_{n=1}^{\infty} c_n + i\mathbb{R}$;

iii. $(\mathbf{x}, \mathbf{y})$ denoting the inner product of the vectors $\mathbf{x}$ and $\mathbf{y}$, if

    a) $\mathbf{x} \overset{\text{def}}{=} (x_1, x_2)$ and $\mathbf{y} \overset{\text{def}}{=} (y_1, y_2)$ are *linearly independent*,

    b) $(\mathbf{x}, \mathbf{S}_\pi)$ converges for every π in Π,

    c) $(\mathbf{y}, \mathbf{S})$ is conditionally convergent,

then $(\mathbf{x}, \mathbf{S}_\pi)$ is a constant K independent of π, each (convergent) $\mathbf{S}_\pi$ is on the line:

$$L \overset{\text{def}}{=} \{\, \mathbf{z} \ : \ (\mathbf{x}, \mathbf{z}) = K \,\},$$

and for each λ in $\mathbb{R}$ there is in Π a π such that $\mathbf{S}_\pi$ converges and is on the line:

$$L_\lambda \overset{\text{def}}{=} \{\, \mathbf{w} \ : \ (\mathbf{w}, \mathbf{y}) = \lambda, \ \lambda \in \mathbb{R} \,\},$$

i.e., $\mathbf{S}_\Pi = L$.

[*Hint:* Show that M in THEOREM **2.1.3.1. 70** is one-dimensional.]

If $a_k = 2^{-k}$, $k \in \mathbb{N}$, then $a_n \le \sum_{k=n+1}^{\infty} a_k$, $n \in \mathbb{N}$, and if $x \in (0, 1]$ there is in $\mathbb{N}$ a subset A_x such that $\sum_{n \in A_x} a_n = x$.

Exercise 2.1.3.10. a) Show that if $0 < a_n \le \sum_{k=n+1}^{\infty} a_k < \infty$ then for every x in $\left(0, \sum_{n=1}^{\infty} a_n \overset{\text{def}}{=} S\right]$ there is in $\mathbb{N}$ a subset A_x for which $\sum_{n \in A_x} a_n = x$. b) Show that if $\infty > S = \sum_{n=1}^{\infty} a_n > \sum_{n=1}^{N} a_n > 0$, $N \in \mathbb{N}$, and for every x in $\left(0, \sum_{n=1}^{\infty} a_n \overset{\text{def}}{=} S\right]$ there is in $\mathbb{N}$ a subset A_x such that $x = \sum_{n \in A_x} a_n$, then for some permutation π of $\mathbb{N}$, $a_{\pi(n)} \le \sum_{k=n+1}^{\infty} a_{\pi(k)}$.

[*Hint:* ad a) By induction define a sequence $\{n_k\}$ so that each n_k is the smallest n such that $a_{n_1} + \cdots + a_{n_{k-1}} + a_n \le x$. ad b) Consider a permutation π such that $a_{\pi(n)} \overset{\text{def}}{=} b_n \downarrow 0$ and show that for each n in $\mathbb{N}$,

$$b_1 + \cdots + b_{n+1} \le \frac{1}{2}\left(S + b_1 + \cdots + b_n\right).]$$

The classical *Cauchy criterion* for convergence of a series $\mathbf{S}(A)$ of vectors in a vector space V complete with respect to the norm $\|\ \|$ is, because it is a criterion, not always simple to apply. Hence, in the study of series in which the terms are constants (real or complex) or not necessarily constant functions, there has evolved a collection of useful tests for convergence. Among the tests most frequently applied is the *comparison test* in the following form:

If $S(A) \stackrel{\text{def}}{=} \sum_{n=1}^{\infty} a_n$ and $S(B) \stackrel{\text{def}}{=} \sum_{n=1}^{\infty} b_n$ are series in which the terms are nonnegative and if $S(A)$ *dominates* $S(B)$, i.e., if $a_n \geq b_n$, then the convergence of $S(A)$ implies the convergence of $S(B)$ (whence the divergence of $S(B)$ implies the divergence of $S(A)$).

For the applicability of comparison test the condition that the terms of the series under study be nonnegative is crucial. If the terms are not necessarily nonnegative, i.e., if some may be negative, the comparison test may fail to be decisive. A series $S(A)$ is said to *dominate* $S(B)$ *absolutely* if $a_n \geq |b_n|$, $n \in \mathbb{N}$, (whence the terms of $S(A)$ are nonnegative, although the terms of $S(B)$ may be positive, zero, or negative).

Exercise 2.1.3.11. Give an example of a divergent series $\sum_{n=1}^{\infty} a_n$ such that $\lim_{n \to \infty} a_n = 0$.

Exercise 2.1.3.12. Give an example of a convergent series $S(A)$ and a divergent series $S(B)$ such that $a_n \geq b_n$.

Exercise 2.1.3.13. Give an example of a convergent series $S(A)$ and a divergent series $S(B)$ for which $|a_n| \geq |b_n|$, $n \in \mathbb{N}$.

[*Hint:* In the last two **Exercises** the solution must involve series in which some terms are negative.]

A series $S(A)$ of positive terms would provide a *universal comparison test* for all series with positive terms if $S(A)$ dominated absolutely every convergent series with positive terms, and $S(A)$ were dominated absolutely by every divergent series with positive terms. However if $S(A)$ diverges then $S(A)$ is not dominated absolutely by the divergent series $\frac{1}{2}S(A)$: $S(A)$ fails to test $\frac{1}{2}S(A)$ by comparison. If $S(A)$ converges and then $S(A)$ does not dominate absolutely the convergent series $2S(A)$: $S(A)$ fails to test $2S(A)$ by comparison.

There is no single series $S(A)$ that can serve as a universal comparison test.

As THEOREM **2.1.3.3** below shows, even more is true.

THEOREM **2.1.3.3.** LET $\mathsf{S} \stackrel{\text{def}}{=} \{\sum_{m=1}^{\infty} a_{mn}\}_{n \in \mathbb{N}} \stackrel{\text{def}}{=} \{S(A_n)\}_{n \in \mathbb{N}}$ BE A SEQUENCE OF CONVERGENT SERIES OF POSITIVE TERMS. THEN THERE IS A CONVERGENT SERIES $S(A)$ OF POSITIVE TERMS THAT IS NOT DOMINATED BY ANY MEMBER OF S.

[**Remark 2.1.3.4:**   The result above may be paraphrased briefly by saying that *there is no universal comparison sequence of positive convergent series.*]

PROOF.   If $S_M(A_n) \overset{\text{def}}{=} \sum_{m=1}^{M} a_{mn}$ and $R_M(A_n) = S(A_n) - S_M(A_n)$ there are natural numbers $M(n)$ such that $1 \le M(1) < M(2) < \cdots$ and

$$\max(R_{M(n)}(A_1), R_{M(n)}(A_2), \ldots, R_{M(n)}(A_n)) < 2^{-n}, \ n \in \mathbb{N}.$$

Let a_m be

$$\begin{cases} 2a_{m1} & \text{if } 1 \le m \le M(2) \\ (k+1)\max(a_{m1}, \ldots, a_{mk}) & \text{if } M(k) < m \le M(k+1), \ k > 1, \ k \in \mathbb{N}. \end{cases}$$

Then $S(A) \overset{\text{def}}{=} \sum_{m=1}^{\infty} a_m$ converges since

$$\sum_{m=M(k)+1}^{M(k+1)} a_m \le \sum_{m=M(k)+1}^{M(k+1)} \left[(k+1)\sum_{n=1}^{k} a_{mn} \right]$$

$$\le (k+1)\sum_{n=1}^{k} \left[\sum_{m=M(k)+1}^{M(k+1)} a_{mn} \right]$$

$$\le (k+1)\sum_{n=1}^{k} R_{M(k)}(A_n)$$

$$\le (k+1)\sum_{n=1}^{k} 2^{-k} < (k+1)^2 2^{-k}.$$

Hence

$$S(A) \overset{\text{def}}{=} \sum_{m=1}^{\infty} a_m = \sum_{m=1}^{M(2)} a_m + \sum_{k=2}^{\infty} \left(\sum_{m=M(k)+1}^{M(k+1)} a_m \right)$$

$$\le S_{M(2)}(A_1) + \sum_{k=2}^{\infty} (k+1)^2 2^{-k} < \infty.$$

Since $a_m/a_{mn} \ge (k+1)$ if $k \ge n$ and $m > M(k)$ it follows that the convergent series $S(A)$ is not dominated by any $KS(A_n)$, $K \in \mathbb{R}$. $\square$

The idea of the preceding proof can be used to show that *there is no sequence of positive divergent series that serves as a universal comparison series sequence for divergence.*

Even when a series diverges some generalized averaging method might lead to a "reasonable" value to assign as the sum of the series. Such a generalized averaging method is often termed a *summability method* that is used to *sum* the series.

Example 2.1.3.1. If $S(A) = \sum_{n=1}^{\infty}(-1)^{n+1}$ then

$$s_n(A) = \begin{cases} 1 & \text{if } n \text{ is odd} \\ 0 & \text{if } n \text{ is even.} \end{cases}$$

It follows that the *average*

$$\sigma_n(A) \overset{\text{def}}{=} \frac{s_1(A) + \cdots + s_n(A)}{n} = \begin{cases} \frac{n+1}{2n} & \text{if } n \text{ is odd} \\ \frac{1}{2} & \text{if } n \text{ is even.} \end{cases}$$

Hence $\lim_{n \to \infty} \sigma_n(A) = \frac{1}{2}$, which is regarded as an acceptable "value" of the divergent series $S(A)$.

Exercise 2.1.3.14. Let f be Lebesgue integrable on $[-\pi, \pi]$, and let the nth Fourier coefficient of f be

$$c_n \overset{\text{def}}{=} \frac{1}{\sqrt{2\pi}} \int_{-\pi}^{\pi} f(x)e^{-inx}\, dx, \ n \in \mathbb{Z}.$$

Let $S_f(x)$ be the (formal) Fourier series

$$\sum_{n=-\infty}^{\infty} c_n \frac{e^{inx}}{\sqrt{2\pi}}$$

and let $s_N(x)$ be $\sum_{n=-N}^{N} c_n \frac{e^{inx}}{\sqrt{2\pi}}$. Show that the average

$$\sigma_N(x) \overset{\text{def}}{=} \frac{1}{N+1} \sum_{n=0}^{N} s_n(x)$$
$$= \int_{-\pi}^{\pi} \frac{1}{2\pi} \frac{\left[\sin \frac{1}{2}(N+1)(x-y)/\sin \frac{1}{2}(x-y)\right]^2}{N+1} f(y)\, dy.$$

The function

$$F_N(x, y) \overset{\text{def}}{=} \frac{1}{2\pi} \frac{\left[\sin \frac{1}{2}(N+1)(x-y)/\sin \frac{1}{2}(x-y)\right]^2}{N+1}$$

is *Fejér's kernel* and the integral in the second line of the display above is the *convolution*, denoted $F_N * f$, of F_N and f.

In discussing convergence resp. uniform convergence it is helpful to use the notation $f_n \to g$ resp. $f_n \overset{u}{\to} g$ to signify that the sequence $\{f_n\}_{n \in \mathbb{N}}$ converges resp. converges uniformly to g as $n \to \infty$.

Among the properties of Fejér's kernel are [Zy]:

i. $F_N \geq 0$;

ii. $\int_{-\pi}^{\pi} F_N(x)\, dx = 1$;

iii. if $0 < \epsilon < \pi$ then $F_N \overset{u}{\to} 0$ in $[-\pi, \pi] \setminus (-\epsilon, \epsilon)$.

Exercise 2.1.3.15. Show:

i. the validity of *i* - *iii* above for F_N;

ii. that if f is continuous on $[-\pi, \pi]$ and $f(-\pi) = f(\pi)$ then

$$|f - F_N * f| \overset{u}{\to} 0$$

(Fejér's theorem);

iii. that if f is Lebesgue integrable on $[-\pi, \pi]$ then

$$\|f - f_N * f\|_1 \overset{\text{def}}{=} \int_{-\pi}^{\pi} |f(x) - F_N * f(x)|\, dx \to 0.$$

[*Hint:* For *iii* use *ii* and the fact that if $\delta > 0$ there is a continuous function g such that $\|f - g\|_1 < \delta$.]

More generally let $T \overset{\text{def}}{=} \{t_{mn}\}_{m,n=1}^{\infty}$ be a (*Toeplitz*) matrix in which each entry is real and for which:

i. there is an M such that $\sum_{n=1}^{\infty} |t_{mn}| \leq M$, $m \in \mathbb{N}$;

ii. $\lim_{m \to \infty} t_{mn} = 0$, $n \in \mathbb{N}$;

iii. $\lim_{m \to \infty} \sum_{n=1}^{\infty} t_{mn} = 1$.

If $\sigma_{m,T}(A) \overset{\text{def}}{=} \sum_{n=1}^{\infty} t_{mn} s_n(A)$ converges for m in $\mathbb{N}$ and if

$$\lim_{m \to \infty} \sigma_{m,T}(A) \overset{\text{def}}{=} \sigma_T(A)$$

exists then $\sigma_T(A)$ is the *T-sum* of $S(A)$. Thus if

$$t_{mn} = \begin{cases} \frac{1}{m} & \text{if } 1 \leq n \leq m \\ 0 & \text{otherwise} \end{cases}$$

then the *T*-sum of $\sum_{n=1}^{\infty}(-1)^{n+1}$ is $\frac{1}{2}$, i.e., *T* sums $S(A)$.

A matrix T is a Toeplitz matrix iff whenever $S(A)$ converges then its *T*-sum is also $S(A)$: $\sigma_T(A) = S(A)$ [**To, Wi**].

Exercise 2.1.3.16. Show that if

$$t_{mn} = \begin{cases} \frac{1}{m} & \text{if } 1 \le n \le m \\ 0 & \text{otherwise} \end{cases}$$

then $T \overset{\text{def}}{=} (t_{mn})_{m,n=1}^{\infty,\infty}$ is a Toeplitz matrix and corresponds to the simple averaging procedure described above, cf. (2.1.3.2), page 79.

There are two large classes of Toeplitz matrices, those derived from *Cesàro summation*, denoted (C, α), and those derived from *Abel summation*. Details about the following statements are discussed in [**Zy**].

i. If

$$\alpha > -1,\ 0 < x < 1,\ A = \{a_n\}_{n=0}^{\infty}$$

$$s_n = \sum_{k=0}^{n} a_k,\ n \in \mathbb{N}$$

$$|s_n| \le M < \infty,\ n \in \mathbb{N}$$

$$S_{x,\alpha}(A) \overset{\text{def}}{=} \frac{\sum_{n=0}^{\infty} a_n x^n}{(1-x)^{\alpha+1}}$$

$$\left(\overset{\text{def}}{=} \frac{\sum_{n=0}^{\infty} s_n^{\alpha} x^n}{(1-x)^{\alpha}} \right)$$

$$\overset{\text{def}}{=} \sum_{n=0}^{\infty} A_n^{\alpha} x^n$$

and if

$$\lim_{n \to \infty} \frac{s_n^{\alpha}}{A_n^{\alpha}} \overset{\text{def}}{=} s_{(C,\alpha)}(A)$$

exists then $S(A)$ is said to be (C, α)-*summable* to $s_{(C,\alpha)}(A)$.

ii. If

$$\lim_{x \to 1} S_{x,0}(A) \overset{\text{def}}{=} s_{Abel}(A)$$

exists then $S(A)$ is *Abel-summable* to $s_{Abel}(A)$.

iii. If $-1 < \alpha < \beta$

    a) and $s_{(C,\alpha)}(A)$ exists then $S(A)$ is (C, β)-summable ($s_{(C,\beta)}(A)$ exists) and $s_{(C,\beta)}(A) = s_{(C,\alpha)}(A)$;

    b) there is an A such that $S(A)$ is (C, β)-summable but is not (C, α)-summable;

    c) and if $S(A)$ is (C, α)-summable it is Abel-summable;

    d) there is an A such that, for each α in $(-1, \infty)$, $S(A)$ is not (C, α)-summable but is Abel-summable.

iv. For each α in $(-1, \infty)$ there is a Toeplitz matrix T_α such that

$$\lim_{m \to \infty} \sigma_{m,T_\alpha}(A) = s_{(C,\alpha)}(A)$$

whenever $s_{(C,\alpha)}(A)$ exists.

$v.$ If $x_n \uparrow 1$ and $\mathbf{x} \overset{\text{def}}{=} \{x_n\}_{n \in \mathbb{N}}$ there is a Toeplitz matrix $T_\mathbf{x}$ such that

$$\lim_{n \to \infty} \sigma_{m, T_\mathbf{x}}(A) = s_{Abel}(A)$$

whenever $s_{Abel}(A)$ exists.

Example 2.1.3.1. 74 shows there are divergent series that can be summed by some Toeplitz matrices.

Exercise 2.1.3.17. Show that the (infinite) identity matrix $I \overset{\text{def}}{=}$ $(\delta_{ij})_{i,j=1}^{\infty}$ is a Toeplitz matrix that sums a series $S(A)$ iff $S(A)$ converges.
    There is no "universal" Toeplitz matrix that sums every series. More emphatic is the next result (cf. THEOREM **2.1.3.3. 72**).

THEOREM **2.1.3.4.** LET $\{T^{(k)}\}_{k \in \mathbb{N}}$ BE A COUNTABLE SET OF TOE-
PLITZ MATRICES. THEN THERE IS A SERIES $S(A)$ SUCH THAT

$$\sigma_{m, T^{(k)}}(A)$$

EXISTS FOR EACH m AND EACH k. YET FOR EACH k IN $\mathbb{N}$,

$$\lim_{m \to \infty} \sigma_{m, T^{(k)}}(A)$$

DOES NOT EXIST, **[GeO, Hab]**.

PROOF. Assume that $T^{(k)} = \{t_{mn}^{(k)}\}_{m,n=1}^{\infty}$. Owing to i - iii in the definition of a Toeplitz matrix, there are in $\mathbb{N}$ two strictly increasing sequences:
$\{m_p\}_{p \in \mathbb{N}}$ and $\{n_p\}_{p \in \mathbb{N}}$ so that:
    if $m \geq m_1$ and $k = 1$,

$$\sum_{n=1}^{\infty} t_{mn}^{(k)} = 1 + \epsilon_{1m}, \ |\epsilon_{1m}| < 0.05,$$

$$\sum_{n=1}^{n_1} t_{m_1 n}^{(k)} = 1 + \delta_1, \ |\delta_1| < 2(0.05), \quad \sum_{n=n_1+1}^{\infty} |t_{m_1 n}^{(k)}| < 0.05;$$

if $m \geq m_2$ and $k = 1, 2$,

$$\sum_{n=1}^{\infty} t_{mn}^{(k)} = 1 + \epsilon_{2m}, \ |\epsilon_{2m}| < (0.05)^2, \quad \sum_{n=1}^{n_1} |t_{mn}^{(k)}| < (0.05)^2,$$

$$\sum_{n=1}^{n_2} t_{m_2 n}^{(k)} = 1 + \delta_2, \ |\delta_2| < 2(0.05)^2, \quad \sum_{n=n_2+1}^{\infty} |t_{m_2 n}^{(k)}| < (0.05)^2;$$

and, in general, if $m \geq m_p$, $k = 1, 2, \ldots, p$, and $p \in \mathbb{N}$,

$$\sum_{n=1}^{\infty} t_{mn}^{(k)} = 1 + \epsilon_{pm}, \quad |\epsilon_{pm}| < (0.05)^p, \quad \sum_{n=1}^{n_p-1} |t_{mn}^{(k)}| < (0.05)^p,$$

$$\sum_{n=1}^{n_p} t_{m_p n}^{(k)} = 1 + \delta_p, \quad |\delta_p| < 2(0.05)^p, \quad \sum_{n=n_p+1}^{\infty} |t_{m_p n}^{(k)}| < (0.05)^p.$$

Let $S(A)$ be such that

$$s_n(A) \stackrel{\text{def}}{=} \begin{cases} 1 & \text{if } 1 \leq n \leq n_1, \ n_2 < n \leq n_3, \ldots \\ -1 & \text{if } n_1 < n \leq n_2, \ n_3 < n \leq n_4, \ldots. \end{cases}$$

(The sequence A itself can be calculated according to the formula

$$a_n = \begin{cases} s_1(A) & \text{if } n = 1 \\ s_n(A) - s_{n-1}(A) & \text{if } 1 < n \in \mathbb{N}.) \end{cases}$$

If p is odd, $p > 1$, and $1 \leq k \leq p$ then

$$\sigma_{m_p, T^{(k)}}(A) = \sum_{n=1}^{n_1} t_{m_p n}^{(k)} - \sum_{n=n_1+1}^{n_2} t_{m_p n}^{(k)} + \cdots$$

$$- \cdots + \sum_{n=n_p+1}^{\infty} t_{m_p n}^{(k)} s_n(A)$$

$$= \sum_{n=1}^{n_1} t_{m_p n}^{(k)} - \left(\sum_{n=1}^{n_2} t_{m_p n}^{(k)} - \sum_{1=n}^{n_1} t_{m_p n}^{(k)} \right) + \cdots$$

$$- \cdots + \left(\sum_{n=1}^{n_p} t_{m_p n}^{(k)} - \sum_{n=1}^{n_p-1} t_{m_p n}^{(k)} \right) + \sum_{n=n_p+1}^{\infty} t_{m_p n}^{(k)} s_n(A).$$

The conditions imposed on the sequences $\{m_p\}_{p \in \mathbb{N}}$ and $\{n_p\}_{p \in \mathbb{N}}$ imply that

$$\sigma_{m_p, T^{(k)}}(A) > 1 - 2(0.05)^{p-1} - 2(p-1)(0.05)^{p-1} - (0.05)^p$$

$$\left(= 1 - [2(p-1) + 2.05] (0.05)^{p-1} \stackrel{\text{def}}{=} f(p) \right).$$

Since $f'(p) = (0.05)^{p-1} [-2 - [2(p-1) + 2.05] \ln 0.05]$ it follows that if p is odd and $p > 1$ then $f'(p) > 0$ and thus on $[3, \infty)$ the minimum value of $f(p)$ is $f(3)$. Hence if p is odd and $p > 1$ then

$$\sigma_{m_p, T^{(k)}}(A) > 0.9.$$

A similar argument shows that if p is even then

$$\sigma_{m_p, T^{(k)}}(A) < -0.7.$$

Therefore for each k in $\mathbb{N}$ the sequence $\{\sigma_{n,T^{(k)}}(A)\}_{n\in\mathbb{N}}$ does not converge.

$\square$

The formula

$$e^z = \lim_{n\to\infty}\left(1+\frac{z}{n}\right)^n$$

may be related to the Toeplitz matrix

$$T_{C,1} \overset{\text{def}}{=} \begin{pmatrix} 1 & 0 & 0 & 0 & \cdots \\ \frac{1}{2} & \frac{1}{2} & 0 & 0 & \cdots \\ \frac{1}{3} & \frac{1}{3} & \frac{1}{3} & 0 & \cdots \\ \vdots & \vdots & \vdots & \ddots & \ddots \\ \vdots & \vdots & \vdots & \vdots & \ddots \end{pmatrix}, \tag{2.1.3.2}$$

cf. **Exercise 2.1.3.16. 76**, corresponding to averaging the terms of a sequence. The formula has a generalization in terms of Toeplitz matrices.

Exercise 2.1.3.18. Assume that $T \overset{\text{def}}{=} (t_{ij})_{i,j=1}^{\infty,\infty}$ is a Toeplitz matrix such that

$$\lim_{i\to\infty}\sum_{j=1}^{\infty}|t_{ij}|^2 = 0,\ i\in\mathbb{N}.$$

Show that if $z\in\mathbb{C}$ then

$$\lim_{i\to\infty}\prod_{j=1}^{\infty}(1+t_{ij}z) = e^z.$$

Give an example of a Toeplitz matrix for which conclusion above is not valid.

[*Hint:* If $0<\delta<\frac{1}{M}$ there is a constant K and a function $\alpha_i(z)$ such that if $|z|<\delta$ then $|\alpha_i(z)|\leq K$ and

$$\exp\left(t_{ij}z - \left(\frac{t_{ij}^2}{2}\right)z^2\alpha_i(z)\right) = (1+t_{ij}z).]$$

The next **Exercises** illustrate some of the unexpected phenomena in the study of series.

Exercise 2.1.3.19. The *alternating series theorem* states that if, for $n\in\mathbb{N}$,

i. $\epsilon_n = (-1)^{n+1}$
ii. $a_n \geq a_{n+1}$
iii. $a_n \downarrow 0$

then $\sum_{n=1}^{\infty} \epsilon_n a_n$ converges. Show that each series below diverges and that for it only the indicated alternating series condition is violated:

$$\sum_{n=1}^{\infty} \frac{1}{n} \quad (i)$$

$$\sum_{n=1}^{\infty} (-1)^n \frac{1}{n^{n \bmod 2}} \quad (ii)$$

$$\sum_{n=1}^{\infty} (-1)^n \quad (iii).$$

Exercise 2.1.3.20. Show that if

$$b_n > 0, \ n \in \mathbb{N}, \text{ and } \liminf_{n \to \infty} b_n = 0$$

there is a divergent series $S(A)$ in which the terms are positive,

$$\lim_{n \to \infty} a_n = 0, \text{ and } \liminf_{n \to \infty} \frac{a_n}{b_n} = 0.$$

[Remark 2.1.3.5: Hence, no matter how rapidly the positive sequence $B \overset{\text{def}}{=} \{b_n\}_{n \in \mathbb{N}}$ converges to 0 there is a positive sequence $A \overset{\text{def}}{=} \{a_n\}_{n \in \mathbb{N}}$ converging to 0 so slowly that $S(A)$ diverges and yet A contains a subsequence converging to 0 more rapidly than the corresponding B-subsequence.]

[Hint: Choose a sequence $\{n_k\}_{k \in \mathbb{N}}$ such that $n_1 < n_1 + 1 < n_2 < n_2 + 1 < \cdots$ and such that

$$\lim_{k \to \infty} b_{n_k} = 0.$$

Choose a_n so that

$$a_n = \begin{cases} b_{n_k}^2 & \text{if } n = n_k, \ k \in \mathbb{N} \\ \frac{1}{j} & \text{if } n = m_j \in \mathbb{N} \setminus \{n_1, n_2, \ldots\} \overset{\text{def}}{=} \{m_1, m_2, \ldots\}. \end{cases}$$

Exercise 2.1.3.21. For a given positive sequence B such that

$$\liminf_{n \to \infty} b_n = 0$$

find a positive sequence A such that $S(A)$ converges and yet

$$\limsup_{n \to \infty} \frac{a_n}{b_n} = \infty.$$

What is the counterpart of **Remark 2.1.3.5. 80**?

[*Hint:* There is a sequence $N \overset{\text{def}}{=} \{n_k\}_{k \in \mathbb{N}}$ such that $b_{n_k} < k^{-3}$. Choose a_n so that

$$a_n = \begin{cases} \frac{1}{k^2} & \text{if } n = n_k \\ \frac{1}{n^2} & \text{if } n \notin N. \end{cases}]$$

In the next **Exercises** dealing with ratio and root tests all series considered are assumed to have positive terms.

Exercise 2.1.3.22. Show that although the limiting ratio exists the ratio test fails for:

$$\sum_{n=1}^{\infty} \frac{1}{n^2} \text{ (convergent) and } \sum_{n=1}^{\infty} \frac{1}{n} \text{ (divergent)}.$$

Exercise 2.1.3.23. Show that the limiting ratio does not exist and the *generalized ratio test* for $S(A)$

  i. $\limsup_{n \to \infty} \frac{a_{n+1}}{a_n} < 1 \Rightarrow$ convergence
  ii. $\limsup_{n \to \infty} \frac{a_{n+1}}{a_n} > 1 \Rightarrow$ divergence

fails for

$$\sum_{n=1}^{\infty} 2^{(-1)^n - n} \text{ (convergent) and } \sum_{n=1}^{\infty} 2^{n - (-1)^n} \text{ (divergent)}.$$

Exercise 2.1.3.24. Show that the *generalized root test* for $S(A)$

  i. $\limsup_{n \to \infty} a_n^{\frac{1}{n}} < 1 \Rightarrow$ convergence
  ii. $\limsup_{n \to \infty} a_n^{\frac{1}{n}} > 1 \Rightarrow$ divergence

fails for

$$\sum_{n=1}^{\infty} \left(\frac{5 + (-1)^n}{2} \right)^{-n} \text{ (convergent) and } \sum_{n=1}^{\infty} \left(\frac{5 + (-1)^n}{2} \right)^{n} \text{ (divergent)}.$$

Exercise 2.1.3.25. For a given positive sequence A show that

$$\liminf_{n\to\infty} \frac{a_{n+1}}{a_n} \leq \liminf_{n\to\infty} a_n^{\frac{1}{n}} \leq \limsup_{n\to\infty} a_n^{\frac{1}{n}} \leq \limsup_{n\to\infty} \frac{a_{n+1}}{a_n}.$$

[**Remark 2.1.3.6:** Hence the (generalized) ratio test can conceivably fail while the (generalized) root test succeeds.]

Exercise 2.1.3.26. Show that the root test succeeds (while the generalized ratio test fails) for

$$\sum_{n=1}^{\infty} 2^{(-1)^n - n} \text{ (convergent) and } \sum_{n=1}^{\infty} 2^{n-(-1)^n} \text{ (divergent)}.$$

The Mertens theorem [O1] states that if one of $S(A)$ and $S(B)$ *converges absolutely* and both converge then their *Cauchy product*

$$S(C) \overset{\text{def}}{=} \sum_{n=1}^{\infty} \left(\sum_{k=1}^{n} a_k b_{n-k+1}\right) \overset{\text{def}}{=} \sum_{n=1}^{\infty} c_n$$

converges to $S(A)S(B)$.

Exercise 2.1.3.27. Show that if $A = B = \{(-1)^n (n+1)^{-\frac{1}{2}}\}_{n\in\mathbb{N}}$ then $S(A)$ (hence $S(B)$) converges but that their Cauchy product does not.
[*Hint:* Show that since $\sqrt{(1+x)(n+1-x)}$ achieves its maximum on $[0,n]$ when $x = \frac{n}{2}$ it follows that

$$|c_n| \geq \frac{2(n+1)}{n+2}.]$$

Exercise 2.1.3.28. Show that the Cauchy product of the divergent series

$$2 + \sum_{n=2}^{\infty} 2^n \text{ and } -1 + \sum_{n=2}^{\infty} 1^n$$

converges.
    Most of the material above deals with series of constants. In the next discussion the emphasis is on series of terms that are not necessarily constants.

Exercise 2.1.3.29. Show that $S(A,x) \overset{\text{def}}{=} \sum_{n=0}^{\infty} e^{-n} \cos n^2 x$ represents a function f in C^{∞}, that the *Maclaurin* series for f consists only of terms of even degree, and the absolute value of the term of degree $2k$ is

$$\left(\sum_{n=0}^{\infty} \frac{e^{-n} n^{4k}}{(2k)!}\right) x^{2k} \left(> \left(\frac{n^2 x}{2k}\right)^{2k} e^{-n}\right), \ n \in \mathbb{N}.$$

Show that if $x \neq 0$ and $k > \frac{e}{2x}$ then the term of degree $2k$ is greater than 1, whence that the Maclaurin series for f converges iff $x = 0$.

Example 2.1.3.2. Assume

$$\phi_{n0}(x) = \begin{cases} ((n-1)!)^2 & \text{if } 0 \leq |x| \leq 2^{-n}(n!)^{-2} \\ 0 & \text{if } |x| \geq 2^{-n+1}(n!)^{-2} \end{cases}$$

and, by means of bridging functions, ϕ_{n0} is made infinitely differentiable everywhere and $0 \leq \phi_{n_0}(x) \leq ((n-1)!)^2$. If

$$\phi_{n1}(x) = \int_0^x \phi_{n0}(t)\, dt,$$

$$\cdots$$

$$f_n(x) \stackrel{\text{def}}{=} \phi_{n,n-1}(x) = \int_0^x \phi_{n,n-2}(t)\, dt, \ n \in \mathbb{N},$$

then direct calculation shows that

$$|f_n^{(k)}(x)| \leq \frac{(2^{-n+1})}{n^2} \frac{|x|^{n-k-2}}{(n-k-2)!}$$

whence the *Weierstraß M-test* shows that for all k in $\{0\} \cup \mathbb{N}$,

$$\sum_{n=1}^{\infty} f_n^{(k)}(x)$$

converges uniformly on every finite interval. It follows that

$$f(x) \stackrel{\text{def}}{=} \sum_{n=1}^{\infty} f_n(x) \in C^{\infty}$$

and that its Maclaurin series is

$$\sum_{n=0}^{\infty} n! x^n,$$

which converges iff $x = 0$. A similar argument shows that an arbitrary series

$$\sum_{n=0}^{\infty} a_n x^n$$

is the Maclaurin series of some function in C^{∞}.

[**Remark 2.1.3.7:** The function f just described is different in character from that of

$$g : \mathbb{R} \ni x \mapsto \begin{cases} \exp\left(-x^{-2}\right) & \text{if } x \neq 0 \\ 0 & \text{otherwise} \end{cases}$$

(cf. (2.1.2.2), page 61). The Maclaurin series for g converges everywhere and represents g only at 0. The Maclaurin series for f converges only at 0 (and represents f only at 0).]

Associated with the power series $P(x) \overset{\text{def}}{=} \sum_{n=0}^{\infty} a_n x^n$ is the number $L \overset{\text{def}}{=} \limsup_{n\to\infty} |a_n|^{\frac{1}{n}}$. The *radius of convergence* R of $P(x)$ is given by the formula:

$$R = \begin{cases} \frac{1}{L} & \text{if } L > 0 \\ \infty & \text{if } L = 0. \end{cases}$$

If $0 \leq r < R$ then $P(x)$ converges uniformly and absolutely in the interval $[-r, r]$. There is no general result about uniformity of convergence in $(-R, R)$ nor about convergence if $x = \pm R$ (when $R < \infty$).

Example 2.1.3.3. The radii of convergence R_1, R_2, R_3 for the power series

$$P_1(x) \overset{\text{def}}{=} \sum_{n=1}^{\infty} x^n$$

$$P_2(x) \overset{\text{def}}{=} \sum_{n=1}^{\infty} \frac{x^n}{n}$$

$$P_3(x) \overset{\text{def}}{=} \sum_{n=1}^{\infty} \frac{x^n}{n^2}$$

are all 1. Yet:

i. $P_1(x)$ converges uniformly in any closed subinterval of $(-1, 1)$, converges but not uniformly in $(-1, 1)$, and diverges if $x = \pm 1$;

ii. $P_2(x)$ behaves just like $P_1(x)$ except that $P_2(x)$ converges if $x = -1$;

iii. $P_3(x)$ converges uniformly on $[-1, 1]$.

Thus the opportunities for finding abnormality of convergence behavior are somewhat limited if the domain of study is the set of power series. On the other hand, *orthonormal series*, in particular *trigonometric series*, and more particularly *Fourier series* provide many examples of unusual convergence phenomena.

[**Remark 2.1.3.8:** It was the study of trigonometric series that gave rise to the proper definition by Riemann of the integral bearing his name. It was the study of the sets of convergence and divergence of trigonometric series that led Cantor to the study of "sets" and thereby opened the field of modern set theory, logic, cardinal and ordinal numbers, etc.]

Let f be integrable on $[-\pi, \pi]$. The *Riemann-Lebesgue* theorem implies that the sequence

$$c_n \stackrel{\text{def}}{=} \frac{1}{\sqrt{2\pi}} \int_{-\pi}^{\pi} f(x) e^{-inx} \, dx, \ n \in \mathbb{Z}$$

of Fourier coefficients of f converges to 0 as $|n| \to \infty$. If the Fourier series for f is written in the form

$$\frac{1}{2} a_0 + \sum_{n=1}^{\infty} (a_n \cos nx + b_n \sin nx)$$

then for n in $\mathbb{N}$, $a_n = \frac{1}{\sqrt{2\pi}} (c_n + c_{-n})$, $b_n = \frac{i}{\sqrt{2\pi}} (c_n - c_{-n})$. Hence

$$\lim_{n \to \infty} (a_n^2 + b_n^2) \stackrel{\text{def}}{=} \lim_{n \to \infty} \rho_n^2 = 0.$$

Cantor, whose research preceded the development of the modern theory of (Lebesgue) integration, showed that if the trigonometric series

$$\frac{1}{2} a_0 + \sum_{n=1}^{\infty} a_n \cos nx + b_n \sin nx$$

(which need not be the Fourier series of an integrable function, cf. **Example 2.1.3.5. 87**) is such that

$$\lim_{n \to \infty} a_n \cos nx + b_n \sin nx \stackrel{\text{def}}{=} \lim_{n \to \infty} \rho_n \cos(nx + \alpha_n) = 0 \qquad (2.1.3.3)$$

everywhere then $\lim_{n \to \infty} \rho_n = 0$.

Lebesgue sharpened Cantor's result as follows.

THEOREM **2.1.3.5.** (CANTOR-LEBESGUE) LET $(2.1.3.3)$ OBTAIN EV-ERYWHERE ON A MEASURABLE SET E OF POSITIVE MEASURE. THEN $\lim_{n \to \infty} \rho_n = 0$.

PROOF. If $\rho_n \not\to 0$ as $n \to \infty$ then there is a sequence $\{n_k\}_{k \in \mathbb{N}}$ and a positive ϵ such that $\rho_{n_k} \geq \epsilon$ for all k. Hence $\lim_{n \to \infty} \cos(n_k x + \alpha_{n_k}) = 0$ a.e. on E and the left member of

$$\int_E \cos^2(n_k x + \alpha_{n_k}) \, dx = \frac{1}{2} \int_E dx + \lim_{k \to \infty} \frac{1}{2} \int_E \cos 2(n_k x + \alpha_{n_k}) \, dx$$

tends to 0 as $k \to \infty$. The Riemann-Lebesgue theorem implies that the second term in the right member above tends to 0 as $k \to \infty$ and it follows that $\frac{1}{2} \lambda(E) = 0$, a contradiction. $\qquad \square$

It should be noted that although the orthogonal set T of trigonometric functions is *complete* in $L^2([0, 2\pi], \mathbb{C})$, the *completeness* of T plays no role in the validity of the last result. Indeed, the argument remains accurate if only a proper but infinite subset of T is at hand. There follow variations inspired by the theme above.

THEOREM **2.1.3.6.** FOR THE MEASURE SITUATION $(X, \mathcal{S}, \mu)$ LET $\{f_n\}_{n \in \mathbb{N}}$ BE AN INFINITE ORTHONORMAL SET CONSISTING OF UNIFORMLY BOUNDED FUNCTIONS: $\|f_n\|_\infty \le M < \infty$, $n \in \mathbb{N}$. IF $a_n f_n(x)$ CONVERGES TO ZERO ALMOST EVERYWHERE THEN $\lim_{n \to \infty} a_n = 0$.

PROOF. If the conclusion is false, via subsequences as needed, it may be assumed that for some positive ϵ and each n in $\mathbb{N}$, $|a_n| \ge \epsilon$. Hence $f_n(x) \overset{\text{a.e.}}{\to} 0$ and $|f_n(x)|^2 \overset{\text{a.e.}}{\to} 0$. The bounded convergence theorem implies the contradiction $1 \equiv \int_X |f_n(x)|^2 \, d\mu(x) \to 0$.

$\square$

On the other hand, absent the condition c) (uniform boundedness of the functions), the conclusion above may fail to obtain.

Example 2.1.3.4. Let E_n be $(\frac{1}{n+1}, \frac{1}{n}]$, $n \in \mathbb{N}$ and for n in $\mathbb{N}$, let f_n be χ_{E_n}. Then $\{g_n \overset{\text{def}}{=} \sqrt{n(n+1)} f_n\}_{n \in \mathbb{N}}$ is an orthonormal set in $L^2([0,1], \lambda)$ and for *every* sequence $\{a_n\}_{n \in \mathbb{N}}$, $\sum_{n=1}^\infty a_n g_n$ converges a.e.

[**Note 2.1.3.2:** The particular "real" form

$$\frac{1}{2} a_0 + \sum_{n=1}^\infty a_n \cos nx + b_n \sin nx$$

of the trigonometric series gives the Cantor-Lebesgue theorem its significance. If a trigonometric series has the form

$$\sum_{n=-\infty}^\infty c_n \frac{e^{inx}}{\sqrt{2\pi}}$$

and if $\lim_{|n| \to \infty} c_n \frac{e^{inx}}{\sqrt{2\pi}} = 0$ for even one value of x then, since $|e^{inx}| \equiv 1$, it follows without further proof that $\lim_{|n| \to \infty} c_n = 0$.]

The next lemma, due to Abel, is useful in many arguments.

LEMMA **2.1.3.1.** IF $S(A)$ IS A SERIES, IF $b_n \downarrow 0$, AND IF THERE IS AN M SUCH THAT $|s_n(A)| \le M$, $n \in \mathbb{N}$ THEN $\sum_{n=1}^\infty b_n a_n$ CONVERGES [**Kno**].

The *Euler formula* ($e^{\pm it} = \cos t \pm i \sin t$) implies that

$$\sum_{n=1}^N \sin nx = \frac{\cos \frac{x}{2} - \cos \left(N + \frac{1}{2}\right) x}{2 \sin \frac{x}{2}}$$

and so

$$\left| \sum_{n=1}^{N} \sin nx \right| \leq \left| \frac{1}{\sin \frac{x}{2}} \right|.$$

If $\sin \frac{x}{2} \neq 0$, i.e., if x is fixed and is not an integral multiple of 2π, then $|\sum_{n=1}^{N} \sin nx| \leq |\frac{1}{\sin \frac{x}{2}}|$ for all N in $\mathbb{N}$. Abel's lemma now implies that if $b_n \downarrow 0$ then $\sum_{n=1}^{\infty} b_n \sin nx$ converges for any x that is not a multiple of 2π. Inspection shows that the series converges if x *is* a multiple of 2π, whence the series converges everywhere.

Example 2.1.3.5. Abel's lemma implies that the trigonometric series

$$\sum_{n=2}^{\infty} \frac{\sin nx}{\ln n} \tag{2.1.3.4}$$

converges for all x.

Example 2.1.3.6. The trigonometric series (2.1.3.4) is *not* the Fourier series of a *Lebesgue integrable* function. Indeed, if

$$f(x) \overset{\text{def}}{=} \sum_{n=2}^{\infty} \frac{\sin nx}{\ln n},$$

if f is Lebesgue integrable, and

$$F(x) \overset{\text{def}}{=} \int_{0}^{x} f(t)\, dt,$$

then F is *absolutely continuous*, periodic (because f is), and *even* (because f is *odd*) whence the Fourier series for F is a cosine series that converges everywhere to F: $F(x) = \sum_{n=0}^{\infty} a_n \cos nx$. Integration by parts shows that if $n \geq 2$ then

$$a_n = -\frac{1}{n \ln n}.$$

Thus if f is Lebesgue integrable there emerges the contradiction that the divergent series

$$\sum_{n=2}^{\infty} \frac{1}{n \ln n} \cos n0 = \sum_{n=2}^{\infty} \frac{1}{n \ln n}$$

converges: f represented by (2.1.3.4) is *not* Lebesgue integrable. In particular, although (2.1.3.4) converges everywhere it does not converge uniformly on $[-\pi, \pi]$.

There remains the question of whether *some* Lebesgue integrable function g is such that its Fourier coefficients are those in (2.1.3.4). However, for such a g it follows from **Exercise 2.1.3.15. 75** that $\|g - F_N * g\|_1 \to 0$

as $n \to \infty$, whence, for some subsequence $\{N_k\}_{k \in \mathbb{N}}$, $F_{N_k} * g \overset{\text{a.e.}}{\to} g$ as $k \to \infty$. Since the functions $F_{N_k} * g$ are, as well, the averages of the partial sums of the series for f it follows that $f = g$ a.e. Since f is not Lebesgue integrable no such g exists.

[**Remark 2.1.3.9:** In particular (2.1.3.4) is not the Fourier series of a Riemann integrable function.]

Exercise 2.1.3.30. Show that

$$\sum_{n=1}^{\infty} \frac{\sin nx}{n} \tag{2.1.3.5}$$

is the Fourier series for $f : [-\pi, \pi] \ni x \mapsto x$, which is of *bounded variation* on $[-\pi, \pi]$: $f \in BV([-\pi, \pi])$. Hence (2.1.3.5) converges uniformly on every closed subinterval of $(-\pi, \pi)$ [**Zy**, I, p. 57]. Show, on the other hand, that the Weierstraß M-test is not effectively applicable: there is no convergent series of positive constants that dominates (2.1.3.5) on $[a, b]$ if $-\pi < a < b < \pi$.

According to the *Riemann-Lebesgue theorem* the *Fourier transform*

$$\hat{f}(t) \overset{\text{def}}{=} \frac{1}{\sqrt{2\pi}} \int_{-\infty}^{\infty} f(x) e^{-itx} \, dx$$

of a function f that is Lebesgue integrable on $\mathbb{R}$ is a continuous function that *vanishes at infinity*: $\lim_{|t| \to \infty} \hat{f}(t) = 0$. The next result shows that not every continuous function vanishing at infinity is a Fourier transform of a Lebesgue integrable function.

Example 2.1.3.7. Let $\sum_{n=-\infty}^{\infty} c_n \exp(inx)/\sqrt{2\pi}$ be the complex exponential form of (2.1.3.4), page 87 or of any trigonometric series converging everywhere to a function that is not Lebesgue integrable. If h is in C^{∞}, $h(x) = 0$, if $x \notin [-\frac{1}{3}, \frac{1}{3}]$, and $\sup_{x \in \mathbb{R}} |h(x)| = h(0) = 2\pi$ then

$$\sum_{n=-\infty}^{\infty} c_n h(x - n) \overset{\text{def}}{=} f(x) \tag{2.1.3.6}$$

is a series in which, for each x, only one term, namely the term for which $x - \frac{1}{3} \le n \le x + \frac{1}{3}$, is nonzero. It follows that the series represents a function f in C^{∞}. For a given x, if $x - \frac{1}{3} \le n \le x + \frac{1}{3}$ then $|f(x)| \le 2\pi |c_n|$ whence $\lim_{|x| \to \infty} f(x) = 0$, i.e., f vanishes at infinity.

On the other hand, if F is Lebesgue integrable on $\mathbb{R}$ then

$$\sum_{m=-\infty}^{\infty} \int_{-\pi}^{\pi} |F(t + 2\pi m)| \, dt \le \int_{-\infty}^{\infty} |F(t)| \, dt < \infty.$$

Hence $\sum_{m=-\infty}^{\infty} |F(x + 2\pi m)| < \infty$ a.e.,

$$g(x) \stackrel{\text{def}}{=} \sum_{m=-\infty}^{\infty} F(x + 2\pi m)$$

is defined a.e., $g(x + 2\pi) = g(x)$ a.e., and

$$\int_{-\pi}^{\pi} |g(t)| \, dt < \infty.$$

If, to boot, $\hat{F} = f$ then a direct calculation shows that nth Fourier coefficient of g is c_n, a contradiction, since the c_n are not the Fourier coefficients of any Lebesgue integrable function.

In [Zy] there is a wealth of information about Fourier series. In particular there are included details about the following counterexamples in Fourier series (cf. also **Note 2.3.1.1. 158**).

  i. [Fejér and Lebesgue, I, pp. 300–1]. If E is a countable subset of $[-\pi, \pi]$ there is a (Lebesgue integrable) function f such that the Fourier series for f diverges on E and converges on $[-\pi, \pi] \setminus E$.
 ii. [Kolmogorov, I, pp. 305, 310]. There is a Lebesgue integrable function f for which the Fourier series diverges everywhere.

The technique of proof for the two results above and for many other theorems in the subject of Fourier series is the definition of a sequence $\{\phi_n\}_{n \in \mathbb{N}}$ of *trigonometric polynomials* and a sequence $\{\alpha_n\}_{n \in \mathbb{N}}$ of constants so that the series

$$\sum_{n=1}^{\infty} \alpha_n \phi_n \qquad\qquad (2.1.3.7)$$

converges in an appropriate sense and defines a function f with the desired properties. For i the two sequences are chosen so that (2.1.3.7) converges uniformly, whence f is continuous. For ii the two sequences are chosen so that the absolute values of the partial sums of (2.1.3.7) are uniformly bounded and the (2.1.3.7) converges a.e., whence f is defined a.e. and is integrable.

[**Note 2.1.3.3:** Despite the result in ii the equation

$$\lim_{N \to \infty} \int_{-\pi}^{\pi} |f(x) - F_N * f(x)| \, dx = 0$$

is valid. Consequently there is a subsequence $\{N_k\}_{k \in \mathbb{N}}$ such that $\lim_{k \to \infty} F_{N_k} * f(x) = f(x)$ a.e.]

Exercise 2.1.3.31. Show that if

$$f_n(x) = \begin{cases} \frac{1}{n} & \text{if } x \in \mathbb{Q} \\ 0 & \text{otherwise} \end{cases}$$

then $f_n \overset{u}{\to} 0$. (The f_n are discontinuous everywhere, their uniform limit is continuous (constant).)

Exercise 2.1.3.32. Show that if $f_n(x) = (\sin nx)/\sqrt{n}$ then $f_n \overset{u}{\to} 0$ while f_n' diverges everywhere.

[*Hint:* If $x \neq 0$ then $|\cos 2nx| = 1 - \cos^2 nx$. Hence if $|\cos nx| < \frac{1}{2}$ then $|\cos 2nx| > \frac{3}{4}$.]

Exercise 2.1.3.33. Show that if

$$f_n(x) = \begin{cases} \inf\left(n, \frac{1}{x}\right) & \text{if } 0 < x \leq 1 \\ 0 & \text{otherwise} \end{cases}$$

then each f_n is bounded on $[0, 1]$ but

$$\lim_{n \to \infty} f_n(x) = \begin{cases} \frac{1}{x} & \text{if } 0 < x \leq 1 \\ 0 & \text{otherwise.} \end{cases}$$

Hence the nonuniform limit of bounded functions can be unbounded.

Exercise 2.1.3.34. Show that if

$$f_n(x) = \begin{cases} \inf(1, nx) & \text{if } x \geq 0 \\ \sup(-1, nx) & \text{if } x < 0 \end{cases}$$

then each f_n is continuous on $\mathbb{R}$ and

$$\lim_{n \to \infty} f_n(x) \overset{\text{def}}{=} E(x) = \begin{cases} 1 & \text{if } x \geq 0 \\ -1 & \text{if } x < 0. \end{cases}$$

Hence the discontinuous function $E(x)$ is the nonuniform limit of the continuous functions f_n. Repeat for $g_n(x) \overset{\text{def}}{=} x^n$, $n \in \mathbb{N}$, $x \in [0, 1]$.

Exercise 2.1.3.35. Show that if

$$f_n(x) \overset{\text{def}}{=} \begin{cases} \max\left(\frac{1}{n}, \frac{1}{q} + 2n^2\left(x - \frac{p}{q}\right)\right) & \text{if } x \in \left(\frac{p}{q} - \frac{1}{2n^2}, \frac{p}{q}\right] \\ \max\left(\frac{1}{n}, \frac{1}{q} - 2n^2\left(x - \frac{p}{q}\right)\right) & \text{if } x \in \left(\frac{p}{q}, \frac{p}{q} + \frac{1}{2n^2}\right) \\ \frac{1}{n} & \text{otherwise} \end{cases}$$

$$1 \leq q < n, \ 0 \leq p \leq q$$

$$f(x) \overset{\text{def}}{=} \begin{cases} \frac{1}{q} & \text{if } x = \frac{p}{q}, \ p, q \in \mathbb{Z}, \ q > 0, \text{ and } (p, q) = 1 \\ 0 & \text{otherwise} \end{cases}$$

(see **Figure 2.1.3.1**) then $f_n \geq f_{n+1}$, each f_n is continuous, and $f_n \to f$. Hence f, which is continuous only on $[0,1] \setminus \mathbb{Q}$ is the nonuniform limit of the continuous functions f_n.

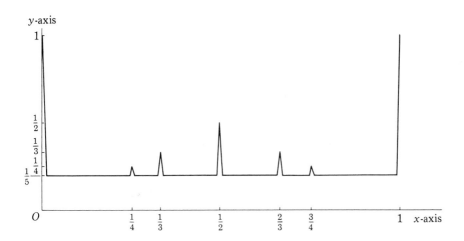

Figure 2.1.3.1. The graph of $y = f_5(x)$ on $[0,1]$.

[**Remark 2.1.3.10:** In [**My**] it is noted that each f_n is continuous, positive everywhere, and if $h_n = \frac{1}{f_n}$ then each h_n is continuous and the limit of $\{h_n(x)\}_{n\in\mathbb{N}}$ on $[0,1]$ is

$$\begin{cases} n & \text{if } m, n \in \mathbb{N}, (m,n) = 1, \text{ and } x = \frac{m}{n} \\ \infty & \text{otherwise.} \end{cases}$$

The limit of $\{h_n\}_{n\in\mathbb{N}}$ is finite only on $(\mathbb{Q} \cap [0,1]) \setminus \{0\}$, an F_σ. In [**My**] there is the following generalization.

Let S be a subset of $[0,1]$. There is a sequence $\{h_n\}_{n\in\mathbb{N}}$ such that each h_n is continuous on $[0,1]$ and

$$\lim_{n\to\infty} h_n(x) \text{ is } \begin{cases} \text{finite} & \text{if } x \in S \\ \text{infinite} & \text{if } x \in ([0,1] \setminus S) \end{cases}$$

iff S is an F_σ.]

Exercise 2.1.3.36. Show that the f_n in **Exercise 2.1.3.34. 90** are such that

$$\lim_{n\to\infty} \int_0^1 f_n(x)\, dx = \int_0^1 \lim_{n\to\infty} f_n(x)\, dx$$

even though the convergence is not uniform.

Exercise 2.1.3.37. Show that if

$$f_n(x) \overset{\text{def}}{=} \begin{cases} 2n^2 x & \text{if } 0 \le x \le \frac{1}{2n} \\ n - 2n^2 \left(x - \frac{1}{2n} \right) & \text{if } \frac{1}{2n} < x \le \frac{1}{n} \\ 0 & \text{if } \frac{1}{n} < x \le 1 \end{cases}$$

then

$$\lim_{n \to \infty} \int_0^1 f_n(x)\, dx = \frac{1}{2}$$

$$\int_0^1 \lim_{n \to \infty} f_n(x)\, dx = 0.$$

Hence the nonuniform limit of Riemann integrable functions can be Riemann integrable but its integral can fail to be the limit of the integrals of the approximating functions. Show that the functions f_n and f of **Exercise 2.1.3.35. 90** exhibit the same failure.

Uniform convergence of a sequence of Riemann integrable functions is a sufficient but not necessary condition for the Riemann integrability of the limit function.

Exercise 2.1.3.38. Let $\{J_n\}_{n \in \mathbb{N}}$ be the set of disjoint open intervals deleted from $[0, 1]$ in the construction of a Cantor-like set C for which $\lambda(C)$ is positive. If $N = 2^n - 1, 2 \le n \in \mathbb{N}$, define f_N so that it is piecewise linear, continuous, and

$$f_N(x) \overset{\text{def}}{=} \begin{cases} 1 & \text{if } x \in \overline{J_n}, 1 \le n \le N \\ 0 & \text{if } x = 0 \text{ or } x = 1 \\ 0 & \text{if } x \text{ is the midpoint of an interval} \\ & \quad \text{separating two adjacent intervals } J_p \text{ and } J_q,\ 1 \le p, q \le N. \end{cases}$$

Show:

i. each f_N is continuous, bounded, and hence Riemann integrable;
ii. $\lim_{N \to \infty} f_N \overset{\text{def}}{=} f$ exists and $\mathrm{Discont}(f) \supset C$;
iii. f is bounded and not Riemann integrable.

Hence on $[0, 1]$, f is the nonuniform limit of uniformly bounded, continuous (hence Riemann and Lebesgue integrable) functions and f is Lebesgue integrable but not Riemann integrable.

Exercise 2.1.3.39. Show that the terms of the series

$$\sum_{n=1}^{\infty} \frac{x^n}{n}$$

converge uniformly to 0 on $[0, 1)$ but that the series itself does not converge uniformly on $[0, 1)$.

Exercise 2.1.3.40. Show that if

$$f_n(x) \stackrel{\text{def}}{=} \begin{cases} \frac{x}{n} & \text{if } n \text{ is odd} \\ \frac{1}{n} & \text{if } n \text{ is even} \end{cases}$$

then $\lim_{n \to \infty} f_n(x) = 0$, $x \in \mathbb{R}$, the convergence to 0 is *not* uniform, and yet $f_{2m} \stackrel{u}{\to} 0$ on $\mathbb{R}$.

Dini's theorem **[HeSt]** states that a *monotone* sequence of *continuous* functions converging to a *continuous* function on a *compact* set converges uniformly.

Exercise 2.1.3.41. Show that each of the following sequences of functions in $\mathbb{R}^{\mathbb{R}}$ fails to conform to one of the *italicized* conditions in Dini's theorem and fails to converge uniformly.

$$f_n(x) \stackrel{\text{def}}{=} \begin{cases} 0 & \text{if } x = 0 \text{ or } \frac{1}{n} \leq x \leq 1 \\ 1 & \text{if } 0 < x < \frac{1}{n} \end{cases}, \ n \in \mathbb{N}$$

$$f_n(x) \stackrel{\text{def}}{=} x^n, \ n \in \mathbb{N}, \ 0 \leq x \leq 1$$

$$f_n(x) \stackrel{\text{def}}{=} \begin{cases} 2n^2 x & \text{if } 0 \leq x \leq \frac{1}{2n} \\ n - 2n^2 \left(x - \frac{1}{2n}\right) & \text{if } \frac{1}{2n} < x \leq \frac{1}{n} \\ 0 & \text{if } \frac{1}{n} < x \leq 1 \end{cases}, \ n \in \mathbb{N}.$$

Exercise 2.1.3.42. Show that if

$$f_n(x) = \frac{x}{(1 + n^2 x^2)}, \ |x| \leq 1, \ n \in \mathbb{N}$$

then $f_n \stackrel{u}{\to} 0$ on $[-1, 1]$ but that iff $x \neq 0$, $f_n'(x) \to 0$ as $n \to \infty$.

Exercise 2.1.3.43. Show that if $f_n(x) = x^n$, $x \in [0, 1)$ $n \in \mathbb{N}$, then $f_n \stackrel{u}{\to} 0$ on every closed subinterval J of $[0, 1)$ but that the convergence $f_n \to 0$ on $[0, 1)$ is *not* uniform.

Exercise 2.1.3.44. Show that if

$$f_n(x) = \begin{cases} \frac{1}{n} & \text{if } 0 \leq x \leq n^2 \\ 0 & \text{if } x > n^2 \end{cases}$$

then $f_n \stackrel{u}{\to} 0$ on $[0, \infty)$ but $\int_0^\infty f_n(x)\, dx \uparrow \infty$ as $n \to \infty$.

Example 2.1.3.8. There is in $\mathbb{Q}$ a sequence $\{a_n\}_{n \in \mathbb{N}}$ such that if

$$f \in C_0([0, 1]) \stackrel{\text{def}}{=} \left\{ f \ : \ f \in \mathbb{R}^{[0,1]}, \ f \in C([0, 1]), \ f(0) = 0 \right\}$$

then there is in $\mathbb{N} \cup \{0\}$ a sequence $\{n_k\}_{k \in \mathbb{N} \cup \{0\}}$ such that

$$0 = n_0 < n_1 < \cdots < n_k < \cdots$$

$$f(x) = \sum_{k=0}^{\infty} \left(\sum_{n=n_k+1}^{n_{k+1}} a_n x^n \right).$$

The *Stone-Weierstraß theorem* [**HeSt, Loo, Sto2, Sto3**] implies that for each m in $\mathbb{N}$ the set $\mathcal{P}_m \stackrel{\text{def}}{=} \mathbb{Q}[x^m]$ of polynomials in x^m and with rational coefficients is $\| \ \|_\infty$-dense in $C_0([0,1])$. If $\{f_n\}_{n \in \mathbb{N}}$ is a countable $\| \ \|_\infty$-dense set in $C_0([0,1])$ there can be defined by induction a sequence $\{P_n\}_{n \in \mathbb{N}}$ and in $\mathbb{N}$ a strictly increasing sequence $\{m_n\}_{n \in \mathbb{N}}$ such that

$$P_n \in \mathcal{P}_{m_n}, \ \deg(P_n) < m_{n+1},$$

$$\left\| f_n - \sum_{i=1}^{n} P_i \right\|_\infty < \frac{1}{n}.$$

If $f \in C_0([0,1])$ and $\epsilon > 0$ let n_k be such that $\frac{1}{n_k} < \frac{\epsilon}{2}$ and $\|f - f_{n_k}\|_\infty < \frac{\epsilon}{2}$. Then $\|f - \sum_{i=1}^{n_k} P_i\|_\infty < \epsilon$. The series $\sum_{m=1}^{\infty} P_m$ may be viewed as a power series $S \stackrel{\text{def}}{=} \sum_{n=1}^{\infty} a_n x^n$ in which the coefficients a_n are in $\mathbb{Q}$. For each f in $C_0([0,1])$ appropriate grouping of the terms in the series S achieves the desired convergence.

[**Note 2.1.3.4:** Actually the grouping is done on the groups formed by the various polynomials P_n: the series $\sum_{n=1}^{\infty} P_n$ is the series in which the terms P_n can be grouped to yield the convergence of the partial sums to a given f in $C_0([0,1])$.]

Compare the following result, a simplified analog of the conclusion above, with the Riemann derangement theorem (**Exercise 2.1.3.8. 69**).

Exercise 2.1.3.45. Show that there is in $\mathbb{R}$ a sequence $\{a_n\}_{n \in \mathbb{N}}$ such that if $t \in \mathbb{R} \cup \{\infty\} \cup \{-\infty\}$ there is a grouping

$$\sum_{p=1}^{\infty} \left(\sum_{n=n_p}^{n_{p+1}} a_n \right) \tag{2.1.3.8}$$

for which the sum of the (possibly divergent) series in (2.1.3.8) is t.

Newton's algorithm for finding the real root(s) of an equation of the form $f(x) = 0$, can fail to produce a convergent sequence of "approximants."

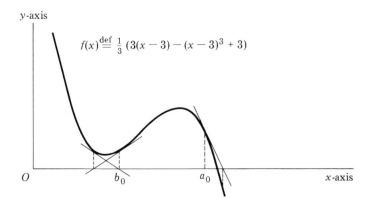

y-axis

$$f(x) \stackrel{\text{def}}{=} \tfrac{1}{3}(3(x-3) - (x-3)^3 + 3)$$

O b_0 a_0 x-axis

Figure 2.1.3.2. Newton's algorithm for the roots of $f(x) = 0$.

Example 2.1.3.9. The curve in **Figure 2.1.3.2** is the graph of the equation $y = f(x)$ and the choice of a_0 as the starting point for the use of Newton's algorithm leads to the real root of $f(x) = 0$. On the other hand, the use of b_0 as the starting point produces a divergent sequence $\{b_n\}_{n=0}^{\infty}$ of "approximants."

2.1.4. $\mathbb{R}^{X \times Y}$

Although the study of $\mathbb{R}^{X}$ subsumes the study of $\mathbb{R}^{X \times Y}$ there are particularities about the latter that deserve special attention. Most of the important phenomena are illustrated in the context of $\mathbb{R}^{\mathbb{R} \times \mathbb{R}}$.

Exercise 2.1.4.1. Let S be a set and let X and Y be topological spaces. Formulate appropriate definitions of uniform convergence and uniform continuity (with respect to S) for a function f in $Y^{X \times S}$. Assume, to boot, there are uniform structures U resp. V for X resp. Y. Formulate an appropriate definition of a Cauchy net, uniform with respect to S, in $Y^{X \times S}$.

In what follows, unless the contrary is stated, the domain of each function is (a subset of) $\mathbb{R}^2 \ (= \mathbb{R} \times \mathbb{R})$.

Exercise 2.1.4.2. Show that if

$$f(x, y) \stackrel{\text{def}}{=} \begin{cases} \frac{xy}{x^2 + y^2} & \text{if } x^2 + y^2 \neq 0 \\ 0 & \text{if } x = y = 0 \end{cases}$$

then f is continuous in each variable separately and yet is not continuous at $(0, 0)$ in $\mathbb{R} \times \mathbb{R}$.

Exercise 2.1.4.3. Show that if

$$f(x, y) \overset{\text{def}}{=} \begin{cases} \frac{x^2 y}{x^4 + y^2} & \text{if } x^2 + y^2 \neq 0 \\ 0 & \text{if } x = y = 0 \end{cases}$$

then f has no limit at $(0,0)$ but that for every straight line L through the origin $\lim_{(x,y)\to(0,0),\ (x,y)\in L} f(x,y) = 0$. Contrast this result with that in **Exercise 2.1.4.2.**

Convergence to $(0,0)$ on a straight line L through the origin can be replaced by convergence on a curve C through the origin if the form of f is appropriately modified.

Exercise 2.1.4.4. Show that if

$$f(x, y) \mapsto \begin{cases} \dfrac{ye^{-\frac{1}{x^2}}}{y^2 + e^{-\frac{2}{x^2}}} & \text{if } x \neq 0 \\ 0 & \text{if } x = 0 \end{cases}$$

and $a \neq 0$ then $f\left(a, e^{-\frac{1}{a^2}}\right) = \frac{1}{2} \neq 0$ as $a \to 0$. If $m, n \in \mathbb{N}$, $(m, n) = 1$, $c \neq 0$, and C is the curve for which the equation is $y = cx^{\frac{m}{n}}$ then

$$\lim_{(x,y)\to 0,\ (x,y)\in C} f(x,y) = 0.$$

[**Note 2.1.4.1:**   Each of the three functions just described is nondifferentiable at the origin but each has first partial derivatives everywhere.]

For X and Y subsets of Hausdorff spaces, for $a \in \overline{X} \setminus X$, for $b \in \overline{Y} \setminus Y$, and for a function f in $\mathbb{R}^{X \times Y}$, there are three limits to be considered at (a, b):

$$\lim_{(x,y)\to(a,b)} f(x,y)$$
$$\lim_{x\to a} \lim_{y\to b} f(x,y)$$
$$\lim_{y\to b} \lim_{x\to a} f(x,y). \tag{2.1.4.1}$$

Exercise 2.1.4.5. Show that for each of the functions described below exactly two of the three limits in (2.1.4.1) exist and are equal at $(0, 0)$:

$$f(x, y) \overset{\text{def}}{=} \begin{cases} \frac{xy}{x^2 + y^2} & \text{if } x^2 + y^2 \neq 0 \\ 0 & \text{if } x = y = 0 \end{cases}$$

$$f(x, y) \overset{\text{def}}{=} \begin{cases} y + x\sin(\frac{1}{y}) & \text{if } y \neq 0 \\ 0 & \text{if } y = 0 \end{cases}$$

$$f(x, y) \overset{\text{def}}{=} \begin{cases} x + y\sin(\frac{1}{x}) & \text{if } x \neq 0 \\ 0 & \text{if } x = 0. \end{cases}$$

Exercise 2.1.4.6. Show that for each of the functions below exactly one of the three limits in (2.1.4.1) exists at $(0,0)$:

$$f(x,y) \stackrel{\text{def}}{=} \begin{cases} x\sin(\frac{1}{y}) + y\sin(\frac{1}{x}) & \text{if } xy \neq 0 \\ 0 & \text{if } xy = 0 \end{cases}$$

$$f(x,y) \stackrel{\text{def}}{=} \begin{cases} \frac{xy}{x^2+y^2} + x\sin(\frac{1}{y}) & \text{if } y \neq 0 \\ 0 & \text{if } y = 0 \end{cases}$$

$$f(x,y) \stackrel{\text{def}}{=} \begin{cases} \frac{xy}{x^2+y^2} + y\sin(\frac{1}{x}) & \text{if } x \neq 0 \\ 0 & \text{if } x = 0. \end{cases}$$

THEOREM 2.1.4.1. (MOORE-OSGOOD) IF

i. X AND Y ARE SUBSETS OF HAUSDORFF SPACES AND $f \in \mathbb{R}^{X \times Y}$;
ii. $a \in \overline{X} \setminus X$, $b \in \overline{Y} \setminus Y$;
iii. $\lim_{x \to a} f(x,y) \stackrel{\text{def}}{=} h(y)$ ON $Y \setminus \{b\}$;
iv. $\lim_{y \to b} f(x,y) \stackrel{\text{def}}{=} g(x)$ UNIFORMLY ON $X \setminus \{a\}$,

THEN ALL THREE LIMITS IN (2.1.4.1) EXIST AND ARE EQUAL [O2].

[**Remark 2.1.4.1:** The result above can be generalized a bit if $\mathbb{R}$ is replaced by a Hausdorff space in which there is defined a *uniformity*.]

Exercise 2.1.4.7. Show that if

$$f(x,y) \stackrel{\text{def}}{=} \begin{cases} \frac{x^2-y^2}{x^2+y^2} & \text{if } x^2 + y^2 \neq 0 \\ 0 & \text{if } x = y = 0 \end{cases}$$

then

$$\lim_{x \to 0} \lim_{y \to 0} f(x,y) = 1$$

$$\lim_{y \to 0} \lim_{x \to 0} f(x,y) = -1$$

(each repeated limit exists but the two are not equal).

The importance of the conditions $a \in \overline{X} \setminus X, \ldots$, is emphasized by **Exercise 2.1.4.8** below.

Exercise 2.1.4.8. Show that if f is defined on $\mathbb{R}^2$ and

$$f(x,y) \stackrel{\text{def}}{=} \begin{cases} 1 & \text{if } xy \neq 0 \\ 0 & \text{if } xy = 0 \end{cases}$$

then

$$\lim_{x \to 0} f(x,y) \overset{\text{def}}{=} h(y) = \begin{cases} 1 & \text{if } y \neq 0 \\ 0 & \text{if } y = 0 \end{cases}$$

$$\lim_{y \to 0} f(x,y) \overset{\text{def}}{=} g(x) = \begin{cases} 1 & \text{if } x \neq 0 \\ 0 & \text{if } x = 0 \end{cases}$$

and both limits above are uniform. Show also that nevertheless

$$\lim_{(x,y) \to (0,0)} f(x,y)$$

does *not* exist. By considering f defined on $\mathbb{R}^2 \setminus \{ (x,y) \ : \ xy = 0 \}$ reconcile the result with THEOREM **2.1.4.1. 97**.

Exercise 2.1.4.9. Show that if

$$f(x,y) \overset{\text{def}}{=} \begin{cases} x^2 \sin(\frac{1}{x}) + y^2 \sin(\frac{1}{y}) & \text{if } xy \neq 0 \\ x^2 \sin(\frac{1}{x}) & \text{if } x \neq 0 \text{ and } y = 0 \\ y^2 \sin(\frac{1}{y}) & \text{if } x = 0 \text{ and } y \neq 0 \\ 0 & \text{if } x = y = 0 \end{cases}$$

then f is differentiable everywhere:

$$f_x(x,y) = \begin{cases} 2x \sin(\frac{1}{x}) - \cos(\frac{1}{x}) & \text{if} x \neq 0 \\ 0 & \text{if } x = 0 \end{cases}$$

$$f_y(x,y) = \begin{cases} 2y \sin(\frac{1}{y}) - \cos(\frac{1}{y}) & \text{if } y \neq 0 \\ 0 & \text{if } y = 0 \end{cases}$$

but neither f_x nor f_y is continuous at $(0,0)$.

[*Hint:* Use the *mean value theorem* in the following form.

There are two functions $\epsilon_1(h,k)$ and $\epsilon_2(h,k)$, defined if $h^2 + k^2 \neq 0$, such that

$$\lim_{(h,k) \to (0,0)} \epsilon_1(h,k) = \lim_{(h,k) \to (0,0)} \epsilon_2(h,k) = 0,$$

and such that

$$f(h,k) - f(0,0) = f_x(0,0)h + f_y(0,0)k + \epsilon_1(h,k) + \epsilon_2(h,k).]$$

Exercise 2.1.4.10. Show that if

$$f(x,y) \overset{\text{def}}{=} \begin{cases} xy \frac{x^2 - y^2}{x^2 + y^2} & \text{if } x^2 + y^2 \neq 0 \\ 0 & \text{if } x = y = 0 \end{cases}$$

then:

i.
$$f_y(x,0) = \begin{cases} x & \text{if } x \neq 0 \\ \lim_{k \to 0} \frac{f(0,k)}{k} = 0 & \text{if } x = 0 \end{cases}$$

$$f_x(0,y) = \begin{cases} -y & \text{if } y \neq 0 \\ \lim_{h \to 0} \frac{f(h,0)}{h} = 0 & \text{if } y = 0 \end{cases}$$

(each first partial derivative exists everywhere);
ii. each first partial derivative is continuous everywhere;
iii.
$$f_{xy}(0,0) = 1$$
$$f_{yx}(0,0) = -1.$$

Exercise 2.1.4.11. Show that if $L = \{ (x,y) \ : \ x \geq 0, \ y = 0 \}$ and f is defined according to

$$f : \mathbb{R}^2 \setminus L \ni (x,y) \mapsto \begin{cases} x^2 & \text{if } x > 0 \text{ and } y > 0 \\ 0 & \text{otherwise} \end{cases}$$

then:

i. f_{xx}, f_{xy}, f_{yx}, and f_{yy} are continuous (whence $f_{xy} = f_{yx}$);
ii. $f_y \equiv 0$ in $\mathbb{R}^2 \setminus L$;
iii. f is not independent of y in $\mathbb{R}^2 \setminus L$.

[**Note 2.1.4.2:**   Thus the implication

$$f_y \equiv 0 \Rightarrow f \text{ is independent of } y$$

is not valid without regard to the nature of the domain of f. The implication *is* valid, e.g., if every vertical line meets the domain of f in an interval.]

Exercise 2.1.4.12. Show that $f(x,y) \overset{\text{def}}{=} (y-x^2)(y-3x^2)$ has no *local extremum* at $(0,0)$. On the other hand, show that if f is confined to any line L through the origin then the resulting function has a local extremum at $(0,0)$ (cf. **Example 2.1.2.3. 60**).

[*Hint:* In every neighborhood of $(0,0)$ there is a point $(0,b)$ where f is positive and there is a point (a,a^2) where f is negative. On the vertical axis f has a minimum at $(0,0)$. On the horizontal axis f has a minimum at $(0,0)$. On the line for which the equation is $y = mx$, $m \neq 0$,

$$f(x,mx) = m^2 x^2 - 4mx^3 + 3x^4 \overset{\text{def}}{=} g(x),$$

$g'(0) = 0$, $g''(0) = 2m^2$, and g has a local minimum at $(0,0)$.]

Exercise 2.1.4.13. Show that if

$$f(x,y) \overset{\text{def}}{=} \begin{cases} \frac{x^3}{y^2} e^{-\frac{x^2}{y}} & \text{if } y > 0 \\ 0 & \text{if } y = 0 \end{cases}$$

and $x \neq 0$ then

$$\int_0^1 f(x,y)\, dy = xe^{-x^2}$$

$$\int_0^1 f_x(x,y)\, dy = e^{-x^2}(1 - 2x^2).$$

Show that

$$\int_0^1 f_x(0,y)\, dy = 0$$

and, consequently,

$$\frac{d}{dx} \int_0^1 f(x,y)\, dy \bigg|_{x=0} = 1 \neq \int_0^1 \frac{\partial f(0,y)}{\partial x}\, dy.$$

Exercise 2.1.4.14. Show that if

$$f(x,y) \overset{\text{def}}{=} \begin{cases} y^{-2} & \text{if } 0 < x < y < 1 \\ -x^{-2} & \text{if } 0 < y < x < 1 \\ 0 & \text{otherwise if } 0 \leq x \leq 1 \text{ and } 0 \leq y \leq 1 \end{cases}$$

then

$$\int_0^1 \int_0^1 f(x,y)\, dx\, dy = 1$$

$$\int_0^1 \int_0^1 f(x,y)\, dy\, dx = -1.$$

Exercise 2.1.4.15. Show that if

$$(a_{mn})_{m,n=1}^{\infty,\infty} \overset{\text{def}}{=} \begin{pmatrix} 0 & \frac{1}{2} & \frac{1}{4} & \cdots & \frac{1}{2^n} & \cdots \\ -\frac{1}{2} & 0 & \frac{1}{2} & \cdots & \frac{1}{2^{n-1}} & \cdots \\ -\frac{1}{2^2} & -\frac{1}{2} & 0 & \cdots & \frac{1}{2^{n-2}} & \cdots \\ \vdots & \vdots & \vdots & \ddots & \vdots & \ddots \end{pmatrix}$$

then

$$\sum_{n=1}^{\infty} a_{mn} = 2^{-m+1}, \; m \in \mathbb{N}$$

$$\sum_{m=1}^{\infty} a_{mn} = -2^{-n+1}, \; n \in \mathbb{N}$$

$$\sum_{m=1}^{\infty} \sum_{n=1}^{\infty} a_{mn} = 2$$

$$\sum_{n=1}^{\infty} \sum_{m=1}^{\infty} a_{mn} = -2.$$

[**Note 2.1.4.3:** In each of the last two instances there is a counterexample to a weakened version of *Fubini's theorem*. The condition $\int_{X \times Y} |g(x,y)| \, d\mu(x,y) < \infty$, for the appropriate interpretation of the *product measure* μ, is violated.]

In 1900 in Paris, Hilbert delivered a lecture on 23 open problems in mathematics [**Hi1**]. Problem 13 posed a question of which a generalized form is:

Can a continuous function f of more than one real variable be expressed as a sum of superpositions [compositions] of continuous functions of one variable?

Example 2.1.4.1. If

$$g(z) \stackrel{\text{def}}{=} \frac{z^2}{4}$$

$$h(z) \stackrel{\text{def}}{=} -\frac{z^2}{4}$$

$$p(z) \stackrel{\text{def}}{=} z$$

$$q(z) \stackrel{\text{def}}{=} -z$$

then

$$f(x,y) \stackrel{\text{def}}{=} xy = g[p(x) - q(y)] + h[p(x) + q(y)]. \tag{2.1.4.2}$$

Despite simple instances such as that in **Example 2.1.4.1** Hilbert conjectured that the answer to his (more restricted) question was "No." In a sequence of papers [**Arn, Ko3, Lor, Sp**] Hilbert's generalized question was discussed, answered "Yes" by Kolmogorov, and given the following dramatic resolution by Sprecher.

THEOREM **2.1.4.2.** ASSUME $2 \leq N \in \mathbb{N}$. THERE IS A MONOTONI-
CALLY INCREASING FUNCTION $\psi_N : [0,1] \mapsto [0,1]$ SUCH THAT $\psi_N([0,1]) = [0,1]$ AND $\psi \in \text{LIP}[\ln 2/\ln(2N+2)]$. IF $\delta > 0$ THERE IS IN $(0, \delta]$ A RA-
TIONAL NUMBER ϵ AND THERE ARE A CONTINUOUS FUNCTION χ AND A
CONSTANT λ SUCH THAT IF $2 \leq n \leq N$ THEN EVERY REAL FUNCTION f OF
n REAL VARIABLES MAY BE EXPRESSED ACCORDING TO THE FOLLOWING
EQUATION:

$$ f(x_1, \ldots, x_n) = \sum_{k=0}^{2n} \chi \left[\sum_{m=1}^{n} \lambda^m \psi_N(x_m + \epsilon k) + k \right]. \qquad (2.1.4.3) $$

THE CONTINUOUS FUNCTION χ DEPENDS ON f, THE CONSTANT λ IS IN-
DEPENDENT OF f, AND ψ_N, WHICH IS A FORTIORI CONTINUOUS, IS INDE-
PENDENT OF f AND n.

[**Remark 2.1.4.2:** The representation (2.1.4.3) is a vast improve-
ment over (2.1.4.2) in that there is only one function χ rather than
two functions g and h and there is only one function ψ_N rather
than two functions p and q.]

The proof of THEOREM **2.1.4.2** is long and detailed and is not repro-
duced here. However, some of its underlying ideas and techniques, namely
coverings and separating functions, are reminiscent of those in the proof of
the *Stone-Weierstraß* theorem [**HeSt, Loo**]. The main ingredients of the
argument may be described as follows.

    i. Assume $2n + 2 \leq \gamma \in \mathbb{N}$. For k in $\mathbb{N}$ there is a finite set Λ_k of indices
such that $S_k \overset{\text{def}}{=} \{S_k^0(i)\}_{i \in \Lambda_k}$ is a set of pairwise disjoint cubes in $\mathbb{R}^n$.
Each cube is of diameter not exceeding γ^{-k-1}. There is a vector $\mathbf{v}$
such that if

$$ S_k^q(i) \overset{\text{def}}{=} S_k^0(i) + q\mathbf{v}, \ q \in \mathbb{N}, \ S_k^q \overset{\text{def}}{=} \{S_k^q(i)\}_{i \in \Lambda_k} $$

then for each k in $\mathbb{N}$ the union $\bigcup_{q=0}^{n+m} S_k^q$ covers the unit cube

$$ \mathcal{E}^n \overset{\text{def}}{=} \{ (x_1, \ldots, x_n) \ : \ 0 \leq x_i \leq 1, \ 1 \leq i \leq n \} $$

$m + 1$ times. Furthermore the labellings are such that if $i_0 \in \Lambda_k$ there
are in Λ_k uniquely determined indices $i_0, \ldots, i_{2n}$ such that

$$ \bigcap_{q=0}^{2n} S_k^q(i_q) \neq \emptyset. $$

    ii. If $\{h^q\}_{1 \leq q \leq n+m}$ is a set of continuous functions that *separate* points of
$\mathcal{E}^n$ and if, for each k,

$$ h^q[S_k^q(i)] \cap h^r[S_k^r(j)] = \emptyset, \ r \neq q, \ i \neq j \qquad (2.1.4.4) $$

then for every continuous function f there is a continuous function χ such that f can be represented in the form

$$f(x_1, \ldots, x_n) = \sum_{q=0}^{m} \chi \left[h^q(x_1, \ldots, x_n) \right].$$

iii. If $\delta > 0$ and $k_0 \in \mathbb{N}$ is such that for $k \geq k_0$, $\epsilon \overset{\text{def}}{=} (\gamma - 1)^{-1} \gamma^{-k} \leq \delta$, there is a monotonically increasing function ψ mapping $\mathcal{E}^1$ on itself and there is a constant λ such that the functions

$$g^q(x_1, \ldots, x_n) \overset{\text{def}}{=} \sum_{m=1}^{n} \lambda^m \psi(x_m + \epsilon q) + q, \ 0 \leq q \leq 2n,$$

behave like the h^q in (2.1.4.4).

iv. If $k \geq k_0$ the sets $S_k^q(i)$ are defined via the parameters γ, q, k, and i restricted to the set $\Lambda_k^q \overset{\text{def}}{=} [(\gamma^k - 1)\epsilon q, \gamma^k + (\gamma^k - 1)\epsilon q] \cap \mathbb{N}$ as follows:

$$e_k(i) \overset{\text{def}}{=} i\gamma^{-k}$$
$$\delta_k \overset{\text{def}}{=} \frac{\gamma - 2}{\gamma - 1} \gamma^{-k}$$
$$E_k(i) \overset{\text{def}}{=} [e_k(i), e_k(i) + \delta_k]$$
$$E_k^q(i) \overset{\text{def}}{=} [e_k(i) - \epsilon q, e_k(i) + \delta_k - \epsilon q].$$

If $i_q \overset{\text{def}}{=} \{i_{1q}, \ldots, i_{nq}\} \subset \Lambda_k^q$ the corresponding Cartesian product

$$\prod_{1 \leq p \leq n} E_k^q(i_{pq})$$

is the cube $S_k^q(i_q)$ in $\mathbb{R}^n$.

v. The construction of the function ψ is based on the intervals $E_k^q(i)$ much as the construction of the Cantor function $\mathcal{C}_0$ is based on the intervals deleted in the formation of the Cantor set C_0.

In [Sp] all the details are given while Lorentz gives a perspicuous presentation for the case in which $N = 2$ [Lor].

2.2. Measure Theory

2.2.1. Measurable and nonmeasurable sets

The setting for discussion of measure theory is a *measure situation* $(X, \mathcal{S}, \mu)$, i.e., a set X, a σ-*ring* $\mathcal{S}$ consisting, by definition, of the *measurable subsets* of X, and a *countably additive* set function μ, here called a *measure*: $\mu : \mathcal{S} \ni A \mapsto \mu(A) \in [0, \infty]$. Very frequently X is $\mathbb{R}^n$ for some n in $\mathbb{N}$, $\mathcal{S}$ is the σ-ring $\mathcal{S}(\mathsf{K})$ generated by the compact sets of $\mathbb{R}^n$ or the σ-ring $\mathcal{L}$ consisting of all Lebesgue measurable sets in $\mathbb{R}^n$, and μ is n-dimensional Lebesgue measure λ_n ($\lambda_1 \overset{\text{def}}{=} \lambda$). In a *locally compact group* G the σ-ring $\mathcal{S}$ is the σ-ring $\mathcal{S}(\mathsf{K})$ generated by the compact sets of G and the measure μ is a (*translation-invariant*) Haar measure:

$$A \in \mathcal{S}, x \in G \Rightarrow xA \in \mathcal{S} \text{ and } \mu(A) = \mu(xA).$$

The facts about measure theory are discussed in some detail in [**Halm, Loo, Rao, Roy, Rud, Sz.-N**]. Important results in measure theory as it applies to Haar measure on locally compact groups, e.g., to Lebesgue measure λ_n on $\mathbb{R}^n$, are:

i.. a set of measure zero, i.e., a null set, contains no nonempty open set;

ii. if A is a set of positive measure then AA^{-1} contains a neighborhood of the identity; since $\mathbb{R}$ is a group in which the binary operation is written additively the set AA^{-1} in $\mathbb{R}$ is written $A - A$.

[**Remark 2.2.1.1:** Although measurable sets and measurable functions are treated in separate **Subsections** of this book, there is no essential distinction between them. If one accepts *measurable set* as a primitive notion, then a measurable function is nothing more than the limit of a sequence of *simple* functions, each of which is a linear combination of characteristic functions of measurable sets. If one accepts, e.g., as in the development of the *Daniell integral* [**Loo, Rao, Roy**], *measurable function* as a primitive notion (derived in turn from an even more elementary notion, that of a nonnegative linear functional defined on a *linear lattice* of extended $\mathbb{R}$-valued functions), then a measurable set is nothing more than a set for which the *characteristic function* is a measurable function.

Thus a result about measurable sets has its counterpart in a result about measurable functions and vice versa. Similar comments apply to sets that have, in $\mathbb{R}^n$, *n-dimensional content* and to functions that are *Riemann integrable* over subsets of $\mathbb{R}^{\mathbb{R}^n}$ [**O3**].

For purposes of illustration, the somewhat artificial distinctions above are useful.]

THEOREM **1.1.4.2. 7** is not an accident. Indeed, Sierpinski [**Si2**] established the following result.

THEOREM **2.2.1.1.** IF $B \overset{\text{def}}{=} \{x_\lambda\}_{\lambda \in \Lambda}$ IS A HAMEL BASIS FOR $\mathbb{R}$ OVER $\mathbb{Q}$ THEN B IS LEBESGUE MEASURABLE IFF $\lambda(B) = 0$.

PROOF. Just the "only if" requires serious attention. Assume B is a measurable Hamel basis and that $\lambda(B) > 0$. It follows that $B - B$ contains a neighborhood of 0, in particular infinitely many rational numbers. Assume that r and s are different nonzero rational numbers in $B - B$. Then $r \neq s$, $rs \neq 0$, and there is in $\mathbb{Q}$ a t and in B elements x_{λ_i}, $1 \leq i \leq 4$, such that

$$r = x_{\lambda_1} - x_{\lambda_2}$$
$$s = x_{\lambda_3} - x_{\lambda_4}$$
$$tr = s = x_{\lambda_3} - x_{\lambda_4} = t(x_{\lambda_1} - x_{\lambda_2}),$$

in contradiction of the linear independence of B over $\mathbb{Q}$.

$\square$

The set of *Borel* sets in $\mathbb{R}^n$ is $\mathcal{S}(\mathsf{O})$, the σ-ring generated by the open sets in $\mathbb{R}^n$. In $\mathbb{R}^n$ the σ-rings $\mathcal{S}(\mathsf{F})$ (generated by the closed sets) and $\mathcal{S}(\mathsf{K})$ (generated by the compact sets) are the same as the set of Borel sets.

[**Note 2.2.1.1:** If $\mathbb{R}^n$ is given the discrete topology so that every set is both open and closed and a set is compact iff it is finite, then

$$\mathcal{S}(\mathsf{O}) = \mathcal{S}(\mathsf{F}) \supsetneq \mathcal{S}(\mathsf{K}).]$$

In [**Si2**] there is also a proof of the next result.

THEOREM **2.2.1.2.** No Hamel basis B can be Borel measurable.

[**Remark 2.2.1.2:**   Hence the Hamel basis B of THEOREM **1.1.4.2. 7** is a non-Borel subset of the Borel set C_0.

The cardinality of the set of all Borel sets is $\#(\mathbb{R})$ whereas the cardinality of $\mathcal{P}(C_0)$, the *power set* of C_0, is $2^{\#(\mathbb{R})}$. It follows, without reference to THEOREM **2.2.1.2**, that there are non-Borel sets of measure zero. Since, for any function f, $\mathrm{Discont}(f)$ is an F_σ it follows that there are null sets that cannot be $\mathrm{Discont}(f)$ for any function f.]

THEOREM **2.2.1.3.** In every nonempty neighborhood U of 0 in $\mathbb{R}$ there is a Hamel basis for $\mathbb{R}$ over $\mathbb{Q}$.

PROOF. Let r be a positive rational number such that $(-r, r) \subset U$ and let H be some Hamel basis for $\mathbb{R}$ over $\mathbb{Q}$. For each h in H there is in $\mathbb{Z}$ a unique m_h such that $m_h r \leq h < (m_h + 1)r$. Let k_h be $h - (m_h + 1)r$. Then $k_h \in (-r, r)$. If K is a maximal linearly independent subset of $\{ k_h : h \in H \} \cup \{\frac{r}{2}\}$ then K is a Hamel basis for $\mathbb{R}$ over $\mathbb{Q}$ and $K \subset (-r, r) \subset U$.

$\square$

The result above is a special case of a more general phenomenon:

In any neighborhood U of the identity in a *Lie group* G there is a *relatively free* subset [Ge5]. In the Lie group $\mathbb{R}$ a maximal relatively free subset of U is perforce a Hamel basis.

The existence of nonmeasurable (Lebesgue) subsets of $\mathbb{R}$ cannot be based on a cardinality argument. The Cantor set C_0 has the cardinality of $\mathbb{R}$. Since $\lambda(C_0) = 0$ every subset of C_0 is Lebesgue measurable it follows that the cardinality of the set $\mathcal{L}$ of all Lebesgue measurable sets is $2^{\#(\mathbb{R})}$, which is also the cardinality of the *power set* $\mathcal{P}(\mathbb{R}) \overset{\text{def}}{=} 2^{\mathbb{R}}$ of $\mathbb{R}$.

Example 2.2.1.1. The map $\theta : \mathbb{R} \ni t \mapsto e^{2\pi i t}$ algebraically and measure-theoretically identifies $\mathbb{R}/\mathbb{Z}$ with the compact multiplicative group $\mathbb{T}$, and identifies $\mathbb{Q}/\mathbb{Z}$ with a countable and infinite subgroup H of $\mathbb{T}$. The Axiom of Choice implies that there is in $\mathbb{T}$ a set S consisting of exactly one element from each of the cosets of H. If r_1 and r_2 are different elements of H and $r_1 S = r_2 S$ then in S there are s_1 and s_2 such that $r_1 s_1 = r_2 s_2$. But then s_1 and s_2 are in the same coset of H, whence the nature of S implies $s_1 = s_2$, i.e., $r_1 = r_2$, a contradiction. Thus

$$\mathbb{T} = \bigcup_{r \in H} rS.$$

If S is measurable then, since λ transferred to $\mathbb{T}$ is again "translation" invariant, S and all the rS have same measure a: $\lambda(rS) \equiv a$. Then

EITHER
$a = 0$, in which case

$$\lambda(\mathbb{T}) = \sum_{r \in H} \lambda(rS) = 0,$$

OR
$a > 0$, in which case

$$\lambda(\mathbb{T}) = \sum_{r \in H} \lambda(rS) = \infty.$$

Since neither conclusion is correct, S is not measurable nor is its counterpart $\theta^{-1}(S) \cap [0, 1)$.

Example 2.2.1.2. Any countable and infinite subgroup G of $\mathbb{T}$ may serve instead of H in the discussion above. In particular, if α is an irrational real number and $\xi \overset{\text{def}}{=} e^{2\pi i \alpha}$ then either of the subgroups

$$A \overset{\text{def}}{=} \{ \xi^n : n \in \mathbb{Z} \}, \ B \overset{\text{def}}{=} \{ \xi^{2n} : n \in \mathbb{Z} \}$$

may be used. Note that:

i. B is a subgroup of index 2 in A;
ii. $B \cap \xi B = \emptyset$ and $A = B \dot\cup \xi B$;
iii. because α is irrational both subgroups A and B are (countably) infinite dense subgroups of the compact group $\mathbb{T}$.

Let P consist of exactly one element of each coset of A and let M be $\mathsf{P}B$. If $MM^{-1} \cap \xi B \neq \emptyset$, i.e., if

$$\rho_i \in \mathsf{P}, b_i \in B, x_i = \rho_i b_i, \ i = 1, 2,$$

and $x_1 x_2^{-1} \in \xi B$, then $\rho_1 \rho_2^{-1} \in \xi B \subset A$ and so, owing to the nature of P, $\rho_1 = \rho_2$. Thus $x_1 x_2^{-1} = b_1 b_2^{-1} \in B$, i.e., $x_1 x_2^{-1} \in \xi B \cap B = \emptyset$, a contradiction whence $MM^{-1} \cap \xi B = \emptyset$. If L is a measurable subset of M and $\lambda(L) > 0$ then $MM^{-1} \supset LL^{-1}$, which contains a $\mathbb{T}$-neighborhood of 1 (cf. THEOREM **1.1.4.1. 5**) and thus an element of the dense set ξB, a contradiction. It follows that the *inner measure* of M is zero: $\lambda_*(M) = 0$.

For x in $\mathbb{T}$ there is in P a ρ such that $x\rho^{-1} \overset{\text{def}}{=} a \in A$. If $x \notin M$ then $a \notin B$ whence for some b in B, $x = \rho\xi b \in \mathsf{P}\xi B = \xi M$. Thus

$$\mathbb{T} \setminus M \overset{\text{def}}{=} M^c \subset \xi M$$

and so $\lambda_* (M^c) = 0$. The inner measure λ_* and *outer measure* λ^* are set functions such that for each measurable set P,

$$\lambda^*(P \cap M) + \lambda_* (P \cap M^c) = \lambda(P),$$

whence $\lambda^*(P \cap M) = \lambda(P)$, in particular, $\lambda^*(M) = 1 > 0 = \lambda_*(M)$.

The set $\theta^{-1}(M) \overset{\text{def}}{=} \widetilde{M}$ in $\mathbb{R}$ has properties analogous to those of M.

i. The set $\widetilde{M}$ is nonmeasurable, $\lambda_* \left(\widetilde{M} \right) = 0$, and $\lambda^* \left(\widetilde{M} \right) = \infty$.
ii. The set $\widetilde{M}$ is *thick* and for every measurable subset P of $\mathbb{R}$,

$$\lambda_*(P \cap M) = 0 \text{ while } \lambda^*(P \cap M) = \lambda(P).$$

Exercise 2.2.1.1. Let G be an infinite subgroup of $\mathbb{T}$. Show:

i. the identity is a limit point of G;
ii. every infinite subgroup of $\mathbb{T}$ is dense in $\mathbb{T}$.
iii. the compact subset $1 \times \mathbb{T}$ is a nowhere dense infinite subgroup of the compact group $\mathbb{T}^2$.

The Cantor set C_0 is one of a family of *nowhere dense perfect* sets. The construction of a typical member of the family is a modification of the construction of the Cantor set C_0.

Example 2.2.1.3. If $\epsilon \in \mathbb{Q} \cap (0,1)$ let α_n be $\epsilon \cdot 2^{-2n+1}$, $n \in \mathbb{N}$. Then $\sum_{n=1}^{\infty} 2^{n-1}\alpha_n = \epsilon$. Let τ_1 and τ_2 be two transcendental numbers such that $\tau_1 < 0$, $\tau_2 > 1$ and $\tau_2 - \tau_1 = 1 + 2\epsilon$ and let $\{r_k\}_{k\in\mathbb{N}}$ be an enumeration of the set $S \overset{\text{def}}{=} [\tau_1, \tau_2] \cap \mathbb{A}$ consisting of the algebraic numbers in $[\tau_1, \tau_2] \overset{\text{def}}{=} I$.

Let $\{I_{mn}, m \in \mathbb{N}, 1 \leq n \leq 2^{m-1}\}$ be the set of open intervals deleted from $[0,1]$ in the construction of the Cantor set C_0. The first open interval I_{11} of length 3^{-1}, the next two I_{21}, I_{22} each of length $3^{-2}, \ldots$ are, for the current construction, replaced by open intervals $J_1, J_2, J_3, \ldots$ so that the endpoints of each J_n are transcendental and:

$$\lambda(J_1) \leq \alpha_1$$
$$\lambda(J_k) \leq \alpha_2, \ k = 2, 3$$
$$\cdots$$
$$\lambda(J_k) \leq \alpha_n, \ 2^{k-1} \leq k \leq 2^k - 1$$
$$\cdots.$$

Furthermore let J_1 be placed to contain r_1: $r_1 \in J_1$. Let r_{k_1} be the first r_k not in J_1. There is a first n, say n_1, such that J_{n_1} may be chosen to contain r_{k_1} and to be disjoint from J_1: $r_{k_1} \in J_{n_1}$, $J_1 \cap J_{n_1} = \emptyset$. By induction one may find a sequence $\{r_{k_p}\}_{p\in\mathbb{N}}$ in S and a sequence $\{J_{n_p}\}_{p\in\mathbb{N}}$ in $\{J_n\}_{n\in\mathbb{N}}$ such that

i. r_{k_p} is the first r_k not in

$$J_1 \cup J_{n_1} \cup \cdots \cup J_{n_{p-1}};$$

ii. $r_{k_p} \in J_{n_p}$;
iii.

$$J_{n_p} \cap \left(J_1 \cup \bigcup_{q=1}^{p-1} J_{n_q}\right) = \emptyset.$$

It follows that:

iv. $J \overset{\text{def}}{=} J_1 \cup \left(\bigcup_{p\in\mathbb{N}} J_{n_p}\right)$ is an open subset of I;
v. $I \setminus J \overset{\text{def}}{=} C_J$ contains no algebraic numbers, i.e., consists entirely of transcendental numbers;
vi. C_J is nowhere dense in $[\tau_1, \tau_2]$ and perfect;
vii. $1 + 2\epsilon \geq \lambda(C_J) \geq 1 + \epsilon$.

Exercise 2.2.1.2. Show that C_J is nowhere dense in $[\tau_1, \tau_2]$ and perfect.

[*Hint:* The complement $I \setminus C_J$ of C_J in I is open and is dense in I. To show C_J is perfect it suffices to prove that each of the countably many endpoints of the intervals J_{n_p} is a *limit point* of C_J.]

Exercise 2.2.1.3. Repeat the construction in **Example 2.2.1.3** with the following modification: $0 < \tau_1 < \tau_2 < 1$, $\tau_2 - \tau_1 > 1 - \epsilon$. The resulting set, say D, should consist entirely of transcendental numbers, be nowhere dense in $[\tau_1, \tau_2]$, be perfect, and have measure greater than $(\tau_2 - \tau_1) - 2\epsilon$. Construct a sequence $\{D_n\}_{n \in \mathbb{N}}$ of sets so that each consists entirely of transcendental numbers, is nowhere dense in $[0, 1]$, and is perfect. Furthermore, the following should obtain:

$$D_n \subset D_{n+1} \subset [0, 1], \ n \in \mathbb{N}, \ \lambda \left(\bigcup_{n \in \mathbb{N}} D_n \right) \stackrel{\text{def}}{=} \lambda(D_\infty) = 1.$$

Hence the set D_∞:

   i. consists of transcendental numbers;
  ii. is dense in $[0, 1]$;
 iii. is an F_σ;
  iv. is a set of the first category in $[0, 1]$;

    Furthermore $E \stackrel{\text{def}}{=} [0, 1] \setminus D_\infty$ is a null set of the second category.

Exercise 2.2.1.4. Assume the endpoints of the closed interval $[a, b]$ are rational. In $[a, b]$ construct a Cantor-like set $C_{a,b}$ such that $\lambda(C_{a,b}) = 0$. Show that the union

$$H \stackrel{\text{def}}{=} \bigcup_{a, b \in \mathbb{Q}, \ a < b} C_{a,b}$$

is a null set and that every x in $\mathbb{R}$ is a point of condensation of H (cf. **Exercise 2.1.1.18. 52**).

    [**Remark 2.2.1.3:** Although every point x of $\mathbb{R}$ is a point of condensation of the set $\mathbb{I}$ of irrational numbers, $\mathbb{I}$ is quite the opposite of a null set since its complement $\mathbb{Q}$ is a null set.]

    When G is a group and A and B are subsets of G the sets AB and AB^{-1} frequently enter arguments about G. When G is abelian and its binary operation is written as $+$ the corresponding sets are the *sum set* $A + B$ and *difference set* $A - B$. When there is a measure situation $(G, \mathcal{S}, \mu)$, in particular when G is a locally compact group and μ is translation invariant (Haar) measure, there arises the question:

    If A and B are measurable is AB measurable?

    A negative answer is provided by the next results.

Example 2.2.1.4. Let G be $\mathbb{R}^2$ and let E be a nonmeasurable subset of $\mathbb{R}^+$. If $A \stackrel{\text{def}}{=} (\{0\} \times E) \cup (E \times \{0\})$ then $\lambda_2(A) = 0$ whence A is measurable (λ_2). Since

$$A + A = [\{0\} \times (E + E)] \cup [(E + E) \times \{0\}] \cup [E \times E] \stackrel{\text{def}}{=} P \cup Q \cup R$$

and since $\lambda_2(P) = \lambda_2(Q) = 0$, if $A + A$ is measurable then so is $E \times E$. However if $x \in E$ the *section* $(E \times E)_x$ is E and so not *almost every* section of $A + A$ is measurable, a contradiction of Fubini's theorem.

Example 2.2.1.5. Let B be a Hamel basis contained in the Cantor set C_0. If $S_1 \stackrel{\text{def}}{=} \bigcup_{r \in \mathbb{Q}}(r + B)$ then $\lambda(S_1) = 0$ and $S_1 = -S_1$. Let S_{n+1} be $S_n + S_n (= S_n - S_n)$, $n \in \mathbb{N}$. If

$$x \stackrel{\text{def}}{=} \sum_{b \in B} a_b b \in S_n, \ a_b \in \mathbb{Q}$$

then x is a sum of not more than 2^{n-1} members of S_1 and hence at most 2^{n-1} of the coefficients a_b are not zero. Since $C_0 + C_0 = [0,2]$ it follows that $\bigcup_{n \in \mathbb{N}} S_n = \mathbb{R}$ and if each S_n is measurable then one of them, say S_{n_0}, has positive Lebesgue measure: $\lambda(S_{n_0}) > 0$. Hence for some M in $\mathbb{N}$

$$S_{n_0} - S_{n_0} (= S_{n_0} + S_{n_0} = S_{n_0+1})$$

contains a neighborhood $U \stackrel{\text{def}}{=} (-2^{-M}, 2^{-M})$. If $K \in \mathbb{N}$ and

$$2^{n_0} < K$$
$$\mathbb{Q} \setminus \{0\} \supset \{r_1, \ldots, r_K\} \text{ (a } K\text{-element set)}$$
$$K \sup_{1 \leq k \leq K} |r_k| < 2^{-M}$$

then whenever $\{b_1, \ldots, b_K\}$ is a K-element subset of B it follows that

$$\sum_{k=1}^{K} r_k b_k \in U \setminus S_{n_0+1} = \emptyset,$$

a contradiction.

Thus some S_n is nonmeasurable and if S_{n_1} is the first nonmeasurable S_n then $n_1 > 1$. Hence S_{n_1-1} is measurable and

$$S_{n_1-1} + S_{n_1-1} (= S_{n_1})$$

is nonmeasurable. (This result was communicated to the writers by Harvey Diamond and Gregory Gellés.)

Example 2.2.1.6. The Cantor-like sets D_0 and D_n in **Example 2.1.1.3. 51** and **Exercise 2.2.1.3. 109** may be chosen so that $\lambda(D_0) = \alpha_0$, $\lambda(D_n) = \alpha_n$, and $\alpha_0 + \sum_{n \in \mathbb{N}} \alpha_n = 1$. In that event D_0 and each D_n is nowhere dense whence the corresponding set D that is the union of them all is of the first category and $\lambda(D) = 1$. The complement $[0,1] \setminus D$ is perforce of the second category and its measure is zero. Relative to $[0,1]$ the set D is *thick*.

Exercise 2.2.1.5. Let A be the (countable) set of endpoints of the intervals deleted in the construction of C_0 and let B be $C_0 \setminus A$. Show that A and B are disjoint nowhere dense sets such that each point of A resp. B is a limit point of B resp. A.

From **Example 2.2.1.6** it follows that category and measure are, at best, loosely related. There are sets of the first category that have measure zero, e.g., $\mathbb{Q}$, and sets of the first category that are thick, e.g., on each interval $[n, n+1]$, $n \in \mathbb{Z}$ construct a set P_n just like D. Then set $P \stackrel{\text{def}}{=} \bigcup_{n \in \mathbb{Z}} P_n$ is such that $\lambda(\mathbb{R} \setminus P) = 0$. There are sets of the second category, e.g., $\mathbb{R} \setminus P$, that have measure zero and sets of the second category, e.g., $\mathbb{R}$, that are thick **[Ox]**.

Category is not preserved under homeomorphism. To see this call a linearly ordered set Λ *ordinally dense* if it has neither first nor last member and between any two members there is a third. For example $\mathbb{Q}$ in its natural order is ordinally dense; $\mathbb{Z}$ is not ordinally dense in its natural order; neither $\mathbb{N}$ in its natural order nor any other well-ordered set in the order of its well-ordering is ordinally dense.

Two ordered sets are *ordinally similar* if there is an order-preserving bijection between them.

THEOREM 2.2.1.4. IF $A \stackrel{\text{def}}{=} \{a_n\}_{n \in \mathbb{N}}$ AND $B \stackrel{\text{def}}{=} \{b_n\}_{n \in \mathbb{N}}$ ARE TWO (COUNTABLE) ORDINALLY DENSE SETS THEY ARE ORDINALLY SIMILAR.

PROOF. The order-preserving bijection is defined by induction:

Let α_1 be a_1 and β_1 be b_1. If $\alpha_1, \ldots, \alpha_{n-1}$ and $\beta_1, \ldots, \beta_{n-1}$ have been chosen so that $\alpha_i \leftrightarrow \beta_i$, $1 \le i \le n-1$, is an order-preserving bijection and n is even let α_n be the first a_m not yet chosen and let β_n be the first b_m not yet chosen and order-related to $\{\beta_1, \ldots, \beta_{n-1}\}$ as α_n is order-related to $\{\alpha_1, \ldots, \alpha_{n-1}\}$. If n is odd let β_n be the first b_m not chosen and let α_n be the first a_m not chosen and order-related to $\{\alpha_1, \ldots, \alpha_{n-1}\}$ as β_n is order-related to $\{\beta_1, \ldots, \beta_{n-1}\}$.

The method of choice is such that $\{\alpha_n\}_{n \in \mathbb{N}} = A$ and $\{\beta_n\}_{n \in \mathbb{N}} = B$ and the bijection $\alpha_n \leftrightarrow \beta_n$ is order-preserving.

$\square$

In particular, the set $\{I_n\}_{n \in \mathbb{N}}$ of intervals deleted in the construction of C_0 is ordinally dense if $I \prec I'$ is taken to mean that I is to the left of I'. Let the set $(0,1) \cap \mathbb{Q} \stackrel{\text{def}}{=} \{r_n\}_{n \in \mathbb{N}} \stackrel{\text{def}}{=} A$, which is also ordinally dense, be in bijective order-preserving correspondence with $\{I_n\}_{n \in \mathbb{N}}$. Define f on $\bigcup_{n \in \mathbb{N}} \overline{I_n}$ so that if $x \in \overline{I_n}$ then $f(x) = r_n$. Thus f is monotone increasing, its range $\{r_n\}_{n \in \mathbb{N}}$ is dense in $[0,1]$, and so f may be extended to a continuous monotone increasing function, again called f, on $[0,1]$. Let B be C_0 shorn of the endpoints of the deleted intervals. Then f maps $[0,1] \setminus B$ onto $\{r_n\}_{n \in \mathbb{N}}$. Owing to the ordinal similarity of $\{r_n\}_{n \in \mathbb{N}}$ and $\{I_n\}_{n \in \mathbb{N}}$, f is increasing on $[0,1]$, strictly increasing and bicontinuous on B, and also $f(B) = [0,1] \setminus \mathbb{Q}$. Thus C_0 (and hence any Cantor-like set) shorn of its endpoints is homeomorphic to the set of $\mathbb{I}_{[0,1]}$ of irrational numbers in $[0,1]$. However B, as a subset of C_0, is nowhere dense and hence of the first category while $\mathbb{I}_{[0,1]}$ is of the second category.

Exercise 2.2.1.6. Show that B above is homeomorphic to $\mathbb{R} \setminus \mathbb{Q}$. Hence a nowhere dense set B is homeomorphic to a dense set $\mathbb{R} \setminus \mathbb{Q}$.

More interesting phenomena in the relationships between measure-theoretic and topological properties arise in the context described below.

The Cantor set C_0, $\{0,1\}^{\mathbb{N}}$, the countable *Cartesian product* of the two-point set $\{0,1\}$ in its discrete topology, is the source of some of these phenomena. More generally, for an arbitrary infinite set M, let $\{0,1\}^{\mathsf{M}}$ be the (possibly uncountable) Cartesian product or *dyadic space* $\mathcal{D}^{\mathsf{M}}$.

The *weight* W of a topological space X is the least of the cardinal numbers W such that the topology of X has a *base* of cardinality W. If $\mathcal{U} \overset{\text{def}}{=} \{U_j\}_{j \in J}$ is a base and $\#(J) = \mathsf{W}$ then $\mathcal{U}$ is a *minimal base* for X.

LEMMA 2.2.1.1.

i. EVERY SEPARABLE METRIC SPACE IS THE CONTINUOUS IMAGE OF A SUBSET OF THE CANTOR SET.

ii. EVERY COMPACT METRIC SPACE IS THE CONTINUOUS IMAGE OF THE CANTOR SET.

ii′ EVERY COMPACT *Hausdorff* SPACE X OF WEIGHT $\#(\mathsf{M})$ IS THE CONTINUOUS IMAGE OF A CLOSED SUBSET OF $\mathcal{D}^{\mathsf{M}}$.

iii. EVERY COMPACT *totally disconnected* METRIC SPACE IS THE HOMEOMORPHIC IMAGE OF A SUBSET OF THE CANTOR SET.

iv. EVERY COMPACT TOTALLY DISCONNECTED *perfect* METRIC SPACE IS THE HOMEOMORPHIC IMAGE OF THE CANTOR SET.

[The fundamental idea behind the proof of *ii′* can be described as follows. If $\#(J) = \#(\mathsf{M})$ and if

$$\xi \overset{\text{def}}{=} \{\xi_j\}_{j \in J} \in \mathcal{D}^{\mathsf{M}}$$

let $\mathcal{U} \overset{\text{def}}{=} \{U_j\}_{j \in J}$ be a minimal base for X. For each j in J there is defined a *dyad* of closed sets:

$$A_j^i \overset{\text{def}}{=} \begin{cases} \overline{U_j} & \text{if } i = 0 \\ X \setminus U_j & \text{if } i = 1. \end{cases}$$

Then $A_\xi \overset{\text{def}}{=} \bigcap_{j \in J} A_j^{\xi_j}$ is either $\emptyset$ or a single point. Let Ξ be $\{\, \xi \; : \; A_\xi \neq \emptyset \,\}$. Then Ξ is a closed subset of $\mathcal{D}^{\mathsf{M}}$, the map

$$F : \Xi \ni \xi \mapsto A_\xi$$

is continuous, and $F(\Xi) = X$. (If $\#\mathsf{M} = \#(\mathbb{N})$ the map F can be extended to a continuous map $\Phi : \mathcal{D}^{\mathsf{M}} \mapsto X$, i.e., ii.)]

See [**AH, Bou, Cs, Eng, HeSt, Kur, Rin**] for detailed proofs of the various parts of LEMMA **2.2.1.1**.

Note that ii' is an imperfect counterpart of ii. In fact, there is no perfect counterpart to ii, as the contents of **Exercises 2.2.1.7** and **2.2.1.8** below show [**Eng**].

Exercise 2.2.1.7. Let X be a set such that $\#(X) > \#(\mathbb{N})$. Fix a point x_0 in X. Define a topology by declaring that a subset A of X is open iff $x_0 \notin A$ or $X \setminus A$ is finite. Show that:

i. X is a compact Hausdorff space;
ii. every one-point set $\{x\}$ other than $\{x_0\}$ is open;
iii. the weight of X is $\#(X)$.

Assume that for some M there is a continuous surjection $f : \mathcal{D}^{\mathsf{M}} \mapsto X$. Then, since each point other than x_0 in X is open, the set

$$\{\, f^{-1}(x) \; : \; x \neq x_0 \,\}$$

consists of uncountably many pairwise disjoint open subsets of $\mathcal{D}^{\mathsf{M}}$.

Exercise 2.2.1.8. Let $\mathcal{U}$ be a set of *basic neighborhoods* for $\mathcal{D}^{\mathsf{M}}$. Show that if the elements of $\mathcal{U}$ are pairwise disjoint then $\mathcal{U}$ is empty, finite, or countable. Show that if $\mathcal{O}$ is a set of pairwise disjoint open subsets of $\mathcal{D}^{\mathsf{M}}$ then $\mathcal{O}$ is empty, finite, or countable. Why do the preceding conclusions show that X in **Exercise 2.2.1.7** is not the continuous image of some dyadic space $\mathcal{D}^{\mathsf{M}}$?

An *arc-image* resp. *open arc-image* γ^* is, for some *arc* resp. *open arc* γ in $C([0,1], X)$ resp. $C((0,1), X)$ the set $\gamma([0,1])$ resp. $\gamma((0,1))$. If γ is injective, the image is *simple*. The *endpoints* of an arc-image γ^* are $\gamma(0)$ and $\gamma(1)$. If $\gamma(0) = \gamma(1)$, γ^* is *closed*; if, to boot, γ is injective on $(0,1)$, γ^* is a *simple closed curve-image* or *Jordan curve-image*.

[**Note 2.2.1.2:** The image γ^*, a subset of a topological space X, is by definition different from γ itself, which is a continuous function. (Nevertheless, by abuse of language, the distinction is occasionally blurred and, e.g., "A Jordan curve in $\mathbb{R}^2$ separates the plane," is an acceptable substitute for the more accurate, "A Jordan curve-image in $\mathbb{R}^2$ separates the plane.")]

For an arc $\gamma : [0, 1] \mapsto (x_1(t), \ldots, x_n(t)) \in \mathbb{R}^n$, the length $\ell(\gamma)$ of γ is defined to be

$$\sup_{0=t_1<\cdots<t_N=1} \sum_{i=2}^{N} \sqrt{\sum_{j=1}^{n} (x_j(t_i) - x_j(t_{i-1}))^2}, \ N \in \mathbb{N}.$$

However, the length $L(\gamma^*)$ of the arc-image γ^* is the infimum, taken over the set P of all *parametric descriptions* s of γ^*, of $\ell(\gamma \circ s)$. Each parametric description is a continuous autojection $s : [0, 1] \mapsto [0, 1]$. Thus

$$L(\gamma^*) \overset{\text{def}}{=} \inf_{\{s \ : \ s \in P\}} \ell(\gamma \circ s).$$

The length of an arc and the length of the corresponding arc-image can be quite different. The length $\ell(\gamma)$ can be infinite while $L(\gamma^*)$ is, in the usual geometric sense, finite.

Example 2.2.1.7.

i. Let γ be
$$\begin{cases} x = \cos 4\pi t \\ y = \sin 4\pi t \end{cases}, \ t \in [0, 1].$$

Then $\ell(\gamma) = 4\pi$, whereas γ^* is a circle of radius 1 and $L(\gamma^*) = 2\pi$.

ii. Let g be a continuous nowhere differentiable function on $[0, 1]$ and let γ be
$$[0, 1] \ni t \mapsto \begin{cases} x = g(t) \\ y = g(t). \end{cases}$$

Then, since g is not of bounded variation on any nondegenerate interval, if $0 \le a < b \le 1$ the arc defined by restricting γ to $[a, b]$ is nonrectifiable: $\ell(\gamma) = \infty$. On the other hand, the arc-image γ^* is a straight line segment and

$$L(\gamma^*) = \sqrt{2}\left(\sup_{t \in [0,1]} g(t) - \inf_{t \in [0,1]} g(t) \right)$$

which, owing to the continuity of g, is finite.

Example 2.2.1.8. The unit n-cube or *parallelotope* $\mathcal{P}^n$ (the topological product of n copies of $[0, 1]$) is the continuous image of C_0. The map $t : C_0 \mapsto \mathcal{P}^n$ may be extended linearly on the *closure* of each interval deleted in the construction of C_0 and the image of the resulting map T is an arc-image $T([0, 1])$ that fills $\mathcal{P}^n$. Since C_0 is *totally disconnected* whereas $\mathcal{P}^n$ is connected neither t nor T is bijective.

When $n = 2$ the continuous map t transforms a set of one-dimensional measure zero onto a set of two-dimensional measure one. Let π_1 be the projection of $\mathbb{R}^2$ onto its first factor: $\pi_1 : \mathbb{R}^2 \ni (x, y) \mapsto x$. If A is a nonmeasurable subset of $[0, 1] \times \{0\}$ then $D \overset{\text{def}}{=} t^{-1}(A)$, as a subset of the null set C_0, is (Lebesgue) measurable whereas $\pi_1 \circ t(D)$ $(= A)$ is a nonmeasurable subset of $[0, 1]$. Since T is an extension of t it follows that $U \overset{\text{def}}{=} \pi_1 \circ T$ is a continuous map of $[0, 1]$ into itself and U maps a null set of $[0, 1]$ onto a nonmeasurable set.

Example 2.2.1.9. Assume $0 \le \alpha < 1$. In each factor of $\mathcal{P}^2$ construct a Cantor set C_α so that $\lambda(C_\alpha) = \alpha$. Then the topological product of the two sets C_α is a compact set $C_{\alpha^2}^2$ such that $\lambda_2(C_{\alpha^2}^2) = \alpha^2$. Each C_α is the intersection of a decreasing sequence $\{K_j\}_{j \in \mathbb{N}}$ of compact sets and each K_j is a finite union of disjoint closed intervals all of the same length: K_1, the complement of the first open interval deleted in the construction of C_α, consists of 2^1 disjoint closed intervals, I_{11}, I_{12}, arranged in natural order from left to right in $[0, 1]$. Similarly K_j, the complement of the union of the first $2^j - 1$ open intervals deleted in the construction of C_α, consists of 2^j disjoint closed intervals, $I_{j1}, \ldots, I_{j2^j}$ arranged in natural order from left to right in $[0, 1]$.

The construction proceeds in sequence of stages of associations between intervals I_{mn} and their Cartesian products $I_{mn} \times I_{m'n'}$.

At stage 1 associate the $2^{2 \times 1} = 4^1$ intervals $I_{21}, \ldots, I_{22^2}$ of K_2 with 4^1 sets in $[0, 1] \times [0, 1]$ as follows:

$$
\begin{array}{cccc}
I_{21} & I_{22} & I_{23} & I_{22^2} \\
\updownarrow & \updownarrow & \updownarrow & \updownarrow \\
I_{21} \times I_{21} & I_{21} \times I_{22} & I_{22} \times I_{22} & I_{22} \times I_{21}
\end{array}
$$

Having completed stages $1, \ldots, j - 1$, at stage j:

i. associate the $2^{2j} = 4^j$ intervals $I_{2j,1}, \ldots, I_{2j,2^{2j}}$ of K_{2j} with the 4^j sets $I_{jp} \times I_{jq}$, $1 \le p, q \le 2^j$ in $[0, 1] \times [0, 1]$;
ii. map each interval deleted from $[0, 1]$ on to a line segment connecting two adjacent *components* of $K_{2j} \times K_{2j}$.

In **Figure 2.2.1.1** there is an indication of the associations made in the first two performances of the procedure just described. Subsequent associations are made similarly, by *inbreeding*, i.e., by repeating in each subinterval and correspondingly in each subsquare the construction just employed in the original square, and by continuing the repetition process endlessly. Although the construction is repeated in each stage, the orientation of the constructions in the subsquares must be such as to permit the connections indicated in **Figure 2.2.1.1**.

Let $\mathcal{K}_j$ be the compact connected set consisting of $K_{2j} \times K_{2j}$ together with the line segments connecting its components (cf. *ii* above). Then

$\mathcal{K}_{j+1} \subset \mathcal{K}_j$ and $\bigcap_{j \in \mathbb{N}} \mathcal{K}_j \overset{\text{def}}{=} \mathcal{K}$ is the homeomorphic image of $[0, 1]$: $\mathcal{K} = \Theta([0, 1])$, i.e., $\mathcal{K}$ is a simple arc-image and since $C_{\alpha^2}^2 \subset \mathcal{K}$ it follows that $\lambda_2(\mathcal{K}) \geq \alpha^2$.

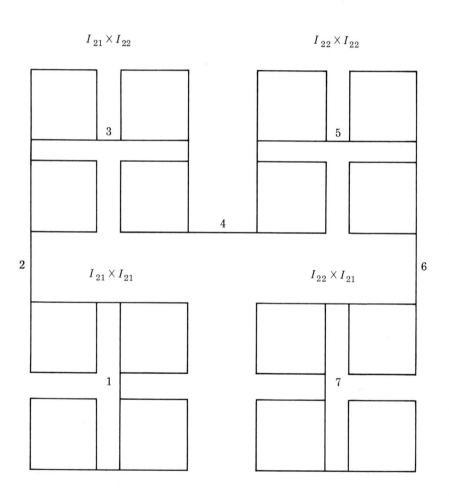

Figure 2.2.1.1. The first steps of the repetition/inbreeding process.

Exercise 2.2.1.9. Show that Θ is a homeomorphism.

[*Hint:* The map Θ is bijective on each of the intervals deleted in the construction of C_α. The set of those intervals is dense in $[0,1]$. Hence if $0 \leq x < y \leq 1$ and one of x and y is not in one of those intervals, then (possibly another) one of those intervals is a proper subset of $[x, y]$. Hence, for some j_0 in $\mathbb{N}$, $\Theta(x)$ and $\Theta(y)$ are in different components of $K_{j_0} \times K_{j_0}$, in particular, $\Theta(x) \neq \Theta(y)$. The continuity of Θ follows because a) the diameters of the components of $K_j \times K_j$ converge to zero as $j \to \infty$ and b) Θ is linear on each of the intervals deleted in the construction of C_α.]

[**Note 2.2.1.3:**   The simple arc-image $\mathcal{K}$ is not *rectifiable*. The very definition of arc-length shows that the arc-image of a rectifiable arc can be covered by rectangles forming a set of arbitrarily small two-dimensional Lebesgue measure.

A similar argument leads to the following conclusion:

For each n in $\mathbb{N}$ and each ϵ in $(0, 1]$ there is in $\mathcal{P}^n$ a (nonrectifiable) simple arc-image $\mathcal{K}$ such that $\lambda_n(\mathcal{K}) \geq 1 - \epsilon$.]

Example 2.2.1.10. When $n = 2$ the simple arc-image $\mathcal{K}$ described above lies in the unit square $[0,1] \times [0,1]$ and the endpoints of $\mathcal{K}$ are $(0,0)$ and $(1,0)$. The union of $\mathcal{K}$ and the simple arc-image

$$B \overset{\text{def}}{=} (\{0\} \times [0, -\delta]) \cup ([0,1] \times \{-\delta\}) \cup (\{1\} \times [0, -\delta])$$

is a *Jordan curve-image* C that is the boundary of a *region* R. Since

$$\lambda_2(C \cup R) = \lambda_2(C) + \lambda_2(R) \leq 1 + \delta$$

it follows that

$$\lambda_2(R) \leq 1 + \delta - \lambda_2(C) = 1 + \delta - \lambda_2(\mathcal{K}) \leq 1 + \delta - (1 - \epsilon) = \delta + \epsilon.$$

Hence $\lambda_2(R) < \lambda_2(\mathcal{K})$ if $\delta < 1 - 2\epsilon$, in which case the measure of R is less than the measure of the Jordan curve-image C that bounds R. In fact, for a positive ϵ there are a Jordan curve C and the region R bounded by C so that

$$0 < \frac{\lambda_2(R)}{\lambda_2(C)} < \epsilon.$$

Exercise 2.2.1.10. Show that a compact convex set in a *separable topological vector space* is an arc-image.

[*Hint:* A separable topological group is metrizable [**Kak1**].]

There are nowhere dense ("thin") sets of positive (Lebesgue) measure, e.g., Cantor-like sets of positive measure. Besicovitch [**Bes2**] used such sets to construct in $\mathbb{R}^3$ a homeomorphic image BES (for Besicovitch) of the surface S_1 of the unit ball

$$B_1 \overset{\text{def}}{=} \left\{ (x,y,z) \ : \ x,y,z \in \mathbb{R}, x^2 + y^2 + z^2 \le 1 \right\}$$

so that $\lambda_3(\text{BES})$ is large while the surface area $A(\text{BES})$ of BES is small.

If $\eta, A > 0$ there is in $\mathbb{R}^3$ a surface BES, homeomorphic to S_1 and such that $\lambda_3(\text{BES}) > A$ while the surface area $A(\text{BES})$ of BES is less than η.

Proceeding by analogy with the definition of arc-length for a curve, one is led to suggest that the area of a surface S in $\mathbb{R}^3$ be defined as the supremum of the set of areas of the polyhedra inscribed in the S. However phenomena such as that in **Exercise 2.2.1.12. 123** below suggest the inadequacy of so simple an approach.

The construction originated by Besicovitch and described below dramatizes even further the need to reformulate a proper theory of surface area. For example, some proper definition of surface area is necessary if there is to be a satisfactory statement, not to mention a satisfactory resolution, of the famous *problem of Plateau*.

For a given Jordan curve-image C in $\mathbb{R}^3$ find in $\mathbb{R}^3$ a surface S bounded by C and of least surface area.

Example 2.2.1.11. Assume $4M^3 > A$, $0 < \alpha < 1$, and $\eta > 2\epsilon > 0$. Let K denote the cube $[-M, M]^3$ in $\mathbb{R}^3$. The cube K is subjected to two operations performed in succession and then repeated endlessly.

   i. Shrinkage by α: replace K by $K_\alpha \overset{\text{def}}{=} [-\alpha M, \alpha M]^3$, $0 < \alpha < 1$, situated cocentrally inside K and with its faces parallel to those of K;
   ii. Subdivision: by passing bisecting planes parallel to the faces of K_α divide it into eight congruent subcubes: $K_\alpha^1, \ldots, K_\alpha^8$.

Inbreed, i.e., repeat the operations *i,ii* above on each of the eight subcubes, on each of the 8^2 subsubcubes, ..., on each of the 8^n subsub...subcubes.

Exercise 2.2.1.11. Let the intersection of the set of all cubes, subcubes, subsubcubes, ... be D. Then D is a dyadic space, a three-dimensional analog and homeomorphic image of the Cantor set. Calculate the measure of D in terms of M and α and thereby show that for some α the three-dimensional measure of D can be made arbitrarily close to but less than $8M^3$, the volume of the original cube.

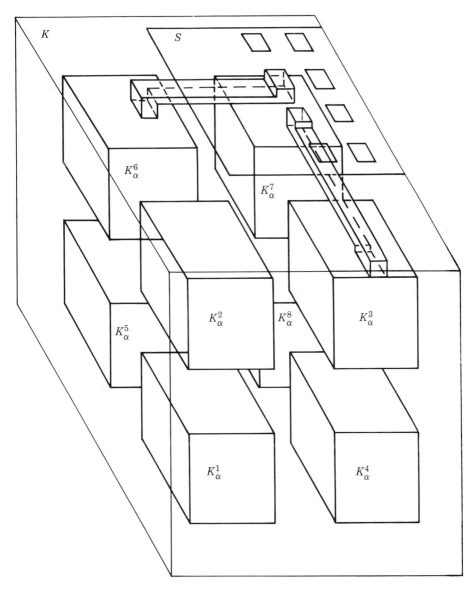

Figure 2.2.1.2. The Besicovitch construction. Only two
of the first eight "ducts" are shown.

The next goal is to construct a polyhedron Π containing (infinitely)
many faces and edges and such that among the vertices of Π are all the
points of D. The procedure given next provides such a polyhedron. As a
polyhedron Π consists of polygonal faces and thus has a well-defined surface
area. The polyhedron constructed below has small area.

On one face of K construct a square S of area not exceeding $\frac{\epsilon}{2}$. Note that S is homeomorphic to a hemisphere. The idea is to distort S in a thorough and systematic manner so that S is formed into a polyhedron of the kind described above.

From S excise eight disjoint pairwise congruent subsquares each of area δ_1 not exceeding $\frac{\epsilon}{32}$ and more narrowly delimited below (cf. **Figure 2.2.1.2** above).

On one face of each of the first eight subcubes construct a square congruent to one of the eight subsquares excised from S.

Again by inbreeding, repeat this construction on each of the subsubcubes, ..., so that on one face of each subsub ... subcube there is a square from which eight congruent subsquares have been excised. In $K \setminus K_\alpha$ run eight tubes, one from each of the eight excised subsquares of S to one of the eight squares on the eight subcubes of K_α. The connected surfaces of the tubes are to be unions of nonoverlapping closed rectangles. The cross-sections of the tubes are rectangles — in short, the tubes are models of heating/air-conditioning ducts. The planar surface area of each tube is proportional to the perimeter of the (rectangular) cross-section. Hence, by a suitable choice of δ_1, the total (planar) surface area of the eight tubes can, be brought below $\frac{\epsilon}{2^2}$.

> The union of S_1, the surfaces of the eight tubes, and the eight squares on the surfaces of the eight subcubes is homeomorphic to S and hence to the surface of a hemisphere.

The process just described is repeated in each of the first eight subcubes, except that a new δ_2 is chosen so that the total surface area of the 64 new tubes does not exceed $\frac{\epsilon}{2^3}$, ...

The basic construction (simplified) is shown in **Figure 2.2.1.2**.

The end-product of the infinite set of tube constructions is a Medusa-like set HEMIBES (hemi+BES) that is homeomorphic to the surface of a hemisphere. As one moves through a first-stage tube, then through one of the second-stage tubes emanating from it, ..., at the "other end" one arrives at precisely one point of D and each point of D is the "other end" of such a trail. Thus D, a dyadic space of positive three-dimensional measure, lies on the surface of a HEMIBES, which is homeomorphic to the *two-dimensional* surface of a hemisphere.

The total surface area of the tubes so traversed is not more than $\frac{\epsilon}{2}$ and so the surface area of HEMIBES does not exceed ϵ whereas HEMIBES contains D and thus the three-dimensional measure of HEMIBES can be made arbitrarily close to $8M^3$. If two "hemispheres" like HEMIBES are conjoined at their "equators" (the perimeters of their squares S) the result BES is homeomorphic to the surface of the ball B_1. The union of BES and the bounded component of its complement is a set $\mathcal{B}$ that is homeomorphic to B_1. The area of the surface of $\mathcal{B}$, i.e., the area of BES, is less than η whereas the three-dimensional measure of $\mathcal{B}$ exceeds A.

[**Remark 2.2.1.4:** Let C be a rectifiable Jordan curve-image in $\mathbb{R}^2$. If R is the bounded component of $\mathbb{R}^2 \setminus C$ and if $\ell(C) = 1$, then

$$\lambda_2(C) = 0, \text{ and } \lambda_2(R \cup C) \leq \frac{1}{4\pi}. \qquad (*)$$

(The second inequality is the famous isoperimetric inequality studied in the calculus of variations.)

The corresponding theorem for $\mathbb{R}^3$ should read: Let Σ be a homeomorphic image of S_1 in $\mathbb{R}^3$ and assume that the surface area of Σ is 1: $A(\Sigma) = 1$. If V is the bounded component of $\mathbb{R}^3 \setminus \Sigma$ then

$$\lambda_3(\Sigma) = 0 \text{ and } \lambda_3(V \cup \Sigma) \leq \frac{1}{6\sqrt{\pi}}. \qquad (**)$$

Whereas $(*)$ is true, owing to BES, $(**)$ is false. The reader is urged to formulate other contrasts stemming from BES.]

[**Note 2.2.1.4:**   The surface BES of $\mathcal{B}$ can be described parametrically by three equations:

$$x = f(u,v), \ y = g(u,v), \ z = h(u,v), \ 0 \leq u, v \leq 1.$$

Since the surface of $\mathcal{B}$ is, for the most part planar, the functions f, g, h are, off a set of two-dimensional measure zero, *linear*, in particular continuously differentiable a.e.

The example BES illuminates not only the problem of Plateau but also the question of *defining* the notion of surface. For example, the parametric description of BES in terms of f, g, h above is qualitatively indistinguishable from that of the surface of a cube or the surface of a cube to which "spines" (closed intervals) or "wings" (closed triangles) have been attached.

In another direction, the ball $\mathcal{B}$ impinges on the circle of ideas under the rubric of *Stokes's theorem*, which is a vast generalization of the FTC. Stokes's theorem and, in particular the FTC, may be written in terms of the symbol ∂ interpreted as a special differentiation operator when ∂ is applied to a (vector-valued) function and as the boundary operator when ∂ is applied to a subset of $\mathbb{R}^n$:

$$\text{Stokes's theorem}: \int_{\partial R} f = \int_R \partial f.$$

(The differential notation in the equation above is omitted deliberately. The integrals are to be interpreted as formed with respect to appropriate measures on ∂R resp. R.) For example, if $R \overset{\text{def}}{=} [a, b]$ and $f \in C^1([a, b], \mathbb{R})$ then $\partial R = \{a, b\}$, $\partial f = f'$, and the FTC reads:

$$\int_{\partial[a,b]} f \overset{\text{def}}{=} f(b) - f(a) = \int_a^b f' \, dx \overset{\text{def}}{=} \int_{[a,b]} \partial f.$$

Similarly in $\mathbb{R}^3$ for a ball

$$B_r \overset{\text{def}}{=} \left\{ (x, y, z) \ : \ x^2 + y^2 + z^2 \le r^2 \right\},$$

its boundary

$$\partial B_r \overset{\text{def}}{=} S_r \overset{\text{def}}{=} \left\{ (x, y, z) \ : \ x^2 + y^2 + z^2 = r^2 \right\},$$

a vector-valued function

$$\mathbf{F}(x, y, z) \overset{\text{def}}{=} (f(x, y, z), g(x, y, z), h(x, y, z)),$$

and

$$\partial \mathbf{F} \overset{\text{def}}{=} \nabla \cdot \mathbf{F} \overset{\text{def}}{=} f_x + g_y + h_z \overset{\text{def}}{=} \text{div}\mathbf{F},$$

the (Gauß) version of Stokes's theorem reads in terms of the (vector) differential $d\mathbf{A}$ of surface area and the (scalar) differential of volume dV:

$$\int_{\partial B_r} \mathbf{F} \overset{\text{def}}{=} \int_{S_r} (f(x, y, z), g(x, y, z), h(x, y, z)) \cdot d\mathbf{A}$$

$$= \int_{B_r} (f_x(x, y, z) + g_y(x, y, z) + h_z(x, y, z)) \, dV$$

$$\overset{\text{def}}{=} \int_{B_r} \text{div}\mathbf{F} \, dV \overset{\text{def}}{=} \int_{B_r} \partial \mathbf{F}.$$

For a smooth $\mathbf{F}$, the theorem fails for the ball $\mathcal{B}$ and its boundary BES.

Similarly, for a surface S in $\mathbb{R}^3$ and bounded by a rectifiable closed curve C: $\partial S = C$, $d\mathbf{s}$ representing the (vector) differential of curve length, there is the formula traditionally named for Stokes:

$$\int_S \partial \mathbf{F} \overset{\text{def}}{=} \int_S \text{curl}\mathbf{F} \overset{\text{def}}{=} \int_S (h_y - g_z, f_z - h_x, g_x - f_y) \cdot d\mathbf{A}$$

$$= \int_C (f, g, h) \cdot d\mathbf{s} \overset{\text{def}}{=} \int_{\partial S} \mathbf{F}.$$

One more comment deserves inclusion. The notion of *Hausdorff dimension* ρ^p, $p \in \mathbb{R}$, $0 < p < \infty$, defined for all subsets of a metric space X, is intimately related to Lebesgue measure when $X = \mathbb{R}^n$. For BES, $0 < \rho^2(BES) < \infty$, whence

$\rho^p(BES) = 0$, $2 < p < \infty$, and $0 < \rho^3(\mathcal{B}) < \infty$, whence $\rho^q(\mathcal{B}) = 0$, $3 < q < \infty$, a result more in harmony with geometric intuition [Ge7].]

The *length* of $\gamma : [0,1] \mapsto \mathbb{R}^n$ is the supremum of the lengths of the polygons inscribed in the curve:

$$\ell(\gamma) \overset{\text{def}}{=} \sup_{0 = t_0 < \cdots < t_n = 1} \sum_{i=1}^{n} \|\gamma(t_i) - \gamma(t_{i-1})\|, \quad n \in \mathbb{N}.$$

Much before Besicovitch produced the construction described above, H. A. Schwarz showed that the corresponding notion of defining the area of a surface by the areas of approximating inscribed polyhedra is useless.

Exercise 2.2.1.12. Let a truncated right circular cylinder Z be of radius r and of height h. For $3 \le m, n < \infty, m, n \in \mathbb{N}$, draw m equally spaced circles on the surface of Z so that their planes are parallel to the base of Z. Inscribe in each circle an n-gon so that the vertices of one n-gon are positioned under or over the midpoints the sides of an adjacent n-gon. Each vertex of a top or bottom n-gon has two *neighboring vertices* on an adjacent n-gon. Each vertex of every other n-gon has four neighboring vertices on two adjacent n-gons. The sides of the n-gons and the line segments connecting neighboring vertices constitute the set of edges of a polyhedron Π_{mn} with vertices lying on the surface of Z: Π_{mn} is a polyhedron *inscribed* in the lateral surface of Z. Each face of Π_{mn} is a triangle. Show that the area of each triangle in Π_{mn} is

$$r \tan \frac{\pi}{n} \sqrt{\left(\frac{h}{m-1}\right)^2 + \left(r - r \cos \frac{\pi}{n}\right)^2}.$$

Find the area A_{mn} of Π_{mn} and show that

$$\liminf_{m,n \to \infty} A_{mn} = 2\pi r h, \quad \limsup_{m,n \to \infty} A_{mn} = \infty$$

whence $\lim_{m,n \to \infty} A_{mn}$ does *not* exist. Show also that if $2\pi r h \le A < \infty$ there is a function $m(n)$ such that $\lim_{n \to \infty} A_{m(n),n} = A$.

Related to the discussions above are the following results.

Example 2.2.1.12. In $\mathbb{R}^2$ the bounded component R of the complement of a Jordan curve-image J is necessarily an open set of positive finite measure. If J is nonrectifiable it follows that for the compact set $K \overset{\text{def}}{=} J \cup R$, $\lambda_2(K) < \infty$ and $L(\partial K) = \infty$ $(= L(R))$. For example, if a) B is the set in **Example 2.2.1.10. 117**, b) f in $C([0,1], \mathbb{R})$ is nowhere differentiable, $f(0) = f(1) = 0$, and c) γ is $\gamma : [0,1] \mapsto (t, f(t)) \in \mathbb{R}^2$, then the Jordan curve-image $J \overset{\text{def}}{=} C \overset{\text{def}}{=} \gamma^* \cup B$ has infinite length. Because γ^*

in $\mathbb{R}^2$ is the graph of a continuous function, $\lambda_2(\gamma^*) = 0$ whence $\lambda_2(C) = 0$. Let R be the bounded component of $\mathbb{R}^2 \setminus J$, let Ω be $R \cup C$, and let V be $\Omega \times [0,1]$, a compact (measurable) subset of $\mathbb{R}^3$. Then $\lambda_3(V)$ is finite, $V^\circ = R \times (0,1)$, and $\lambda_3(V^\circ)$ is positive (and finite). Owing to the nonrectifiability of the curve C, the area of the boundary $V \setminus V^\circ$ of V is necessarily infinite.

The set V° may be viewed as the inside of a container that can be filled with paint but the (inside) surface $V \setminus V^\circ$ of the container cannot be painted.

Working in a different part of the domain of "sizes" of figures in two- and three-dimensional Euclidean space, Besicovitch [**Bes1, Bes3**] proved THEOREMs **2.2.1.5** and **2.2.1.6** below. The former provided a surprising solution to a problem posed by Kakeya.

THEOREM **2.2.1.5**. LET $\mathcal{K}$ BE THE SET OF INTERIORS K DETERMINED BY PLANE CLOSED POLYGONS (JORDAN CURVE-IMAGES) P AND CONTAINING AT LEAST ONE UNIT LINE SEGMENT (OF LENGTH ONE). IF $\epsilon > 0$ THERE IS IN $\mathcal{K}$ A K_ϵ SUCH THAT THE AREA OF K_ϵ DOES NOT EXCEED ϵ AND THERE IS IN K_ϵ A DIRECTED UNIT LINE SEGMENT S THAT CAN BE MOVED CONTINUOUSLY IN K_ϵ AND, IN THE COURSE OF ITS MOTION, S POINTS IN ALL DIRECTIONS: "A UNIT LINE SEGMENT CAN BE ROTATED THROUGH 360° WITHIN AN ARBITRARILY SMALL POLYGONAL AREA."

PROOF. Since $\mathcal{K}$ contains the interiors of all squares of side 2, as in **Figure 2.2.1.3** below, $\mathcal{K}$ is not empty. In what follows the basic ideas go back to Besicovitch and to a device due to J. Pal [**Bes 1, Bes3**]. In a square of side 2 the diagonals form four isosceles right triangles $T_1, \ldots, T_4$ having a common vertex at the center of the square (see **Figure 2.2.1.3**). For $p \geq 3$ let each side of the square be divided into $n \overset{\text{def}}{=} 2^{p-2}$ equal subintervals, each the base of a triangle having its third vertex at the center of the square. Assume translations parallel to the base of T_1 can bring its n subtriangles $\tau_1, \ldots, \tau_n$ into overlapping triangles $\tau_1', \ldots, \tau_n'$ so that the area of their union, the *Perron tree*, is not more than $\frac{2}{p}$. Those vertices formerly at the center of the square are now the tips of the branches of the Perron tree. Let an interval of length 1 be pivoted at the tip of τ_1', rotated counterclockwise through the vertex angle of τ_1', lifted and translated to the tip of τ_2', pivoted at the tip of τ_2', rotated through the vertex angle of τ_2', ... The result is a (discontinuous) movement that keeps the interval always within the Perron tree when rotation takes place and effects a counterclockwise rotation of the interval through an angle of 90°. Repeated for $T_2, \ldots, T_4$ the procedure effects a rotation of S through 360° within a figure of total area not exceeding four times the area of a Perron tree.

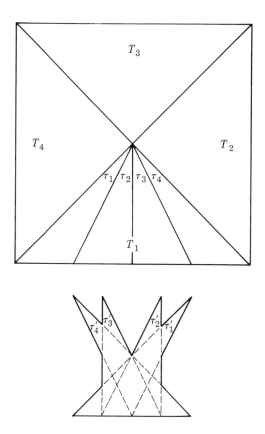

Figure 2.2.1.3. The building of a Perron tree.

As in **Figure 2.2.1.4** the *Pal join*

$$GL \cup LMN \cup DL$$

permits the replacement of the discontinuous movement just described by a continuous movement. The triangles DEF and GHI are translates of typical adjacent subtriangles in the original square. Thus DF and GH are parallel. If $\epsilon > 0$ the point K is chosen so that

$$\lambda_2(GHK) \; [= \lambda_2(LMN)] < \frac{\epsilon}{8}\lambda_2(GHI) \; \left[= \frac{\epsilon}{8n}\right].$$

Then the interval S, say XY, lying on DE and with Y at D is:

i. pivoted at D, rotated until XY lies on DF ($DF \subset DL$);

ii. slid along DL until X is at L;

iii. pivoted at L, rotated until XY lies on LN;

iv. slid along GL until Y is at G;

v. pivoted at G, and rotated until XY lies on GI.

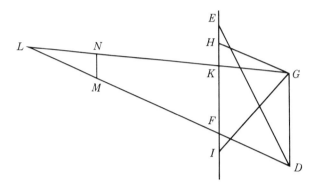

Figure 2.2.1.4. The Pal join.

Thus S, lying in $DEF \cup \text{Pal join} \cup GHI$, is rotated through an angle equal to $\angle EDI + \angle HGI$, the sum of the angles of two adjacent subtriangles. Since the area of the Pal join is less than $\frac{\epsilon}{8n}$ the area of the union of all Pal joins and of all Perron trees is not more than

$$\frac{\epsilon}{2} + 4\lambda_2 \,(\text{Perron tree})\,.$$

Hence if $p > \frac{16}{\epsilon}$ it follows that carrying out this process in each of the large triangles $T_1, \ldots, T_4$ leads to a continuous $360°$ rotation of the interval in a figure of area not more than ϵ.

Although the Pal joins permit continuous motions and do not significantly add to the area of the Perron trees, the *diameter* of the polygon produced via the Pal joins is significantly larger than the diameter of the original square T.

The core of the Besicovitch solution is a systematic device for constructing a Perron tree of arbitrarily small area. The description that follows is drawn not from [**Bes1**], in which the solution of the Kakeya problem first appeared, but from [**Bes3**], where the author's expository skill, accumulated over 35 years, is plainly evident.

For p at least 3, lines parallel to and at heights $\frac{1}{p}, \frac{2}{p}, \ldots, 1$ above the base of T_1 are drawn. By *recomposition-compression* the decomposition of T_1 into 2^{p-2} subtriangles is successively reversed while compression is applied to yield versions of T_1 that are similar but of heights $\frac{p-1}{p}, \ldots, \frac{1}{p}$. Furthermore, each recomposition-compression *halves* the number of subtriangles, cf. **Figure 2.2.1.5**, where the relation between the recomposed-compressed triangles and the decomposed original triangle T_1 and its subtriangles $\tau_1, \ldots, \tau_n$ is shown. The purpose of recomposition-compression is simply to reduce to its most primitive form the operation of translating subtriangles for optimal overlap.

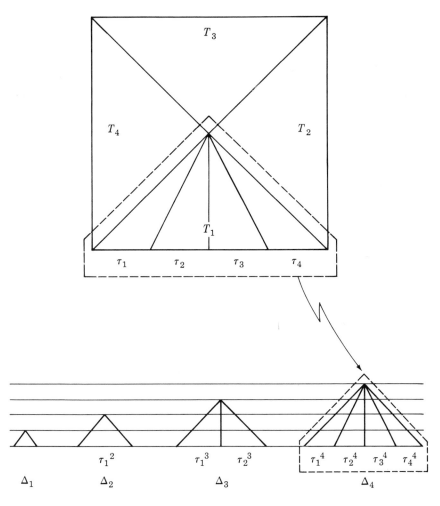

Figure 2.2.1.5. Recomposition-compression ($p = 4$, $n = 2^{4-2} = 4$).

Then the procedure *bisection-expansion* illustrated in **Figure 2.2.1.6** undoes recomposition-compression and leads to the creation of the Perron tree, a union of translates of the triangles τ_i. Notice that the area of the top end of each triangle Δ_k is always the area of Δ_1, i.e., $\frac{1}{p^2}$. In **Figure 2.2.1.6** bisection-expansion is shown for the triangle τ_1^2 of **Figure 2.2.1.5**. The "sapling" that "grows" into the Perron tree of **Figure 2.2.1.3** is shown for the parameter values $p = 4$, $n = 2^{p-2} = 4$.

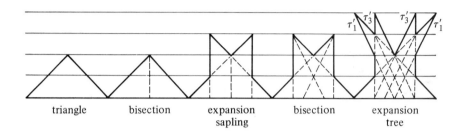

triangle   bisection   expansion   bisection   expansion
                                      sapling   tree

Figure 2.2.1.6. Bisection-expansion ($p = 4$, $n = 2^{p-2} = 4$).

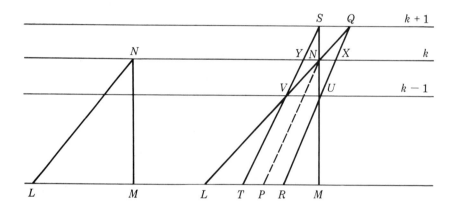

Figure 2.2.1.7. The basis of the area calculation.

Because the quadrangle $SQUV$ in **Figure 2.2.1.7** is a parallelogram it follows that

$$\lambda_2 \left(\triangle SNV \right) = \lambda_2 \left(\triangle QNU \right) = \lambda_2 \left(\triangle NUV \right).$$

Thus there obtains the equality

$$\lambda_2 \left(\triangle LQR \cup \triangle TSM \right) = \lambda_2 \left(\triangle LMN \right) + 2\lambda_2 \left(\triangle NVU \right).$$

Note also that in **Figure 2.2.1.5** the area of the part (the "top end") of Δ_4 between levels 3 and 4 is the area of Δ_1. Hence in the bisection-expansion illustrated above, the area of Δ_1 is increased by no more than $2\lambda_2(\Delta_1)$. When bisection-expansion is applied $p - 2$ times to Δ_2 to produce the full Perron tree for the parameter values p and $n = 2^{p-2}$, the area of that Perron tree does not exceed

$$\lambda_2(\Delta_2) + 2(p - 2)\lambda_2(\Delta_1) = \frac{4}{p^2} + 2(p - 2)\lambda_2(\Delta_1) = \frac{2}{p}.$$

Since Perron trees of arbitrarily small area can be constructed and since Pal joins of arbitrarily small area can be used, the Besicovitch solution is validated.

$\square$

Exercise 2.2.1.13. Show that when $p = 3$ the bisection-expansion procedure yields the optimal overlap of τ_1' and τ_2', i.e., the Perron tree when $n = 2$ has the minimal area achievable by overlapping translates of τ_1 and τ_2.

In [**Bes1**] Besicovitch shows how his solution of the Kakeya problem yields as well the next result.

THEOREM **2.2.1.6.** THERE IS IN $\mathbb{R}^2$ A SET OF INTERVALS, ORIENTED IN ALL POSSIBLE DIRECTIONS, EACH OF LENGTH 1, AND THE UNION $\mathcal{S}$ OF THOSE INTERVALS IS A NULL SET: $\lambda_2(\mathcal{S}) = 0$.

PROOF. When $p \geq 3$ the corresponding Perron tree is the union of translates τ_i' of the constituent triangles τ_i in T_1. If τ_i is to the right resp. left of the midpoint M of the base of T_1 then τ_i is translated to the left resp. right. Each line segment L from the vertex of T_1 to a point X on the base of T_1 is translated to a line segment within the Perron tree. Using $p + 1$ rather than p, Perronize the translate τ_i' of τ_i, ...

Each line segment L is translated to a sequence $\{L_p\}_{3 \leq p}$ and the sequence $\{X_p\}_{3 \leq p}$ of base points converges to a point X_∞ on the base of T_1. Because the line segments L_p are translates of L the sequence $\{L_p\}_{3 \leq p}$ converges to a line segment L_∞. The set $\mathcal{S}_1$ of all such limiting line segments is a subset of each (multiply) Perronized figure and hence $\lambda_2(\mathcal{S}_1) = 0$. On the other hand, because all Perronizations involve only translation, the set $\mathcal{S}_1$ consists of line segments oriented like the original line segments L of T_1. If a similar process is applied to T_2, T_3, T_4 the result is sets $\mathcal{S}_2, \mathcal{S}_3, \mathcal{S}_4$ and

$$\mathcal{S} \stackrel{\text{def}}{=} \mathcal{S}_1 \cup \mathcal{S}_2 \cup \mathcal{S}_3 \cup \mathcal{S}_4$$

is a set of line segments, each of length not less than 1 and $\lambda_2(\mathcal{S}) = 0$. For each direction θ there is in $\mathcal{S}$ a unique line segment oriented in the direction θ.

$\square$

If R is a Lebesgue measurable subset of $\mathbb{R}^2$ and if every line in $\mathbb{R}^2$ meets R in at most two points then, owing to *Fubini's theorem,* $\lambda_2(R) = 0$. On the other hand, in **Example 2.2.1.13** below there is defined a nonmeasurable set N meeting each line in at most two points. The idea is due to Sierpinski [**Si3**].

Example 2.2.1.13. Let Ψ be the first ordinal number corresponding to a well-ordered set of cardinality $\#(\mathbb{R})$. Then

$$\#[\{\alpha \,:\, \alpha < \Psi\}] = \#(\mathbb{R}).$$

Let O be the set of all open subsets of $\mathbb{R}$. Then, because $\mathbb{R}^2$ is separable in its Euclidean topology, $\#(\mathsf{O}) = \#(\mathbb{R})$ and so the set

$$\mathsf{F}_{pos} \stackrel{\text{def}}{=} \left\{ F \,:\, F \subset \mathbb{R},\ F = \overline{F},\ \lambda_2(F) > 0 \right\}$$

of closed subsets of positive measure in $\mathbb{R}^2$ and $\mathcal{S} \stackrel{\text{def}}{=} \{\alpha \,:\, \alpha < \Psi\}$ are of the same cardinality, i.e., the sets of F_{pos} may be indexed by the elements in $\mathcal{S}$: $\alpha \leftrightarrow F_\alpha$. In the set of all maps p of some *initial segment*

$$\{\alpha \,:\, 1 \le \alpha < \beta \le \Psi\} \stackrel{\text{def}}{=} [1, \beta)$$

of $\mathcal{S}$ into the *power set* $2^{\mathbb{R}}$ of $\mathbb{R}^2$ let $\mathcal{P}$ consist of those maps such that:

i. $p(\alpha) \in F_\alpha$;
ii. no three points in the range of p are collinear.

Then $\mathcal{P}$ is nonempty, e.g., if $\beta = 2$ and $p(\{1\}) \subset F_1$ then $p \in \mathcal{P}$. The set $\mathcal{R}$ of ranges of maps in $\mathcal{P}$ may be partially ordered by inclusion. Zorn's lemma implies there is a maximal element R in $\mathcal{R}$ and for some initial segment $[1, \beta)$ and some q in $\mathcal{P}$: $q\{[1, \beta)\} = R$.

If $\beta < \Psi$ then $\#(R) < \#(\mathbb{R})$ and there is a direction θ different from that determined by every pair of points in R. Since $F_\beta \in \mathsf{F}_{pos}$, Fubini's theorem implies that some line in the direction θ meets F_β in a set A of positive measure. Hence there is in A a point P_β not collinear with any pair of points in R. Define q' according to:

$$q'(\alpha) = \begin{cases} q(\alpha) & \text{if } \alpha < \beta \\ P_\beta & \text{if } \alpha = \beta. \end{cases}$$

Then q' maps the initial segment $[1, \beta+1)$ into a set R' properly containing R, in contradiction of the maximality of R. Hence $\beta = \Psi$ and

iii. R meets every set F_α;
iv. no three points in R are collinear.

The set $R^c \stackrel{\text{def}}{=} \mathbb{R}^2 \setminus R$ contains no set B of positive measure since such a set B must contain some F_α, hence must meet R. Fubini's theorem implies that if R is measurable then $\lambda_2(R) = 0$. Hence if R is measurable so is R^c and since R^c contains no set of positive measure, $\lambda_2(R_2) = 0$, whence $\lambda_2(\mathbb{R}^2) = 0$, a contradiction. In other words, R is a nonmeasurable subset of $\mathbb{R}^2$ and R meets every line in at most two points.

Exercise 2.2.1.14. Let R be the set of **Example 2.2.1.13.** If $x \in \mathbb{R}$ and the vertical line V_x through x meets R in one point (x, y) let $f(x)$ be y. If V_x meets R in two points let $f(x)$ be the larger of the corresponding ordinates. If V_x does not meet R let $f(x)$ be 0, i.e.,

$$f(x) \overset{\text{def}}{=} \begin{cases} \max\{y \ : \ (x,y) \in R\} & \text{if } \{y \ : \ (x,y) \in R\} \neq \emptyset \\ 0 & \text{otherwise.} \end{cases}$$

Let G be the graph of f. Show that at least one of G and $R \setminus G$ is a nonmeasurable subset of $\mathbb{R}^2$. If G is measurable subset of $\mathbb{R}^2$ let h be such that

$$h(x) = \begin{cases} \min\{y \ : \ (x,y) \in R\} & \text{if } \{y \ : \ (x,y) \in R\} \neq \emptyset \\ 0 & \text{otherwise.} \end{cases}$$

Show that either the graph of f or the graph of h is a nonmeasurable subset of $\mathbb{R}^2$.

Example 2.2.1.14. For n in $\mathbb{N}$ and R a *region* in $\mathbb{R}^n$ the set $R \cap \mathbb{Q}^n$ does not have content.

Example 2.2.1.15. The region R bounded by the Jordan curve-image of **Example 2.2.1.10. 117** does not have content since the measure of the boundary ∂R of R is positive.

Example 2.2.1.16. For α positive, the compact set $S \overset{\text{def}}{=} C_{\alpha^2}^2$ (cf. **Example 2.2.1.9. 115**) does not have content since $\lambda_2(\partial S) > 0$.

If f is nonnegative and Riemann resp. Lebesgue integrable on $[0, 1]$ then

$$S \overset{\text{def}}{=} \{(x, y) \ : \ 0 \leq y \leq f(x), \ x \in [0, 1]\}$$

has (two-dimensional) content resp. is a Lebesgue measurable subset of $\mathbb{R}^2$ and the two-dimensional content resp. two-dimensional Lebesgue measure of S is

$$\int_0^1 f(x)\, dx.$$

By contrast there are the phenomena illustrated in **Exercises 2.2.1.15, 2.2.1.16**.

Exercise 2.2.1.15. Let ϕ be $\chi_{\mathbb{Q}\cap[0,1]}$ the and let ψ be $\phi + 1$. Show that:

i. for x in $\mathbb{R}$, $\phi(x) < \psi(x)$;

ii. $\psi - \phi$ is Riemann integrable on $[0, 1]$ and the Riemann integral

$$\int_0^1 (\psi(x) - \phi(x))\ dx$$

is 1;

iii. the set

$$S \overset{\text{def}}{=} \{ (x, y)\ :\ \phi(x) \le y \le \psi(x),\ x \in [0, 1] \}$$

does not have two-dimensional content.

Exercise 2.2.1.16. Let E be a nonmeasurable subset of $[0, 1]$. Show that if $\phi \overset{\text{def}}{=} \chi_E$ and $\psi \overset{\text{def}}{=} \phi + 1$ then:

i. for x in $\mathbb{R}$ $\phi(x) < \psi(x)$;

ii. $\psi - \phi$ is Lebesgue integrable on $[0, 1]$ and the Lebesgue integral

$$\int_0^1 (\psi(x) - \phi(x))\ dx$$

is 1;

iii. the set

$$S \overset{\text{def}}{=} \{ (x, y)\ :\ \phi(x) \le y \le \psi(x),\ x \in [0, 1] \}$$

is not a Lebesgue measurable subset of $\mathbb{R}^2$.

2.2.2. Measurable and nonmeasurable functions

Example 2.2.2.1. The Cantor function $\mathcal{C}_0$ permits the definition of a continuous bijection $\Psi : [0, 1] \ni x \mapsto x + \mathcal{C}_0(x) \in [0, 2]$ (hence Ψ^{-1} is also a continuous bijection) that maps a Lebesgue measurable set of measure zero into a nonmeasurable set.

Indeed, $\lambda\left[\Psi\left([0, 1] \setminus \mathcal{C}_0\right)\right] = 1$ whence $\lambda\left[\Psi\left(\mathcal{C}_0\right)\right] = 1$ and so $\Psi\left(\mathcal{C}_0\right)$ contains a nonmeasurable set E. On the other hand:

i. $A \overset{\text{def}}{=} \Psi^{-1}(E) \subset \mathcal{C}_0$ and so A is Lebesgue measurable;

ii. $\lambda(A) = 0$;

iii. $\Psi(A)\ (= E)$ is not measurable;

iv. since the continuous image of a Borel set is a Borel set it follows that A is a non-Borel subset of the Lebesgue measurable set $\mathcal{C}_0$ of measure zero;

v. in particular A is not an F_σ;

vi. there is no function f such that $\mathrm{Discont}(f) = A$.

[**Remark 2.2.2.1:**   Any two closed Cantor-like sets are homeo-morphic (cf. LEMMA **2.2.1.1. 112**). One may have measure zero and the other may have positive measure, cf. **Example 2.2.1.3. 108.**]

If f is a bounded measurable function and p is a polynomial then $p \circ f$ is measurable. The *Stone-Weierstraß theorem* implies that if g is continuous (on a domain containing the range of f) then $g \circ f$ is measurable.

Exercise 2.2.2.1. Let the notation be that used in **Example 2.2.2.1**. Show that although the *characteristic function* χ_A is measurable yet the composition $\chi_A \circ \Psi^{-1}$ is not measurable.

A measurable function of a continuous function need not be mea-surable.

Exercise 2.2.2.2. Show that if $f : \mathbb{R} \mapsto \mathbb{R}$ is monotone and $g : \mathbb{R} \mapsto \mathbb{R}$ is measurable then both f and $f \circ g$ are measurable. (Hence if h is a function of bounded variation then both h and $h \circ g$ are measurable.)

The function χ_A resp. Ψ^{-1} of **Exercise 2.2.2.1** is measurable resp. monotone but the composition $\chi_A \circ \Psi^{-1}$ is not measurable.

A measurable function of a monotone function is not necessarily measurable.

The following result is used often in a measure situation $(X, \mathcal{S}, \mu)$.

THEOREM **2.2.2.1.**   (EGOROFF) IF $E \in \mathcal{S}$, IF $\mu(E) < \infty$, AND IF $\{f_n\}_{n\in\mathbb{N}}$ IS A SEQUENCE OF MEASURABLE FUNCTIONS CONVERGING TO f A.E. ON E THEN FOR EACH POSITIVE δ THERE IS A SUBSEQUENCE $\{f_{n_k}\}_{k\in\mathbb{N}}$ AND A MEASURABLE SUBSET D OF E SUCH THAT

$$\mu(D) < \delta$$
$$f_{n_k} \xrightarrow{\text{u}} f \text{ ON } E \setminus D$$

[**Halm, Roy, Rud**].

If the sequence $\{f_n\}_{n\in\mathbb{N}}$ in THEOREM **2.2.2.1** is replaced by an un-countable net $\{f_\lambda\}_{\lambda\in\Lambda}$ the conclusion need not obtain.

Example 2.2.2.2. Let Λ be $[0, 1)$ ordered so that $t_1 \succeq t_2$ iff $t_1 \leq t_2$. Then a net ν_t converges to ν iff $\lim_{t\to 0} \nu_t = \nu$.

Let S be the nonmeasurable set in **Example 2.2.1.1. 106**. Let $\{r_n\}_{n\in\mathbb{N}}$ be an enumeration of $\mathbb{Q} \cap [0, 1)$ and define S_n to be $(S + r_n)/\mathbb{Z}$. Then there is a positive δ such that

$$\lambda^*(S) = \delta \ (= \lambda^*(S_n), \ n \in \mathbb{N}).$$

For t in $J_n \overset{\text{def}}{=} [2^{-n-1}, 2^{-n})$ let f_t be defined by the equation

$$f_t(x) = \begin{cases} 1 & \text{if } x \in S_n \text{ and } x = 2^{n+1}t - 1 \\ 0 & \text{otherwise.} \end{cases}$$

Since $[0,1) = \bigcup_{n \in \mathbb{N}} S_n$, $(0,1) = \bigcup_{n \in \mathbb{N}} J_n$, if $t \in (0,1)$ then

$$\#(\{x : f_t(x) \neq 0\}) = \begin{cases} 1 & \text{if } 2^{n+1}t - 1 \in S_n \\ 0 & \text{otherwise.} \end{cases}$$

It follows that each f_t is a bounded measurable function different from zero for at most one x in $[0,1)$ and that if $x \in (0,1)$ then $\lim_{t \to 0} f_t(x) = 0$. In short, $f_t \overset{\text{a.e.}}{\to} 0$.

If $\lambda^*(D) < \delta$ then for each n in $\mathbb{N}$, $S_n \setminus D \neq \emptyset$. Choose x_n in $S_n \setminus D$. As t traverses J_n, $2^{n+1}t - 1$ traverses $[0,1)$ and there is in J_n a t_n such that $2^{n+1}t_n - 1 = x_n$, whence $f_{t_n}(x_n) = 1$. As $n \to \infty$, $t_n \to 0$ and thus off D, $f_t \overset{\text{u}}{\not\to} 0$

Although $\lim_{t \to 0} f_t(x) = 0$ for each x in $(0,1)$ there is in $(0,1)$ no set D such that $\lambda^*(D) < \delta$ and as $t \to 0$, $f_t(x) \overset{\text{u}}{\to} 0$ off D.

If the hypothesis $\mu(E) < \infty$ is dropped from Egoroff's THEOREM, again the conclusion fails to obtain.

Example 2.2.2.3. Consider the measure situation $(\mathbb{N}, 2^{\mathbb{N}}, \mu)$ in which μ is *counting measure*. If for n in $\mathbb{N}$, f_n is the characteristic function of the set $\{1, 2, \dots, n\}$ then on $E \overset{\text{def}}{=} \mathbb{N}$, $\lim_{n \to \infty} f_n$ is the constant 1. If $0 < \epsilon < 1$ and $\mu(D) < \epsilon$ then $D = \emptyset$. However, $\{f_n\}_{n \in \mathbb{N}}$ does not converge uniformly to 1 on $\mathbb{N} \setminus D$ $(= \mathbb{N})$.

Let $(X, \mathcal{S}, \mu)$ be a measure situation and let $\{f_n\}_{n \in \mathbb{N}}$ be a sequence of measurable functions. There are defined several modes in which the sequence might converge to a function f.

Convergence a.e.:

$$f_n \overset{\text{a.e.}}{\to} f \Leftrightarrow \mu[\{x : f_n(x) \not\to f(x)\}] = 0.$$

Convergence in measure:

$$f_n \overset{\text{meas}}{\to} f \Leftrightarrow$$
$$\{\epsilon > 0 \Rightarrow \lim_{n \to \infty} \mu[\{x : |f_n(x) - f(x)| > \epsilon\}] = 0\}.$$

Convergence in p-mean (when f_n, f are in $L^p(X)$):

$$f_n \overset{\| \ \|_p}{\to} f \Leftrightarrow \lim_{n \to \infty} \int_X |f_n(x) - f(x)|^p \, d\mu = 0.$$

Dominated convergence: (when f_n, $n \in \mathbb{N}$ and f are in $L^p(X)$, $1 \le p < \infty$):

$$f_n \overset{\text{dom}}{\to} f \Leftrightarrow$$
$$f_n \overset{\text{a.e.}}{\to} f \text{ and there is in } L^p(X) \text{ a } g \text{ such that } |f_n| \le |g|.$$

If $\mu(X) < \infty$ then

$$\overset{\text{dom}}{\to} \Rightarrow \left\{ \begin{matrix} \overset{\text{a.e.}}{\to} \\ \underset{\to}{\| \ \|_p} \end{matrix} \right\} \Rightarrow \overset{\text{meas}}{\to} .$$

If $\mu(X) = \infty$ then

$$\overset{\text{dom}}{\to} \Rightarrow \left\{ \begin{matrix} \overset{\text{a.e.}}{\to} \\ \underset{\to}{\| \ \|_p} \end{matrix} \right. \underset{\Rightarrow}{} \overset{\text{meas}}{\to} \cdot$$

Exercise 2.2.2.3. Show that if $f_n \overset{\text{meas}}{\to} f$ then there is a subsequence $\{f_{n_k}\}_{k\in\mathbb{N}}$ such that $f_{n_k} \overset{\text{a.e.}}{\to} f$.

The **Exercises** that follow are designed to show that the implications above are the only valid ones relating the different modes.

Exercise 2.2.2.4. Show that if

$$f_n(x) \overset{\text{def}}{=} \begin{cases} n & \text{if } 0 < x < \frac{1}{n} \\ 0 & \text{if } x \in \mathbb{R} \setminus \left(0, \frac{1}{n}\right) \end{cases} , \quad n \in \mathbb{N}$$

then $f_n \overset{\text{a.e.}}{\to} 0$ but $f_n \overset{\| \ \|_1}{\not\to} 0$, i.e., $\overset{\text{a.e.}}{\to} \not\Rightarrow \overset{\| \ \|_1}{\to}$ and $\overset{\text{a.e.}}{\to} \not\Rightarrow \overset{\text{dom}}{\to}$.

Exercise 2.2.2.5. Show that if $n = 2^k + m$, $0 \le m < 2^k$ then the *marching sequence* $\{f_n\}_{n\in\mathbb{N}}$ given by

$$f_n(x) \overset{\text{def}}{=} \begin{cases} 1 & \text{if } \frac{m}{2^k} \le x \le \frac{m+1}{2^k} \\ 0 & \text{otherwise} \end{cases}$$

is such that $f_n \overset{\| \ \|_1}{\to} 0$ and $f_n \overset{\text{meas}}{\to} 0$ but for all x in $[0,1]$, $f_n(x) \not\to 0$, i.e., $\overset{\| \ \|_1}{\to} \not\Rightarrow \overset{\text{a.e.}}{\to}$, $\overset{\| \ \|_p}{\to} \not\Rightarrow \overset{\text{dom}}{\to}$, and $\overset{\text{meas}}{\to} \not\Rightarrow \overset{\text{a.e.}}{\to}$.

Exercise 2.2.2.6. Show that, in the notation above, if

$$f_n(x) \overset{\text{def}}{=} \begin{cases} 2^k & \text{if } \frac{m}{2^k} \le x \le \frac{m+1}{2^k} \\ 0 & \text{otherwise} \end{cases}$$

then $f_n \overset{\text{meas}}{\to} 0$ but $f_n \overset{\| \ \|_1}{\not\to} 0$, i.e., $\overset{\text{meas}}{\to} \not\Rightarrow \overset{\| \ \|_1}{\to}$ and $\overset{\text{meas}}{\to} \not\Rightarrow \overset{\text{dom}}{\to}$.

Exercise 2.2.2.7. Show that if

$$f_n(x) \overset{\text{def}}{=} \begin{cases} 1 & \text{if } n \le x \le n+1 \\ 0 & \text{otherwise} \end{cases}, \ n \in \mathbb{N}$$

then $f_n \overset{\text{a.e.}}{\to} 0$ but $f_n \overset{\text{meas}}{\not\to} 0$: if $\mu(X) = \infty$ then $\overset{\text{a.e.}}{\to} \not\Rightarrow \overset{\text{meas}}{\to}$.

Exercise 2.2.2.8. Assume a) $\{f_n\}_{n=0}^{\infty} \subset L^1([0,1], \lambda)$, b) $f_n \ge 0$, $n \in \mathbb{N}$, c) $f_n \overset{\text{meas}}{\to} f_0$, and d)

$$\int_0^1 f_n(x)\, dx \to \int_0^1 f_0(x)\, dx \text{ as } n \to \infty.$$

Show that $f_0 \in L^1([0,1], \lambda)$ and that if E is a Lebesgue measurable subset of $[0,1]$ then

$$\lim_{n \to \infty} \int_E f_n(x)\, dx = \int_E f_0(x)\, dx.$$

[*Hint:* **Exercise 2.2.2.3. 135** and **Theorem 2.2.2.1. 133** imply that if $\epsilon > 0$ then E contains a measurable subset E_ϵ such that $\lambda(E \setminus E_\epsilon) < \epsilon$ and

$$\lim_{k \to \infty} \int_{E_\epsilon} f_{n_k}(x)\, dx = \int_{E_\epsilon} f_0(x)\, dx.$$

Hence $\limsup_k \int_E f_{n_k}(x)\, dx \le \int_E f_0(x)\, dx$. Fatou's lemma implies

$$\int_E f_0(x)\, dx \le \liminf_k \int_E f_{n_k}(x)\, dx \le \limsup_k \int_E f_{n_k}(x)\, dx \le \int_E f_0(x)\, dx.$$

If $\lim_{n \to \infty} \int_E f_n(x)\, dx \ne \int_E f_0(x)\, dx$ there is a subsequence $\{f_{n'_k}\}_{k \in \mathbb{N}}$ and a positive η such that $|\int_E f_{n'_k}(x)\, dx - \int_E f_0(x)\, dx| \ge \eta$. The preceding argument applied to $\{f_{n'_k}\}_{k \in \mathbb{N}}$ yields a contradiction.]

Example 2.2.2.4. Show that if

$$g_0(x) = \begin{cases} \frac{1}{x} & \text{if } x \ne 0 \\ 0 & \text{if } x = 0 \end{cases}, \ g_n(x) = \begin{cases} n & \text{if } 0 \le x \le \frac{1}{n} \\ \frac{1}{x} & \text{if } \frac{1}{n} \le x \end{cases}, \ n \in \mathbb{N}$$

then for the "marching sequence" in **Exercise 2.2.2.5. 135** and for properly chosen constants c_n, the sequence $\{g_n + c_n g_n f_n \overset{\text{def}}{=} h_n\}_{n \in \mathbb{N}}$ is such that a) $\{h_n\}_{n \in \mathbb{N}} \subset L^1([0,1], \lambda)$, b) $h_n \ge 0$, $n \in \mathbb{N}$, c) $h_n \overset{\text{meas}}{\to} g_0$, and d)

$$\int_0^1 h_n(x)\, dx \to \infty = \int_0^1 g_0(x)\, dx$$

whereas if $\epsilon \in (0, 1)$ then

$$\limsup_{n \to \infty} \int_\epsilon^1 h_n(x)\, dx = \infty \neq \int_\epsilon^1 g_0(x)\, dx.$$

For contrast cf. **Exercise 2.2.2.8. 136**.

The *Radon-Nikodým theorem* states that if X in $(X, \mathcal{S}, \mu)$ is σ-*finite* and if ν is a σ-finite measure *absolutely continuous* with respect to μ ($\nu \ll \mu$) then there is a nonnegative measurable function f such that for every E in $\mathcal{S}$: $\nu(E) = \int_E f(x)\, d\mu$.

Exercise 2.2.2.9. Show that if $X \overset{\text{def}}{=} \mathbb{R}$, $\mathcal{S} \overset{\text{def}}{=} 2^\mathbb{R}$ and

$$\nu(E) \overset{\text{def}}{=} \begin{cases} 0 & \text{if } E \text{ is countable, finite, or empty} \\ \infty & \text{otherwise} \end{cases}$$
$$\mu(E) \overset{\text{def}}{=} \begin{cases} n & \text{if } \#(E) = n, \ n \in \mathbb{N} \\ \infty & \text{otherwise} \end{cases}$$

(μ is *counting measure*) then $\nu \ll \mu$. Show there is no nonnegative measurable function f such that for every subset E of X

$$\nu(E) = \int_E f(x)\, d\mu.$$

Let $(X, \mathcal{S}, \mu)$ and $(Y, \mathcal{R}, \nu)$ be two measure situations and let $H : X \mapsto Y$ be a *measurable map*, i.e.,

$$B \in \mathcal{R} \Rightarrow H^{-1}(B) \in \mathcal{S}.$$

Then on $\mathcal{R}$ there can be defined an *image measure* $H\mu$ according to the formula:

$$H\mu : \mathcal{R} \ni B \mapsto \mu\left(H^{-1}(B)\right).$$

Example 2.2.2.5. Assume $\mathcal{S}$ is the set $\mathcal{L}$ of Lebesgue measurable subsets of $[0, 1]$, that $\mathcal{R} = \{\emptyset, [0, 1]\}$, and that H is the identity map. Then $[.25, .75] \notin \mathcal{R}$, $H^{-1}([.25, .75]) = [.25, .75] \in \mathcal{S}$ but

$$H([.25, .75]) = H\left[H^{-1}([.25, .75])\right] = [.25, .75] \notin \mathcal{R}.$$

In other words, $H\mu$ is not definable for all images of sets in $\mathcal{S}$. However, since

$$H^{-1}\{[0, 1] \setminus H([0, 1])\} = H^{-1}(\emptyset)$$

it follows that

$$H\mu\{[0, 1] \setminus H([0, 1])\} = \mu(\emptyset) = 0,$$

i.e., $[0,1] \setminus H([0,1])$ is measurable and is a null set:

The image measure $H\mu$ is *concentrated* on $\text{im}(H)$.

Associated with a measure situation $(Z, \mathcal{T}, \rho)$ are the *outer measure* ρ^* and *inner measure* ρ_*. Each is defined on 2^Z. In terms of the inner measure and outer measure it is possible to illustrate the *image measure catastrophe* in which the image measure $H\mu$ is *not* concentrated on $\text{im}(H)$.

Example 2.2.2.6. Let a subset X of $[0,1]$ be nonmeasurable and such that $\lambda_*(X) = 0 = 1 - \lambda^*(X)$. The *inclusion map*

$$H : X \ni x \hookrightarrow x \in [0,1]$$

permits the definition in X of a σ-algebra $\mathcal{S} \overset{\text{def}}{=} H^{-1}(X \cap \mathcal{L})$ and on $\mathcal{S}$ a measure μ: if $A \in \mathcal{S}$ there is in $\mathcal{L}$ an $\widetilde{A}$ such that $A = X \cap \widetilde{A}$ and $\mu(A) \overset{\text{def}}{=} \lambda\left(\widetilde{A}\right)$. (If $B \in \mathcal{L}$ and $A = X \cap B$ then $(B \setminus A) \cup (A \setminus B) \subset [0,1] \setminus X$ and so $\lambda(B) = \lambda(A)$, i.e., $\mu(A)$ is well-defined and the image measure $H\mu$ is definable.)

However, although $H^{-1}\{[0,1] \setminus H([0,1])\}$ is empty, $[0,1] \setminus H([0,1])$ is *not* empty: $H^{-1}\{[0,1] \setminus H([0,1])\} = \emptyset$ and $[0,1] \setminus H([0,1]) \neq \emptyset$. Furthermore, $[0,1] \setminus H([0,1])$ is *not* measurable, and

$$\lambda_* \{[0,1] \setminus H([0,1])\} = 0 = 1 - \lambda^* \{[0,1] \setminus H([0,1])\} :$$

The *image measure catastrophe* has occurred: the image measure $H\mu$ is *not* concentrated on $\text{im}(H)$.

Exercise 2.2.2.10. Let $\{I_n\}$ be the sequence of open intervals deleted from $[0,1]$ in the construction of a Cantor set C of positive measure. For any open interval $J \overset{\text{def}}{=} (a,b)$, $0 \leq a < b \leq 1$ and n in $\mathbb{N}$, let $g_{n,J}$ be continuous, piecewise linear, and such that :

$$g_{n,J}(x) = \begin{cases} 1 & \text{if } 0 \leq x \leq a \\ 0 & \text{if } a + \frac{b-a}{2^n} \leq x \leq b - \frac{b-a}{2^n} \\ 1 & \text{if } b \leq x \leq 1. \end{cases}$$

For n in $\mathbb{N}$ let f_n be the product

$$g_{n,I_1} \cdot g_{n,I_2} \cdot \cdots \cdot g_{n,I_n}.$$

Show that:

i. $f_n \downarrow \chi_C$;
ii. χ_C is a bounded *semicontinuous* function that is not Riemann integrable;
iii. there is no Riemann integrable function h such that $h = \chi_C$ a.e.

[**Note 2.2.2.1:** By means of *bridging functions* (cf. **Exercise 2.1.2.6. 62**) the functions $g_{n,J}$, and hence also the functions f_n, may be replaced by infinitely differentiable functions while the conclusions above remain valid.]

Exercise 2.2.2.11. Show that if C in $[0, 1]$ is a Cantor set of positive measure, (a, b) is the generic notation for an interval deleted from $[0, 1]$ in the construction of C, and

$$f(x) \overset{\text{def}}{=} \begin{cases} 0 & \text{if } 0 \leq x < 1 \\ 1 & \text{if } x = 1 \end{cases}$$

$$g(x) \overset{\text{def}}{=} \begin{cases} 1 & \text{if } x \in C \\ 1 - \frac{1}{2}(b - a) + |x - \frac{1}{2}(a + b)| & \text{if } x \in (a, b) \end{cases}$$

then f is Riemann integrable, g is continuous, and there is no Riemann integrable function h such that $h = f \circ g$ a.e.

Exercise 2.2.2.12. Show that, in contrast to the results above, a continuous function of a Riemann integrable function is Riemann integrable: if f is continuous and g is Riemann integrable then $f \circ g$ is Riemann integrable. Show also that if f is continuous and g is measurable then $f \circ g$ is measurable (cf. **Exercise 2.2.2.1. 133**).

[*Hint:* Assume first that f is a polynomial. Then use the *Weierstraß approximation theorem*.]

Example 2.2.2.7. Assume

$$g(x) \overset{\text{def}}{=} \begin{cases} x^2 \sin(\frac{1}{x}) & \text{if } x \neq 0 \\ 0 & \text{otherwise.} \end{cases}$$

If $c > 0$ let x_c be $\sup \{ x : 0 < x \leq c, g'(x) = 0 \}$. Then define g_c according to the rule

$$g_c(x) = \begin{cases} g(x) & \text{if } 0 \leq x \leq x_c \\ g(x_c) & \text{if } x_c < x \leq c. \end{cases}$$

Let C in $[0, 1]$ be a Cantor set of positive measure, let (a, b) denote the generic interval deleted from $[0, 1]$ in the construction of C, and let c be $\frac{1}{2}(b - a)$. For x in $[a, b]$ define f according to the rule:

$$f(x) = \begin{cases} g_c(x - a) & \text{if } a \leq x \leq \frac{1}{2}(a + b) \\ g_c(-x + b) & \text{if } \frac{1}{2}(a + b) \leq x \leq b. \end{cases}$$

A direct calculation shows that f is differentiable on $[0, 1]$ and that $f'(x) = 0$ on C. Since $g'(0) = 0$ and

$$g'(x) = 2x \sin \left(\frac{1}{x} \right) - \cos \left(\frac{1}{x} \right) \text{ if } x \neq 0,$$

it follows that $|f'(x)| \leq 3$ on $[0,1]$ and f' is discontinuous on C, whence f' is not Riemann integrable:

> The function f' is bounded, has a bounded *primitive* f and yet f' is not Riemann integrable nor is there a Riemann integrable function h such that $h = f'$ a.e.

Exercise 2.2.2.13. Show that if

$$f(x) = \begin{cases} \frac{\sin x}{x} & \text{if } x \neq 0 \\ 1 & \text{if } x = 0 \end{cases}$$

then f is continuous on $\mathbb{R}$,

$$\lim_{R \to \infty} \int_0^R f(x)\, dx$$

exists, but f is not Lebesgue integrable on $[0, \infty)$.

If, in $(X, \mathcal{S}, \mu)$, $\mu(X)$ is not finite, the following pathology may arise.

Exercise 2.2.2.14. Assume $R_n \uparrow \infty$. Find in $\mathbb{R}$ sequences $\{a_n\}_{n\in\mathbb{N}}$ and $\{\alpha_n\}_{n\in\mathbb{N}}$ of positive numbers such that $a_n \uparrow \infty$, $\sum_{n=1}^{\infty} \alpha_n < \infty$,

$$L^1(\mathbb{R}, \mathbb{R}) \ni f_n \stackrel{\text{def}}{=} \sum_{k=n}^{\infty} \alpha_k \chi_{[-a_k, a_k]}, \text{ and } R_n < \frac{\|f_n(x)\|_1}{\|f_n\|_\infty}.$$

Note that $f_n \downarrow 0$, $f_n \stackrel{u}{\to} 0$, and $\|f_n\|_1 \downarrow 0$ but the convergence to zero of $\|f_n\|_1$ can be arbitrarily slow compared to the convergence of $\|f_n\|_\infty$.

For the *product measure situation* $(X \times Y, \mathcal{S} \times \mathcal{T}, \mu \times \nu)$ there are two important theorems, due to *Fubini* and *Tonelli*.

THEOREM (FUBINI). IF $(X, \mathcal{S}, \mu)$ AND $(Y, \mathcal{T}, \nu)$ ARE MEASURE SITUATIONS AND IF $f : X \times Y \mapsto \mathbb{R}$ IS MEASURABLE AND $|f|$ IS INTEGRABLE WITH RESPECT TO $\mu \times \nu$ THEN FOR ALMOST EVERY x RESP. y THE RESTRICTED MAP

(WHEN x IS HELD CONSTANT) $f_{(x)} : Y \mapsto \mathbb{R}$

RESP.

(WHEN y IS HELD CONSTANT) $f^{(y)} : X \mapsto \mathbb{R}$

IS MEASURABLE AND

$$\int_{X \times Y} f(x, y)\, d(\mu \times \nu) = \int_X \left(\int_Y f_{(x)}(y)\, d\nu \right) d\mu$$
$$= \int_Y \left(\int_X f^{(y)}(x)\, d\mu \right) d\nu.$$

THEOREM (TONELLI). IF f IS NONNEGATIVE AND MEASURABLE, IF $(X, \mathcal{S}, \mu)$ AND $(Y, \mathcal{T}, \nu)$ ARE σ-FINITE, AND IF EITHER OF THE ITERATED INTEGRALS

$$\int_X \left(\int_Y f_{(x)}(y) \, d\nu \right) d\mu$$

$$\int_Y \left(\int_X f^{(y)}(x) \, d\mu \right) d\nu$$

IS FINITE THEN f IS INTEGRABLE (WHENCE, BY FUBINI'S THEOREM, THE TWO ITERATED INTEGRALS ARE EQUAL AND THEIR COMMON VALUE IS

$$\int_{X \times Y} f(x, y) \, d(\mu \times \nu).)$$

The importance of the integrability of f in Fubini's theorem and the importance of the σ-finiteness assumption in Tonelli's theorem are revealed in the two parts of **Example 2.2.2.8.**

Example 2.2.2.8.

i. For the measure situation $([0,1]^2, \mathcal{L} \times \mathcal{L}, \lambda \times \lambda)$ and for n in $\mathbb{N}$, let h_n be a continuous nonnegative function such that the support of h_n is contained in $I_n \overset{\text{def}}{=} \left(\frac{1}{n+1}, \frac{1}{n} \right)$ and $\int_{[0,1]} h_n(x) \, dx = 1$. Then for each (x, y) in $[0, 1]^2$, at most one term of the series

$$\sum_{n=1}^{\infty} (h_n(x) - h_{n+1}(x)) \, h_n(y) \overset{\text{def}}{=} f(x, y)$$

is not zero, $|f(x, y)| = \sum_{n=1}^{\infty} |h_n(x) - h_{n+1}(x)| \, |h_n(y)|$, and f is continuous except at $(0, 0)$. Hence f is measurable. Furthermore,

$$\int_{I_n \times I_n} |f(x, y)| \, d(\lambda \times \lambda) = 1, \quad \int_{[0,1] \times [0,1]} |f(x, y)| \, d(\lambda \times \lambda) = \infty$$

$$\int_{[0,1]} \left(\int_{[0,1]} f(x, y) \, dy \right) dx = 1 \neq 0 = \int_{[0,1]} \left(\int_{[0,1]} f(x, y) \, dx \right) dy.$$

Thus, absent the integrability of $|f|$, the conclusion of Fubini's theorem cannot be drawn.

ii. For a measure situation $(X, 2^X, \mu)$, μ is *counting measure* iff whenever $S \subset X$ then

$$\mu(S) = \begin{cases} \#(S) & \text{if } S \text{ is finite} \\ \infty & \text{otherwise.} \end{cases}$$

Assume that in the measure situation $([0,1], 2^{[0,1]}, \mu)$ μ is counting measure (whence $[0, 1]$ is not σ-finite) and consider the measure situation $([0, 1] \times [0, 1], 2^{[0,1]} \times (\mathcal{L} \cap [0, 1]), \mu \times \lambda)$. Assume

$$B \overset{\text{def}}{=} \{ (a, a) \, : \, a \in [0, 1] \}.$$

Then, $B_{(x)}$ resp. $B^{(y)}$ denoting the set of y resp. x such that $(x, y) \in B$,

$$\int_0^1 \mu\left(B_{(x)}\right) d\lambda = \int_0^1 1\, dx = 1$$

$$\int_{[0,1]} \lambda\left(B^{(y)}\right) d\mu = \int_{[0,1]} 0\, d\mu = 0.$$

In other words,

$$\int_0^1 \left[\int_{[0,1]} \chi_B(x, y)\, d\mu \right] d\lambda = 0$$

and

$$\int_{[0,1]} \left[\int_0^1 \chi_B(x, y)\, d\lambda \right] d\mu = 1,$$

i.e., both iterated integrals exist but are unequal even though χ_B is a bounded nonnegative $2^{[0,1]} \times (\mathcal{L} \cap [0,1])$-measurable function. Thus, absent the σ-finiteness condition, the conclusion of Tonelli's theorem cannot be drawn.

Exercise 2.2.2.15. In the context of **Example 2.2.2.8** above, B is the *graph* of the measurable function $f : [0,1] \ni x \mapsto x$ whence B is $\mu \times \lambda$-measurable. Show

$$\mu \times \lambda(B) = \int_{[0,1] \times [0,1]} \chi_B(x, y)\, d(\mu \times \lambda) = \infty.$$

Example 2.2.2.9. Let R in $\mathbb{R}^2$ be a nonmeasurable subset that meets every line in at most two points (cf. **Example 2.2.1.13. 130**). Then χ_R is nonnegative and not measurable whence

$$\int_{\mathbb{R}^2} \chi_R(x, y)\, d\lambda_2$$

does not exist but

$$\int_{\mathbb{R}} \left(\int_{\mathbb{R}} \chi_R(x, y)\, dx \right) dy = \int_{\mathbb{R}} \left(\int_{\mathbb{R}} \chi_R(x, y)\, dy \right) dx = 0.$$

[**Remark 2.2.2.2:** In **Example 3.1.2.5. 193** there is described a set Γ that is dense in $\mathbb{R}^2$ and meets every horizontal resp. vertical line in exactly one point. Let Γ_1 be $\Gamma \cap [0,1]^2$. Then the Riemann double integral $\int_{[0,1]^2} \chi_{\Gamma_1}(x,y)\, dA$ does not exist although

$$\int_{[0,1]} \left(\int_{[0,1]} \chi_{\Gamma_1}(x,y)\, dx \right) dy = \int_{[0,1]} \left(\int_{[0,1]} \chi_{\Gamma_1}(x,y)\, dy \right) dx = 0.]$$

Exercise 2.2.2.16. For f in $\mathbb{R}^{\mathbb{R}}$ and α in $\mathbb{R}$ define the sets

$$S_{\leq\alpha} \stackrel{\text{def}}{=} f^{-1}\left((-\infty, \alpha]\right)$$
$$S_{<\alpha} \stackrel{\text{def}}{=} f^{-1}\left((-\infty, \alpha)\right)$$
$$S_{\geq\alpha} \stackrel{\text{def}}{=} f^{-1}\left([\alpha, \infty)\right)$$
$$S_{>\alpha} \stackrel{\text{def}}{=} f^{-1}\left((\alpha, \infty)\right).$$

Show that f is (Lebesgue or Borel) measurable iff for all α, $S_{\leq\alpha}$ is measurable, iff for all α, $S_{<\alpha}$ is measurable, iff for all α, $S_{\geq\alpha}$ is measurable, and iff for all α, $S_{>\alpha}$ is measurable.

Exercise 2.2.2.17. Let E be a nonmeasurable subset of $\mathbb{R}$. Show that if id denotes the map $\mathbb{R} \ni x \mapsto x \in \mathbb{R}$ and

$$f \stackrel{\text{def}}{=} \chi_E \cdot \text{id} - \chi_{\mathbb{R} \setminus E} \cdot \text{id} \quad \left(\in \mathbb{R}^{\mathbb{R}} \right)$$

then f is nonmeasurable although for every α in $\mathbb{R}$

$$S_{=\alpha} \stackrel{\text{def}}{=} f^{-1}\left(\{\alpha\}\right)$$

consists of at most two points and hence is measurable.

2.2.3. Group-invariant measures

Let S be a set, let S_0 be a fixed nonempty subset of S, and let G be a group of *autojections*, i.e., bijections of S onto itself. The problem to be considered is that of determining whether, on the power set $\mathcal{S} \stackrel{\text{def}}{=} 2^S$, there exists a finitely additive measure μ such that:

i. $\mu(S_0) = 1$ (μ is *normalized*);
ii. if $g \in G$ and $A \in \mathcal{S}$ then $\mu(g(A)) = \mu(A)$.

Such a μ is called an $[S, S_0, G]$-measure and is an instance of a *group-invariant measure*.

Example 2.2.3.1. For S an arbitrary set, S_0 a finite subset of S, and G the set of all bijections of S onto itself, assume $\#(S_0) = n$ $(\in \mathbb{N})$. Define μ as follows:

$$\mu(A) \overset{\text{def}}{=} \begin{cases} \frac{1}{n}\#(A) & \text{if } \#(A) \in \mathbb{N} \cup \{0\} \\ \infty & \text{otherwise.} \end{cases}$$

Then μ is an $[S, S_0, G]$-measure. In other words, if S_0 is finite and μ is an S_0-normalized counting measure then, for any group G of autojections of S, i–ii are satisfied.

Let a group G be called *measurable* if there is a $[G, G, G]$-measure. In [**N1**] von Neumann showed that:

i. every abelian group G is measurable;

ii. if H is a normal subgroup of G and if both H and G/H are measurable then G is measurable. Thus "measurability" is, in the current context, a QL property.

In particular, for n in $\mathbb{N}$, $\mathbb{Q}^n$, $\mathbb{R}^n$, and $\mathbb{T}^n$ are measurable groups.

[**Note 2.2.3.1:** It must be noted that the measures with respect to which abelian groups are measurable are not necessarily countably additive. On the one hand, counting measure, which is a countably additive measure, is, for any countable group, abelian or not, automatically a measure with respect to which the group is a measurable group. However, if $G = \mathbb{T}$ then the group-invariant measure, say μ, derivable from von Neumann's result cannot be countably additive. Indeed, if μ is countably additive it is, in particular, a nontrivial translation-invariant countably additive measure on the σ-ring $\mathcal{S}(\mathsf{K})$ generated by the compact subsets of $\mathbb{T}$. Thus μ is Haar measure and, according to the results in **Subsections 1.1.4** and **2.2.1**, there is in $\mathbb{T}$ a set S such that $\mu(S) = 0 = \infty$, a contradiction. Similar observations apply to $\mathbb{R}^n$.]

On the other hand, Hausdorff [**Hau**] showed that if G_{rigid} is the group of rigid motions of $\mathbb{R}^3$ and

$$B_1 \overset{\text{def}}{=} \left\{ (x, y, z) \ : \ x^2 + y^2 + z^2 \leq 1 \right\}$$
$$S_1 \overset{\text{def}}{=} \left\{ (x, y, z) \ : \ x^2 + y^2 + z^2 = 1 \right\} \ (= \partial B_1)$$

are the unit ball and the surface of the unit ball of $\mathbb{R}^3$ then there is no $\left[\mathbb{R}^3, S_1, G_{rigid}\right]$-measure. Consideration of unions of spherical shells reveals that there is no $\left[\mathbb{R}^3, B_1, G_{rigid}\right]$-measure. Hausdorff's result is consonant with von Neumann's because the group G_{rigid} of rigid motions of $\mathbb{R}^3$ contains the subgroup $SO(3)$ of all rotations about axes through the origin O of $\mathbb{R}^3$, and as the next lines show among other things, $SO(3)$ is not abelian.

The group $SO(3)$ is isomorphic, according to the maps described next, to the multiplicative group $\mathbb{H}_1$ of quaternions of norm 1.

Example 2.2.3.2. The correspondence

$$\mathbf{q} \overset{\text{def}}{=} a\mathbf{1} + b\mathbf{i} + c\mathbf{j} + d\mathbf{k}$$

$$\leftrightarrow a \begin{pmatrix} 1 & 0 \\ 0 & 1 \end{pmatrix} + b \begin{pmatrix} 0 & i \\ i & 0 \end{pmatrix} + c \begin{pmatrix} 0 & -1 \\ 1 & 0 \end{pmatrix} + d \begin{pmatrix} i & 0 \\ 0 & -i \end{pmatrix}$$

$$= \begin{pmatrix} a + di & bi - c \\ bi + c & a - di \end{pmatrix} \overset{\text{def}}{=} \begin{pmatrix} \alpha & -\overline{\beta} \\ \beta & \overline{\alpha} \end{pmatrix} \overset{\text{def}}{=} A_{\mathbf{q}}.$$

is an isomorphism between $\mathbb{H}$ and a subalgebra of the algebra Mat_{22} of 2×2 matrices over $\mathbb{C}$. Furthermore, if $\mathbf{q} \in \mathbb{H}_1$ then $A_{\mathbf{q}}$ is a unitary matrix. Let $\mathbb{C}_\infty$ denote the extended complex plane with the "point at infinity" ∞ adjoined. The map

$$T_{A_{\mathbf{q}}} : \mathbb{C}_\infty \ni z \mapsto \begin{cases} \frac{\alpha z - \overline{\beta}}{\beta z + \overline{\alpha}} & \text{if } \beta z + \overline{\alpha} \neq 0,\ z \neq \infty \\ \infty & \text{if } \beta z + \overline{\alpha} = 0,\ z \neq \infty \\ \frac{\alpha}{\beta} & \text{if } z = \infty,\ \beta \neq 0 \\ \infty & \text{if } z = \infty,\ \beta = 0 \end{cases}$$

is an auteomorphism of the extended plane $\mathbb{C}_\infty$. The association $T_{A_{\mathbf{q}}} \leftrightarrow A_{\mathbf{q}}$ is a group isomorphism. The standard *stereographic projection* of $\mathbb{C}_\infty$ onto the *Riemann sphere* $S_{\frac{1}{2}}$ converts the map $T_{A_{\mathbf{q}}}$ into an auteomorphism $\widetilde{T_{A_{\mathbf{q}}}}$ of $S_{\frac{1}{2}}$. Every auteomorphism T of S_R has a fixed point. [PROOF: If $\mathbf{x} \in S_R$ the sequence $\{T^n\mathbf{x}\}_{n \in \mathbb{N}}$ has a limit point $\mathbf{y}$ and $T\mathbf{y} = \mathbf{y}$.]

The corresponding fixed point of $T_{A_{\mathbf{q}}}$ is a solution ζ of the equation

$$\beta z^2 + (\overline{\alpha} - \alpha)z + \overline{\beta} = 0.$$

Furthermore, $\overline{\zeta}^{-1}$ is a second solution corresponding to a fixed point. The stereographic images of these fixed points are diametrically opposite points of $S_{\frac{1}{2}}$ and $\widetilde{T_{A_{\mathbf{q}}}}$ is a rotation about the axis through them. In this way $\mathbb{H}_1$ is isomorphic to the set of rotations of $S_{\frac{1}{2}}$, i.e., to the set of rotations of S_1 or of $\mathbb{R}^3$. Since $\mathbb{H}_1$ is not abelian neither is $SO(3)$ nor G_{rigid} abelian.

An important consequence of **Remark 1.1.5.1. 17** is that $\mathbb{H}_1$, i.e., $SO(3)$, contains a free set of cardinality $\#(\mathbb{R})$. This fact is basic to the derivation of the Banach-Tarski "paradox" to which the remainder of this **Section** is devoted.

Call two subsets A and B of $\mathbb{R}^3$ *congruent* if there is a rigid motion U such that $U(A) = B$. Hausdorff's idea was exploited by Banach and Tarski to show that, $\cong$ denoting "*congruent*," the ball B_1 in $\mathbb{R}^3$ can be decomposed into m pieces $C_1, \ldots, C_m$ such that

$$B_1 \cong C_1 \uplus \cdots \uplus C_m,$$

into n pieces $C_{m+1}, \ldots, C_{m+n}$ such that

$$B_1 \cong C_{m+1} \uplus \cdots \uplus C_{m+n},$$

into $m + n$ pieces $A_1, \ldots, A_{m+n}$ such that

$$B_1 = A_1 \uplus \cdots \uplus A_{m+n},$$

and such that

$$A_i \cong C_i, \ 1 \le i \le m + n$$

[BanT].

Thus B_1 can be decomposed and the pieces can then be reassembled via rigid motions to form two balls, each congruent to B_1.

This theorem was polished and refined by Sierpinski, von Neumann, and finally by Robinson to yield the following result.

THEOREM **2.2.3.1.** IN THE UNIT BALL B_1 OF $\mathbb{R}^3$ THERE ARE FIVE PAIRWISE DISJOINT SETS, $\mathbf{A}_1, \ldots, \mathbf{A}_5$, THE LAST A SINGLE POINT, AND

$$B_1 = \mathbf{A}_1 \uplus \cdots \uplus \mathbf{A}_5$$
$$B_1 \cong \mathbf{A}_1 \uplus \mathbf{A}_3 \cong \mathbf{A}_2 \uplus \mathbf{A}_4 \uplus \mathbf{A}_5.$$

The ingredients of the proof of THEOREM **2.2.3.1** are straightforward and are assembled below in a pattern based on Robinson's development [Robi]. Except at the very end, where a single translation is invoked, only rotations of $\mathbb{R}^3$ are used for the rigid motions that establish the relevant congruences. Reflections are not used.

At first the focus is on the decomposition of the surface $S_1 \overset{\text{def}}{=} \partial B_1$ of the unit ball B_1 and in that discussion the only rigid motions used are rotations. The goal is to show that there are two different decompositions of S_1:

$$S_1 = A_1 \uplus A_2 \uplus A_3 \uplus A_4$$
$$S_1 = C_1 \uplus C_2 \uplus C_3 \uplus C_4 \uplus C_5 \tag{2.2.3.1}$$

such that

$$A_1 \cong A_2 \cong A_1 \uplus A_2$$
$$A_3 \cong A_4 \cong A_3 \uplus A_4 \tag{2.2.3.2}$$
$$C_1 \cong C_2 \cong C_1 \uplus C_2 \uplus C_5$$
$$C_3 \cong C_4 \cong C_3 \uplus C_4. \tag{2.2.3.3}$$

In the second decomposition, C_5 is a single point P. Furthermore, there are for S_R, $0 < R < 1$, decompositions analogous to that in (2.2.3.1) and with properties analogous to those in (2.2.3.2)–(2.2.3.3).

Associated with a finite decomposition $\{A_1, \ldots, A_n\}$ of S_1 and a congruence

$$A_{k_1} \uplus \cdots \uplus A_{k_r} \cong A_{l_1} \uplus \cdots \uplus A_{l_s}$$

is a *canonical* relation R having domain and range $N \overset{\text{def}}{=} \{1, \ldots, n\}$ and such that iRj iff $i \in K \overset{\text{def}}{=} \{k_1, \ldots, k_r\}$ and $j \in L \overset{\text{def}}{=} \{l_1, \ldots, l_s\}$. A rotation U is *compatible* with the congruence if no point of $U(A_i)$ lies in A_j unless $i \in K$ and $j \in L$, i.e., U is compatible with the corresponding relation R if no point of $U(A_i)$ lies in A_j unless iRj: $U(A_i) \cap A_j \neq \emptyset \Rightarrow iRj$.

Any relation in N is defined by a subset $\mathcal{R}$ of $N \times N$: $iRj \Leftrightarrow (i,j) \in \mathcal{R}$. Hence without regard to congruence, one may speak of a relation R and its corresponding subset $\mathcal{R}$ of $N \times N$: $R \sim \mathcal{R}$. The discussion below is confined to those relations R having domain and range N. In other words each image of the two projections of $\mathcal{R}$ onto the factors of $N \times N$ is N. If $S_1 = \bigcup_{i=1}^{n} A_i$ then for any relation R and any rotation U the notion of their compatibility remains unchanged.

If R_1 and R_2 are relations their product $R_1 R_2$ is the relation R_3 such that $iR_3 k$ iff there is a j such that $iR_1 j$ and $jR_2 k$. The inverse R^{-1} of a relation R is characterized by the statement: $iR^{-1}j$ iff jRi. If iRi then i is a fixed point for R. The identity relation R_{id} corresponds to the "diagonal" $\triangle \overset{\text{def}}{=} \{(i,j) : i = j\}$: $iR_{id}j \Leftrightarrow i = j$.

If $R \sim N \times N$ then $R = R^{-1}$ and so $RR^{-1} = R \neq R_{id}$, i.e., the product of a relation and its inverse need not be the identity relation.

Exercise 2.2.3.1. Show that if U and R are compatible then (since U has a fixed point) R has a *fixed point*.

Exercise 2.2.3.2. Show that if U_i and R_i are compatible, $1 \leq i \leq m$, then $U_1 \cdot \cdots \cdot U_m$ is compatible with $R_m \cdot \cdots \cdot R_1$.

[*Hint:* Note the reversal of order in the product of relations. Use induction.]

For a free set $\{U_1, \ldots, U_m\}$ of rotations of S_1 let G be the group generated by them. Then each element of G is uniquely representable as a *reduced* word $U_{i_1}^{\epsilon_1} \cdots U_{i_j}^{\epsilon_j}$, i.e., a word that does not simplify (cf. **Exercise 1.1.5.1. 9**). If $\mathbf{x} \in S_1$ then $G\mathbf{x} \overset{\text{def}}{=} \{U(\mathbf{x}) : U \in G\}$ is the *orbit* or *trajectory* of $\mathbf{x}$. A point $\mathbf{x}$ in S_1 is a *fixed point* if, for some U in G and not the identity id of G, $U(\mathbf{x}) = \mathbf{x}$. As an auteomorphism of S_1 each U in G has a fixed point $\mathbf{z}$. Since $U \in [\mathbb{R}^3]$ it follows that the antipodal point $-\mathbf{z}$ is also a fixed point for U, just as physical intuition suggests. Since $\det(U) = 1$ and all the eigenvalues of U are in $\mathbb{T}$, it follows that if $U \neq \text{id}$ then the eigenvalues of U are, for some ζ in $\mathbb{T}$, $\{1, \zeta, \overline{\zeta}\}$.

Exercise 2.2.3.3. Show: a) that a trajectory consists entirely of fixed points or contains no fixed points; b) two trajectories are either disjoint or coincide; c) S_1 is the (disjoint) union of the trajectories.

[*Hint:* If $\mathbf{x}$ is fixed for U then $V(\mathbf{x})$ is fixed for VUV^{-1}.]

Exercise 2.2.3.4. Show that if τ is a trajectory without fixed points and $\mathbf{x} \in \tau$ then for each $\mathbf{y}$ in τ there is in G a U such that $\mathbf{y} = U(\mathbf{x})$.

A trajectory consisting of fixed points may be described in a manner similar to that in **Exercise 2.2.3.4** although the details of the description, given next, are more complex.

Let τ consist entirely of fixed points. Among all rotations having fixed points in τ there is at least one, say W, for which the corresponding reduced word is shortest. Assume $W(\mathbf{x}) = \mathbf{x}$.

Exercise 2.2.3.5. Show that the first and last factors of W are not inverses of each other. Thus W and W^{-1} do not begin with the same factor nor end with the same factor.

[*Hint:* Otherwise, for some rotation V, $V^{-1}WV$ has a fixed point in τ and $V^{-1}WV$, reduced, is shorter than W.]

LEMMA **2.2.3.1.** IF $V(\mathbf{x}) = \mathbf{x}$ THEN FOR SOME n IN $\mathbb{Z}$, $V = W^n$.

PROOF. Since W and V have the same fixed point, they are rotations around the axis through $\mathbf{x}$ and hence they commute: $WV = VW$. Hence $V = WVW^{-1}$.

If WV does not simplify, then the unique representation of V begins with the block W. Hence for some n in $\mathbb{N}$, $V = W^n Z$ and Z does not begin with W. However, $V = W^n V W^{-n} = W^{2n} Z W^{-n}$ whence $W^n Z W^{-n} = Z$, and so $V = W^n Z = Z W^n$, which does not begin with W. If $Z W^n$ simplifies then, since V begins with W, $V = W^{n-k}$, $k > 0$, a contradiction. Hence $Z W^n$ does not simplify and so $Z = \mathrm{id}$ and $V = W^n$.

If WV does simplify then, owing to **Exercise 2.2.3.5**, $W^{-1}V$ does not simplify and the previous argument shows that for some n in $\mathbb{N}$, $V = W^{-n}$. □

Exercise 2.2.3.6. Show that if $\mathbf{y} \in \tau$ then for some X that does not end with W nor with the inverse of the first factor of W, $\mathbf{y} = X(\mathbf{x})$. Show also that such an X is unique.

[*Hint:* For some Z, $\mathbf{y} = Z(\mathbf{x})$ and if Z ends with W, then $\mathbf{y} = Z(\mathbf{x}) = YW(\mathbf{x}) = Y(\mathbf{x})$. After finitely many steps, $\mathbf{y} = X(\mathbf{x})$ and X does not end with W. If X ends with the inverse of the first factor of W, consider XW^n for large enough n.]

If $\mathbf{y} = X(\mathbf{x}) = X'(\mathbf{x})$ while X and X' are as described, then $X^{-1}X'$ fixes $\mathbf{x}$ and so $X^{-1}X' = W^n$, $n \in \mathbb{Z}$. If $n > 0$ then X' ends with W. If $n < 0$ then reverse the rôles of X and X'. Hence $n = 0$.]

The next step in the argument is the derivation of the connection between a set of relations and the possibility of decomposing S_1 in a manner associated to the relations. For this purpose the algebra described above for relations is quite useful.

THEOREM **2.2.3.2.** LET $R_1, \ldots, R_m$ BE RELATIONS FOR WHICH N IS BOTH DOMAIN AND RANGE. THEN S_1 CAN BE DECOMPOSED INTO n PIECES $A_1, \ldots, A_n$ AND FOR THIS DECOMPOSITION THERE ARE ROTATIONS $U_1, \ldots, U_m$ COMPATIBLE RESPECTIVELY WITH $R_1, \ldots, R_m$ IFF EACH PRODUCT OF FACTORS OF THE FORM R_i^ϵ, $\epsilon = \pm 1$, HAS A FIXED POINT. FURTHERMORE, IF SUCH ROTATIONS EXIST THEY MAY BE CHOSEN TO BE A FREE SET IN $SO(3)$.

PROOF. If $S_1 = \bigcup_{i=1}^n A_i$ and if rotations U_i as described exist and $R \overset{\text{def}}{=} R_{i_1}^{\epsilon_1} \cdots R_{i_s}^{\epsilon_s}$ is given then $U \overset{\text{def}}{=} U_{i_s}^{\epsilon_s} \cdots U_{i_1}^{\epsilon_1}$ is compatible with R and since U has a fixed point so does R, cf. **Exercise 2.2.3.3. 148.**

Conversely, assume every R as described has a fixed point. Choose m free rotations, $U_1, \ldots, U_m$. The next argument uses the results in **Exercises 2.2.3.4. 148** and **2.2.3.6. 148.**

The task is to define a decomposition $\{A_1, \ldots, A_n\}$ of S_1 so that, for the free set $\mathcal{U} \overset{\text{def}}{=} \{U_1, \ldots, U_m\}$ of rotations, each U_i is compatible with the corresponding R_i. Since the group G generated by $\mathcal{U}$ is countable G may be enumerated systematically so that first only rotations (*reduced* words) that have exactly one factor are listed, then those having only two factors, $\ldots$. Let V_0 be id and let the enumeration of $G \setminus \{\text{id}\}$ be V_n, $n \in \mathbb{N}$. Throughout what follows the fundamental assumption that the domain and range of each R_i is N proves essential.

Case 1. Assume the trajectory τ has no fixed points. Let $\mathbf{x}$ be a point in τ. Start the construction of A_1 by the declaration: $\mathbf{x} \in A_1$. If $V_1 = U_i^{\epsilon_i}$, then since there is an l such that $1R_i^{\epsilon_i}l$, start the construction of A_l by the declaration: $V_1(\mathbf{x}) \in A_l$. Note that $A_1 \cap A_l = \emptyset$. Having constructed or made assignments to pairwise disjoint sets already constructed for all *reduced* words having at most n factors, assume $V_{M+1} \overset{\text{def}}{=} U_j^{\epsilon_j} V_M$ is the first word having $n + 1$ factors. If $V_M(\mathbf{x}) \in A_k$, there is a p such that $kR_j^{\epsilon_j}p$. If A_p has been constructed, assign $V_{M+1}(\mathbf{x})$ to A_p. Otherwise construct A_p by the declaration: $V_{M+1}(\mathbf{x}) \in A_p$. By definition, A_p is disjoint from all sets A_i already in existence.

The inductive *procedure* described above defines pairwise disjoint sets $A_1, \ldots$ for a given trajectory without fixed points. The *procedure* is independent of the trajectory and thus the sets $A_1, \ldots$ are defined for all trajectories having no fixed points.

Case 2. Assume the trajectory τ consists of fixed points. According to the earlier discussion, for a $\mathbf{x}$ in τ, there is a rotation X such that every $\mathbf{y}$ in τ is uniquely of the form $X(\mathbf{x})$, and the rotation X ends neither with W nor with the inverse of the first factor of W. Let the reduced form of W be $\prod_{i=1}^{s} U_{k_i}^{\epsilon k_i}$. Thus the points

$$\mathbf{x}, U_{k_s}^{\epsilon k_s}(\mathbf{x}), \ldots, \prod_{i=1}^{s} U_{k_i}^{\epsilon k_i}(\mathbf{x}) = \mathbf{x}$$

form a closed cycle. Once the points of the cycle have led to constructions or assignments to sets A_q the other points of τ lead to constructions or assignments following the procedure in *Case 1.*

Note that the hypothesis concerning the existence of a fixed point for every product of factors R_i^{ϵ} has not yet been invoked. Now the hypothesis is used to conclude that $\prod_{i=1}^{s} R_{k_i}^{\epsilon_i} \stackrel{\text{def}}{=} R$ has a fixed point. Thus there are integers $k_0, \ldots, k_s$ such that $k_{r-1} R_{k_r}^{\epsilon_r} k_r$, $1 \le r \le s$, and $k_0 = k_s \stackrel{\text{def}}{=} k$. If A_k exists, assign $\mathbf{x}$ to A_k. Otherwise declare A_k to consist of $\mathbf{x}$. Similarly, for the other points of the cycle assign them to, or construct by declaration for them, sets A_{k_r}.

Since the sets $A_1, \ldots$ are pairwise disjoint and since every point of S_1 is on some trajectory, it follows that $S_1 = \bigcup_{i=1}^{?} A_i$. Since the domain and range of each relation is N it follows that $? = n$.

$\square$

LEMMA **2.2.3.2.** IF $S_1 = A_1 \uplus A_2$ AND U IS A ROTATION THEN $U(A_1) \ne A_2$.

PROOF. As a rotation, U is an auteomorphism of S_1 and has a fixed point $\mathbf{v}$. Assume $\mathbf{v} \in A_1$. Then $U(\mathbf{v}) \in A_1 \setminus A_2$.

$\square$

Let $A_1, \ldots, A_n$ be pairwise disjoint subsets of S_1 and assume $S_1 = \bigcup_{i=1}^{n} A_i$. Then $\{A_1, \ldots, A_n\}$ is a *finite decomposition* of S_1. The set of all congruences, of which a typical one is

$$A_{k_1} \uplus \cdots \uplus A_{k_r} \cong A_{l_1} \uplus \cdots \uplus A_{l_s}, \tag{2.2.3.4}$$

is decomposable with respect to an equivalence relation $\equiv$ defined as follows. Let K be $\{k_1, \ldots, k_r\}$, L be $\{l_1, \ldots, l_s\}$, and denote a congruence such as (2.2.3.4) by C_L^K. Then:

  i. $C_L^K \equiv C_{N \setminus L}^{N \setminus K}$;
  ii. $C_L^K \equiv C_{L \setminus K}^{K \setminus L}$;

iii. $M \subset N \Rightarrow C_L^K \equiv C_{L \cup M}^{K \cup M}$.

Furthermore $C_L^K \equiv C_{L'}^{K'}$ if there is a finite chain of congruences linked by $\equiv$ and of which C_L^K is the first and $C_{L'}^{K'}$ is the last. An equivalence such as *i* is an *equivalence by complementation* and an equivalence such as *ii* or *iii* is an *equivalence by transitivity*.

For the most part, the argument below is concerned with canonical relations tied to congruences, but the intermediate results are more easily described with respect to relations that are not necessarily canonical.

If R is a relation on N, if $k \in N$, and if

$$\mathcal{R} \supset \{(1, k), \ldots, (n, k)\}$$

then, by abuse of language, R is said to *contain a constant* (the constant relation R_k, by further abuse of language, the constant k).

Exercise 2.2.3.7. Show that:

i. if R contains the constant k and if $\{(k, l)\} \subset S$ then RS contains the constant l;

ii. if R contains a constant then R has a fixed point;

iii. if $n \geq 3$ there are two canonical relations R and S such that RS is not canonical (hence there are noncanonical relations);

iv. if R and S are canonical relations then RS contains a constant or RS is itself canonical;

v. if $R \sim (K, L)$ then R has a fixed point iff

$$[K \cap L] \cup [(N \setminus K) \cap (N \setminus L)] \neq \emptyset.$$

[*Hint:* Ad *iv*: It suffices to consider the product of two canonical relations and then to proceed by induction. Assume

$$kR_1 s \Leftrightarrow (k \in K_1 \Leftrightarrow s \in L_1)$$
$$sR_2 l \Leftrightarrow (s \in K_2 \Leftrightarrow l \in L_2).$$

Show that if $K_1 = L_1$ or $K_1 = L_2$ then $R_1 R_2$ is canonical. Show that if $K_2 \cap L_1 \neq \emptyset$ and $K_2 \cap (N \setminus L_2) \neq \emptyset$ then $R_1 R_2$ contains a constant. Argue similarly if $K_2 \cap L_1 \neq \emptyset$ or $K_2 \cap (N \setminus L_1) \neq \emptyset$.]

The contents of THEOREM **2.2.3.2. 149** can be translated into a statement about congruences, complementary congruences, and congruences arising from transitivity.

THEOREM **2.2.3.3.** THE SURFACE S_1 MAY BE DECOMPOSED INTO n PIECES SATISFYING A GIVEN SYSTEM $\mathcal{C}$ OF CONGRUENCES IFF:

i. NONE OF THE CONGRUENCES IN $\mathcal{C}$ IS A CONGRUENCE OF TWO COMPLEMENTARY SUBSETS OF S_1;

ii. NONE OF THE CONGRUENCES IN $\mathcal{C}$ IS EQUIVALENT ($\equiv$) TO A CONGRUENCE OF TWO COMPLEMENTARY SUBSETS OF S_1.

PROOF. Since complementary subsets of S_1 cannot be congruent (cf. LEMMA **2.2.3.2. 150**) the necessity of i–ii follows.

The proof of sufficiency of the conditions rests on the conclusion of **Exercise 2.2.3.7iv. 151**:

EITHER

the product R of two canonical relations contains a constant, whence R has a fixed point,

OR

R is itself canonical.

When a product of two canonical relations is itself canonical, say

$$R_1 \sim (K_1, L_1), \ R_2 \sim (K_2, L_2), \ R_1 R_2 \overset{\text{def}}{=} R \sim (K, L).$$

let the superscript * on a subset A of N denote either A itself or $N \setminus A$. Then kRl means there is in N an s such that

$$
\begin{aligned}
(k, s) &\in K_1^* \times L_1^* \\
(s, l) &\in K_2^* \times L_2^* \\
(k, l) &\in K^* \times L^* \\
L_1^* \cap K_2^* &\neq \emptyset.
\end{aligned}
\tag{2.2.3.5}
$$

(Note that there are *sixteen* such sets of conditions.) Each corresponds to the equivalence ($\equiv$) of the congruence corresponding to R and the congruence corresponding to R_1 or to R_2. One of the conditions (2.2.3.5) serves as the transitivity or complementation from which the cited equivalence can be inferred.

When the product of canonical relations is itself canonical its associated congruence is equivalent ($\equiv$) to the congruence associated to one of the factors in the product.

If the product R contains no fixed point then R does not contain a constant and hence R is canonical. Thus, in the notation used above,

$$[K \cap L] \cup [(N \setminus K) \cap (N \setminus L)] = \emptyset$$

whence $K = N \setminus L$ and so R corresponds to a congruence of complementary subsets of S_1, i.e., R corresponds to a congruence equivalent to one of the congruences in the original system, contrary to the hypothesis of THEOREM **2.2.3.3. 151**.

$\square$

Example 2.2.3.3. Let n be 4 and let $\mathcal{C}$ be the system

$$
\begin{aligned}
A_1 &\cong A_2 \cong A_1 \cup A_2 \\
A_3 &\cong A_4 \cong A_3 \cup A_4.
\end{aligned}
$$

Then the only congruences equivalent via complementation and/or transitivity are the following:

$$A_1 \cong A_2 \cong A_1 \cup A_2 \cong A_1 \cup A_2 \cup A_3 \cong A_1 \cup A_2 \cup A_4$$
$$A_3 \cong A_4 \cong A_3 \cup A_4 \cong A_1 \cup A_3 \cup A_4 \cong A_2 \cup A_3 \cup A_4.$$

Hence there exist rotations $U_i, 1 \leq i \leq 4$, such that

$$U_1(A_1) \cup U_3(A_3) \cong A_1 \cup \cdots \cup A_4 = S_1$$
$$U_2(A_2) \cup U_4(A_4) \cong A_1 \cup \cdots \cup A_4 = S_1,$$

i.e., two copies of S_1 can be made from S_1 itself.

Example 2.2.3.4. In **Example 2.2.3.3** choose a trajectory τ consisting of nonfixed points and choose a point P in τ. Define a new decomposition of S_1 by assigning P to $C_5 \overset{\text{def}}{=} \{P\}$ and assigning $U_i(P)$, $1 \leq i \leq 4$, according to the following pattern:

$$U_1(P) \mapsto C_3 \text{ or } C_4, \; U_1^{-1}(P) \mapsto C_1$$
$$U_2(P) \mapsto C_3 \text{ or } C_4, \; U_2^{-1}(P) \mapsto C_2$$
$$U_3(P) \mapsto C_1 \text{ or } C_2, \; U_3^{-1}(P) \mapsto C_1 \text{ or } C_2 \text{ or } C_4$$
$$U_4(P) \mapsto C_1 \text{ or } C_2, \; U_4^{-1}(P) \mapsto C_1 \text{ or } C_2 \text{ or } C_3$$
$$(S_1 = C_1 \cup C_2 \cup C_3 \cup C_4 \cup C_5).$$

(Notice the considerable flexibility in the assignments above.) For any other point Q in τ make assignments according to the algorithm in *Case 1* of the proof of THEOREM **2.2.3.2. 149**. The (canonical) relations to be observed are precisely those listed next:

$$R_1 \sim \mathcal{R}_1 \overset{\text{def}}{=} \{1\} \times \{1, 2\}$$
$$R_2 \sim \mathcal{R}_2 \overset{\text{def}}{=} \{2\} \times \{1, 2\}$$
$$R_3 \sim \mathcal{R}_3 \overset{\text{def}}{=} \{3\} \times \{3, 4\}$$
$$R_4 \sim \mathcal{R}_4 \overset{\text{def}}{=} \{4\} \times \{3, 4\}.$$

The corresponding congruences are

$$C_1 \cong C_2 \cong C_1 \cup C_2 \cup C_5$$
$$C_3 \cong C_4 \cong C_3 \cup C_4$$

and then

$$U_1(C_1) \cup U_3(C_3) \cong S_1$$
$$U_2(C_2) \cup U_4(C_4) \cong S_1.$$

If $0 < r < 1$ let $S_1(r)$ be $\{ (x, y, z) \ : \ x^2 + y^2 + z^2 = r^2 \}$ and, following the patterns in **Examples 2.2.3.3** and **2.2.3.4**, decompose them as follows:

$$S_1(r) = A_1(r) \uplus A_2(r) \uplus A_3(r) \uplus A_4(r), \ 0 < r < 1$$
$$S_1 = C_1 \uplus C_2 \uplus C_3 \uplus C_4 \uplus C_5$$
$$(= C_1 \uplus C_2 \uplus C_3 \uplus C_4 \uplus \{P\}).$$

Let $\mathbf{A}'_k$ be $C_k \uplus \bigcup_{0 < r < 1} A_k(r)$, $1 \le k \le 4$, and let $\mathbf{A}'_5$ be C_5. Then, **O** denoting $(0,0,0)$,

$$U_1\left(\mathbf{A}'_1\right) = U_2\left(\mathbf{A}'_2\right) = \mathbf{A}'_1 \uplus \mathbf{A}'_2 \uplus \mathbf{A}'_5$$
$$U_3\left(\mathbf{A}'_3\right) = U_4\left(\mathbf{A}'_4\right) = \mathbf{A}'_3 \uplus \mathbf{A}'_4$$
$$U_3\left(\mathbf{A}'_3 \uplus \{\mathbf{O}\}\right) = U_4\left(\mathbf{A}'_4 \uplus \{\mathbf{O}\}\right) = \mathbf{A}'_3 \uplus \{\mathbf{O}\} \uplus \mathbf{A}'_4.$$

It follows that B_1 is the union of five disjoint subsets:

$$B_1 = \mathbf{A}'_1 \uplus \mathbf{A}'_2 \uplus \left(\mathbf{A}'_3 \uplus \{\mathbf{O}\}\right) \uplus \mathbf{A}'_4 \uplus \mathbf{A}'_5 \qquad (2.2.3.6)$$

and if $\mathbf{A}_i \overset{\text{def}}{=} \mathbf{A}'_i$, $i = 1, 2, 4, 5$, $\mathbf{A}_3 \overset{\text{def}}{=} \mathbf{A}'_3 \uplus \{\mathbf{O}\}$, then

$$U_1(\mathbf{A}_1) = U_2(\mathbf{A}_2) = \mathbf{A}_1 \uplus \mathbf{A}_2 \uplus \mathbf{A}_5 \qquad (2.2.3.7)$$
$$U_3(\mathbf{A}_3) = U_4(\mathbf{A}_4 \uplus \{\mathbf{O}\}) = \mathbf{A}_3 \uplus \mathbf{A}_4. \qquad (2.2.3.8)$$

From (2.2.3.7) and (2.2.3.8) it follows that

$$\mathbf{A}_1 \cong \mathbf{A}_1 \uplus \mathbf{A}_2 \uplus \mathbf{A}_5 \qquad (2.2.3.9)$$
$$\mathbf{A}_3 \cong \mathbf{A}_3 \uplus \mathbf{A}_4 \qquad (2.2.3.10)$$

whence

$$\mathbf{A}_1 \uplus \mathbf{A}_3 \cong \biguplus_{i=1}^{5} \mathbf{A}_i = B_1.$$

From (2.2.3.7), (2.2.3.8), and the fact that $\mathbf{A}_5 \cong \{\mathbf{O}\}$ it follows as well that

$$\mathbf{A}_2 \uplus \mathbf{A}_4 \uplus \mathbf{A}_5 \cong \mathbf{A}_2 \uplus \left(\mathbf{A}_4 \uplus \{\mathbf{O}\}\right) \cong B_1.$$

$\square$

[**Note 2.2.3.2:** One of the pieces of B_1 is $\mathbf{A}_5$, i.e., the set $\{P\}$ consisting of the single point P. Since the congruence $\{\mathbf{O}\} \cong \mathbf{A}_5$ cannot be achieved by a rotation around **O** but can be achieved by a translation, one of the rigid motions employed is a translation, not a rotation.]

THEOREM **2.2.3.4.** IT IS IMPOSSIBLE TO FIND FOUR PAIRWISE DIS-JOINT SETS

$$K_1, \ldots, K_4$$

AND FOUR *isometric* MAPS

$$T_i : \mathbb{R}^3 \mapsto \mathbb{R}^3, \ 1 \le i \le 4,$$

SUCH THAT

$$B_1 = K_1 \cup \cdots \cup K_4$$
$$B_1 = T_1(K_1) \cup T_2(K_2) \tag{2.2.3.11}$$
$$B_1 = T_3(K_3) \cup T_4(K_4). \tag{2.2.3.12}$$

PROOF. (Note that the sets $T_i(K_i)$ are not assumed to be pairwise disjoint and that the T_i are not assumed to be sense preserving.) If each T_i does not move **O**, assume **O** $\in K_1$. Then $T_1(\mathbf{O}) = \mathbf{O}$ and, since each T_i, as an isometry, is perforce one-one, **O** $\notin T_3(K_3) \cup T_4(K_4) = B_1$. In short, if no T_i moves **O** then B_1 in (2.2.3.11) or in (2.2.3.12) lacks a center, a contradiction. Thus it may be assumed that $T_4(\mathbf{O}) \neq \mathbf{O}$.

Since $B_1 \setminus T_4(B_1)$ covers more than a hemisphere of S_1 (cf. **Exercise 2.2.3.8** below) and since $T_3(K_3)$ covers $B_1 \setminus T_4(B_1)$ it follows that $T_3(K_3)$ covers more than a hemisphere of S_1.

Furthermore, $T_3(\mathbf{O}) = \mathbf{O}$ since otherwise, each of $T_3(B_1)$ and $T_4(B_1)$ failing to cover a hemisphere of S_1 (cf. **Exercise 2.2.3.8** below), their union cannot cover S_1; a fortiori, $T_3(K_3) \cup T_4(K_4) \subsetneq B_1$, a contradiction.

Since T_3^{-1} is an isometry and $T_3^{-1}(\mathbf{O}) = \mathbf{O}$ it follows that $T_3^{-1}(S_1) = S_1$. If H is a hemisphere of S_1, let P be its center, e.g., the "North Pole." Since T_3^{-1} is an isometry it carries every spherical triangle having a vertex at P and contained in H into a congruent spherical triangle. Hence $K_3 \ (= T_3^{-1}(T_3(K_3)))$ itself covers more than a hemisphere of S_1. Hence $K_1 \cup K_2 \cup K_4$ and, a fortiori, $K_1 \cup K_2$ covers less than a hemisphere of S_1. It follows that $T_1(K_1) \cup T_2(K_2)$ cannot cover S_1, much less B_1.

$\square$

Exercise 2.2.3.8. Show that $B_1 \setminus T_4(B_1)$ covers more than a hemisphere of S_1.

[*Hint:* It may be assumed that for some positive a, $T_4(\mathbf{O}) = (a, 0, 0)$. Hence if $(x, y, z) \in T_4(S_1) \cap S_1$ then $x = \frac{a}{2}$.]

For an extensive treatment of measure-theoretic paradoxes see [**Wag**] where there is a treatment of the next generalization, proved also in [**BanT, Str**], of THEOREM **2.2.3.1. 146.**

Let $A \cong' B$ mean that A and B are equivalent under the group of rigid motions *and reflections*. If X and Y are two bounded subsets of $\mathbb{R}^3$ and if both their interiors are nonempty: X°, $Y^\circ \neq \emptyset$, then they are equivalent by finite decomposition, i.e.,

$$X = X_1 \uplus \cdots \uplus X_m$$
$$Y = Y_1 \uplus \cdots \uplus Y_m$$
$$X_i \cong' Y_i, \ 1 \leq i \leq m.$$

Thus a pea can be decomposed into finitely many pieces that can be reassembled to form a body as large as the Sun.

2.3. Topological Vector Spaces

2.3.1. Bases

Zorn's lemma implies that for every *Banach space*, indeed for every vector space, there is a Hamel basis. On the other hand, a *complete orthonormal set* (CON) in a separable *Hilbert space* is a *Schauder basis* or simply a *basis* in the sense given next.

A subset $\mathcal{B} \overset{\text{def}}{=} \{\mathbf{b}_\lambda\}_{\lambda \in \Lambda}$ of a Banach space V (endowed with a norm $\| \ \|$) is a (Schauder) basis (for V) iff for every $\mathbf{x}$ in V there is in $\mathbb{C}^\Lambda$ a unique set $\{x_\lambda\}_{\lambda \in \Lambda}$ such that $\mathbf{x} = \sum_{\lambda \in \Lambda} x_\lambda \mathbf{b}_\lambda$, i.e., there is a countable or finite subset S of Λ and if $\epsilon > 0$ there is in S a finite subset $S(\epsilon)$ such that $\|\mathbf{x} - \sum_{\lambda \in S(\epsilon)} x_\lambda \mathbf{b}_\lambda\| < \epsilon$.

Note that $\mathcal{B}$ need not be countable. For example, if

$$V \overset{\text{def}}{=} \left\{ \{a_t\} \ : \ a_t \in \mathbb{C}, \sum_{t \in \mathbb{R}} |a_t|^2 < \infty \right\},$$

i.e., if V is a nonseparable Hilbert space, and if, for t in $\mathbb{R}$,

$$\mathbf{b}_t \overset{\text{def}}{=} \{\delta_{ts}\}_{s \in \mathbb{R}}$$

then the set of all $\mathbf{b}_t$ is a complete orthonormal set for V and is a (Schauder) basis for V. If V is finite-dimensional the notions of Hamel basis and Schauder basis for V coincide.

Exercise 2.3.1.1. Show that if V is infinite-dimensional *no* Hamel basis is a Schauder basis.

[*Hint:* It may be assumed that each element $\mathbf{h}_\mu$ of a Hamel basis H (in a Banach space) is of norm one. Consider

$$\sum_{n=1}^{\infty} 2^{-n} \mathbf{h}_{\mu_n} \overset{\text{def}}{=} \mathbf{x}$$

for which there must be a finite representation $\sum_\mu x_\mu \mathbf{h}_\mu$.]

From this point on, *basis* means *Schauder basis*. In every infinite-dimensional Banach space V there is an infinite-dimensional closed subspace U for which there is a basis [**Ban, Day, Ge3**]. If a Banach space V has a countable basis then V is separable. Only after many years of study by many mathematicians did Davie and Enflo [**Dav, Enf**] show that the converse statement is false. They constructed in $C([0,1],\mathbb{R})$, the set of $\mathbb{R}$-valued functions continuous on $[0,1]$ and topologized by the norm,

$$\| \ \|_\infty : C([0,1],\mathbb{R}) \ni f \mapsto \sup_{x \in [0,1]} |f(x)|,$$

a closed subspace for which there is no basis.

A basis $\mathcal{S} \stackrel{\text{def}}{=} \{s_n\}_{n \in \mathbb{N}}$ for $C([0,1],\mathbb{R})$ was constructed by Schauder. The functions in $\mathcal{S}$ are piecewise linear and $\text{var}(s_n)$, the *variation* of each s_n, is finite. Indeed, $\text{var}(s_n) \leq 2$, $n \in \mathbb{N}$, i.e., the Schauder system is of uniformly bounded variation. Subsequently Franklin designed a sequence $\mathcal{F} \stackrel{\text{def}}{=} \{F_n\}_{n \in \mathbb{N}}$ of functions constituting a basis for $C([0,1],\mathbb{R})$ [**KacSt**]. With respect to the standard inner product, $(f,g) \stackrel{\text{def}}{=} \int_0^1 f(x)\overline{g(x)}\,dx$, the functions in $\mathcal{F}$ are orthonormal. Since there is a basis for $C([0,1],\mathbb{R})$, the Davie/Enflo result might appear surprising. On the other hand, since every separable Banach space is isometrically isomorphic to some closed subspace of $C([0,1],\mathbb{R})$ [**Ban**], the Davie/Enflo search among the closed subspaces of $C([0,1],\mathbb{R})$ for a separable Banach space for which there is no basis is most reasonable. Were there a basis for every closed subspace of $C([0,1],\mathbb{R})$, there would be a basis for every separable Banach space.

There is an interesting history behind the "basis problem" for Banach spaces. Between WWI and WWII the Polish school of functional analysis flourished. Some of its members started the *"Scottish book"* [**Mau**] in which problems in analysis were proposed and rewards for some solutions were offered. Most rewards were free drinks. One prize was a live goose, for the solution of **Problem 153**, posed by Mazur in the following form.

If $f \in C([0,1]^2,\mathbb{R})$ and $\epsilon > 0$ are there in $[0,1]$ numbers $x_j, y_i, 1 \leq j \leq n, 1 \leq i \leq m$, and is there a matrix $(t_{ij})_{i,j=1}^{m,n}$ such that $\sup_{(x,y) \in [0,1]^2} \left| f(x,y) - \sum_{i,j} f(x,y_i) t_{ij} f(x_j,y) \right| < \epsilon$?

Mazur, loosely paraphrased, asks: When f is regarded as a "matrix" $(f_{xy})_{(x,y) \in [0,1]^2}$ does f have an approximate Moore-Penrose inverse $f^+ \stackrel{\text{def}}{=} (t_{ij})_{i,j=1}^{m,n}$? (Cf. **Exercise 1.3.1.4. 28.**)

The formulation in [**Mau**] of Mazur's question appears to be different from what is given above. The two are actually equivalent. The one offered here provides a useful and motivating connection with the Moore-Penrose inverse.

Grothendieck [**Groth**] vastly extended the problem and constructed an edifice of statements, each equivalent to the affirmative solution of the problem cited above. One formulation is the following.

Let a linear transformation T be said to have *finite rank* iff the dimension of $\text{im}(T)$ is finite. A Banach space V is said to have the *bounded approximation property (BAP)* iff a) there is a sequence $\{F_n\}_{n\in\mathbb{N}}$ of endomorphisms, each with finite rank and such that $\sup_n \|F_n\| < \infty$ and b) for every $\mathbf{x}$ in V

$$\lim_{n\to\infty} \|\mathbf{x} - F_n(\mathbf{x})\| = 0.$$

Prove that every Banach space has the BAP.

Enflo proved more than the existence of a separable Banach without a basis. He constructed a separable reflexive Banach space V in which the BAP fails to hold. Since the existence of a countable basis in a Banach space implies the validity of the BAP in that space, Enflo cooked the goose. (Enflo received a live goose in a special ceremony conducted in Warsaw by some of the contributors to the Scottish book. Regrettably, other contributors, having died in the Holocaust, were absent.)

[**Note 2.3.1.1:** Despite the importance of the set

$$T \overset{\text{def}}{=} \{[0, 2\pi) \ni \theta \mapsto \cos n\theta, \ \theta \mapsto \sin n\theta\}_{n\in\mathbb{N}\cup\{0\}}$$

of trigonometric functions, it is *not* a basis for $C(\mathbb{T}, \mathbb{R})$. Indeed, if T were a basis for $C(\mathbb{T}, \mathbb{R})$ the Fourier series for every function f in $C(\mathbb{T}, \mathbb{R})$ would converge uniformly to f. On the other hand, there is in $C(\mathbb{T}, \mathbb{R})$ a set E that is a dense G_δ and for each f in E there is in $\mathbb{T}$ a set D_f that is a dense G_δ on which the Fourier series for f diverges [**Rud, Zy**].

The set T is an (orthogonal) basis for $L^2([0, 1], \mathbb{C})$, the Hilbert space of (equivalence classes) of measurable functions g defined on $[0, 1]$ and such that

$$\|g\|_2^2 \overset{\text{def}}{=} \int_0^1 |g(x)|^2 \, dx < \infty.$$

For a long time there were two open questions about Fourier series:

    i. If $f \in C([0, 1], \mathbb{R})$ does the Fourier series for f converge anywhere?

    ii. If $g \in L^2([0, 1], \mathbb{C})$ does the Fourier series for g converge a.e.?

Many years after the questions were asked, Carleson [**Ca**], in a spectacular and difficult paper, answered *ii* (and hence also *i*) affirmatively. Subsequently Hunt [**Hu**] extended the result to $L^p([0, 1], \mathbb{C})$, $p > 1$.]

If $\mathcal{B} \overset{\text{def}}{=} \{\mathbf{b}_n\}_{n\in\mathbb{N}}$ is a basis for a Banach space V then in the *dual space* V^* the set $\mathcal{B}^* \overset{\text{def}}{=} \{\mathbf{b}_n^*\}_{n\in\mathbb{N}}$ of *coefficient functionals* such that

$$\mathbf{b}_n^* : V \ni \mathbf{x} \overset{\text{def}}{=} \sum_{n=1}^{\infty} x_n \mathbf{b}_n \mapsto x_n$$

forms with $\mathcal{B}$ a *biorthogonal* set $\{\mathcal{B}, \mathcal{B}^*\}$: $\mathbf{b}_m^*(\mathbf{b}_n) = \delta_{mn}$.

If V is the dual space of a Banach space W $(V = W^*)$ and if $\mathcal{B}$ is a basis for V the set $\mathcal{B}^*$ can lie in W, e.g., if V is a *Hilbert space* or a *reflexive Banach space*. Furthermore if $\mathcal{B}^* \subset W$ then $\mathcal{B}^*$ is a basis for W, [**Day**]. However there are Banach spaces V for which $\mathcal{B}^*$ fails to lie in W [**Ge1**].

Example 2.3.1.1. For p in $[1,\infty)$ the set

$$l^p \overset{\text{def}}{=} \left\{ \{a_n\}_{n\in\mathbb{N}} \ : \ a_n \in \mathbb{R}, \ \sum_{n=1}^{\infty} |a_n|^p < \infty \right\} \overset{\text{def}}{=} V$$

is a Banach space with respect to the norm

$$\|\ \|_p : l^p \ni \{a_n\}_{n\in\mathbb{N}} \mapsto \left(\sum_{n=1}^{\infty} |a_n|^p \right)^{\frac{1}{p}}.$$

If $p = 1$ then

$$V^* = \mathbf{m} \overset{\text{def}}{=} \left\{ \{c_n\}_{n\in\mathbb{N}} \ : \ c_n \in \mathbb{R}, \ \sup_{n\in\mathbb{N}} |c_n| < \infty \right\}$$

and V is itself the dual of

$$W \overset{\text{def}}{=} \mathbf{c}_0 \overset{\text{def}}{=} \left\{ \{d_n\}_{n\in\mathbb{N}} \ : \ d_n \in \mathbb{R}, \ \lim_{n\to\infty} d_n = 0 \right\}.$$

For n in $\mathbb{N}$ let $\mathbf{b}_n$ be the sequence $\{a_{nk}\}_{k\in\mathbb{N}}$ such that

$$a_{1k} = \delta_{1k}, \ a_{nk} = \begin{cases} (-1)^n & \text{if } k = 1 \\ 1 & \text{if } k = n \\ 0 & \text{otherwise} \end{cases} , \ n > 1.$$

Then $\{\mathbf{b}_n\}_{n\in\mathbb{N}}$ is a basis for l^1 and yet $\mathbf{b}_1^*$ is the sequence $\{(-1)^{n+1}\}_{n\in\mathbb{N}}$, which is in $\mathbf{m} \setminus \mathbf{c}_0$: $\{\mathbf{b}_n\}_{n\in\mathbb{N}}$ is not a *retrobasis* for l^1. Furthermore, since $\mathbf{m}$ is not separable, $\{\mathbf{b}_n^*\}_{n\in\mathbb{N}}$ cannot be a basis for $\mathbf{m}$.

Every *orthonormal* basis $\Phi \overset{\text{def}}{=} \{\phi_n\}_{n\in\mathbb{N}}$ for a Hilbert space $\mathcal{H}$ is an *unconditional basis* in the sense that whenever $\sum_{n=1}^{\infty} a_n \phi_n$ converges and $\epsilon_n = \pm 1$, $n \in \mathbb{N}$, then $\sum_{n=1}^{\infty} \epsilon_n a_n \phi_n$ also converges. On the other hand, there are in $\mathcal{H}$ *conditional* bases $\mathcal{B} \overset{\text{def}}{=} \{\mathbf{b}_n\}_{n \in \mathbb{N}}$ such that $\sum_{n=1}^{\infty} a_n \mathbf{b}_n$ converges while for some sequence $\{\epsilon_n = \pm 1 \ : \ n \in \mathbb{N}\}$, $\sum_{n=1}^{\infty} \epsilon_n a_n \mathbf{b}_n$ does *not* converge.

Example 2.3.1.2. If $0 < \alpha < \frac{1}{2}$ and

$$\mathbf{b}_0 = \frac{|x|^\alpha}{(2\pi)^{\frac{1}{2}}}$$

$$\mathbf{b}_{2n+1} = |x|^\alpha \frac{\cos(n+1)x}{(\pi)^{\frac{1}{2}}}$$

$$\mathbf{b}_{2n+2} = |x|^{-\alpha} \frac{\sin(n+1)x}{(\pi)^{\frac{1}{2}}}$$

$$n = 0, 1, \ldots$$

then

$$\mathcal{B} \overset{\text{def}}{=} \{\mathbf{b}_n\}_{n=0,1,\ldots}$$

is a basis but not an unconditional basis for $L^2\left([0,1], \mathbb{C}\right)$ **[Ge9]**.

Karlin **[Day, Kar1, Kar2]** showed that in $C\left([0,1], \mathbb{R}\right)$ *no* basis can be unconditional.

The total variation of each function in Schauder's construction of a basis for $C\left([0,1], \mathbb{R}\right)$ is not more than two, i.e., the Schauder system of functions is of uniformly bounded variation. Each of the Franklin functions is of bounded variation but the Franklin system is not of uniformly bounded variation. COROLLARY **2.3.1.1** to THEOREM **2.3.1.1** below provides an explanation for this phenomenon.

THEOREM **2.3.1.1.** LET $(X, \mathcal{S}, \mu)$ BE A MEASURE SITUATION AND ASSUME $\{\phi_n\}_{n \in \mathbb{N}}$ IS AN INFINITE ORTHONORMAL SYSTEM OF FUNCTIONS DEFINED ON X. ON THE SET

$$E \overset{\text{def}}{=} \left\{ x \ : \ \lim_{n \to \infty} \phi_n(x) \overset{\text{def}}{=} \phi(x) \text{ exists} \right\}$$

$\phi(x) = 0$ A.E.

PROOF. If $E = \emptyset$ the conclusion follows vacuously. Since each ϕ_n is in $L^2(X, \mu)$ it follows that $C \overset{\text{def}}{=} \bigcup_{n=1}^{\infty} \{x \ : \ \phi_n(x) \neq 0\}$ is σ-finite and $\phi = 0$ on $X \setminus C$. In the argument that follows it may be assumed that $E \subset C$ and that $\mu(E) > 0$.

Egoroff's theorem implies that if $\mathcal{S} \ni A \subset E$ and $0 < \epsilon < \mu(A) < \infty$ there is in A a subset A_ϵ such that $\mu\left(A \setminus A_\epsilon\right) < \epsilon$ and $\phi_n \overset{u}{\to} \phi$ on A_ϵ. Since

$\{\phi_n\}_{n\in\mathbb{N}}$ is orthonormal,

$$\int_{A_\epsilon} |\phi(x)|\, d\mu \le \int_{A_\epsilon} |\phi(x) - \phi_n(x)|\, d\mu + \int_{A_\epsilon} |\phi_n(x)|\, d\mu$$

$$\le \|\phi - \phi_n\|_\infty \mu(A_\epsilon) + \sqrt{\mu(A_\epsilon)}$$

whence ϕ is integrable. Note that $\overline{\operatorname{sgn}(\phi)}\phi_n \overset{u}{\to} |\phi|$ on A_ϵ and that *Bessel's inequality* implies $\lim_{n\to\infty} \int_{A_\epsilon} \overline{\operatorname{sgn}(\phi(x))}\phi_n(x)\, d\mu = 0$. Thus $\int_{A_\epsilon} |\phi(x)|\, d\mu = 0$ and $\phi = 0$ a.e. on A_ϵ. Since $\mu(A \setminus A_\epsilon)$ can be made arbitrarily small, it follows that $\phi = 0$ a.e. on A. Moreover, E is σ-finite and so there are measurable sets A_n of finite measure and such that $E = \bigcup_{n\in\mathbb{N}} A_n$, whence $\phi = 0$ a.e. on E.

$\square$

COROLLARY **2.3.1.1.** IF $-\infty < a < b < \infty$, $X = [a,b]$, $\mathcal{S}=\mathcal{L}$, AND $\mu = \lambda$, THEN $\limsup_{n\in\mathbb{N}} \operatorname{var}(\phi_n) = \infty$.

PROOF. If $\limsup_{n\in\mathbb{N}} \operatorname{var}(\phi_n) < \infty$ then, since $\{\phi_n\}_{n\in\mathbb{N}}$ is orthonormal, it may be assumed that $\sup_{n\in\mathbb{N}} |\phi_n(a)| < \infty$. The *Helly selection theorem* **[Wi]** implies the existence of a subsequence, for convenience denoted again $\{\phi_n\}_{n\in\mathbb{N}}$, such that $\lim_{n\to\infty} \phi_n(x) \overset{\text{def}}{=} \phi(x)$ exists everywhere. Hence $\phi(x) = 0$ a.e. and the argument used above implies the contradiction

$$0 = \int_{[a,b]} |\phi(x)|^2\, dx = \lim_{n\to\infty} \int_{[a,b]} |\phi_n(x)|^2\, dx = 1.$$

$\square$

For a continuous function f defined on a metric space with metric d the *modulus of continuity* $\omega(f, x_0, \epsilon)$ at a point x_0 is

$$\sup\{\delta \ : \ d(y, x_0) < \delta \Rightarrow |f(y) - f(x_0)| < \epsilon\}.$$

The *uniform modulus of continuity* $\omega(f, \epsilon)$ is

$$\sup\{\delta \ : \ d(y, x) < \delta \Rightarrow |f(y) - f(x)| < \epsilon\}.$$

Exercise 2.3.1.2. Show that if (X, d) is compact and $\{\phi_n\}_{n\in\mathbb{N}}$ is, for $(X, \mathcal{S}(\mathsf{K}), \mu)$, $(\mu(X) < \infty)$, an infinite, uniformly bounded, and orthonormal system in $L^2(X, \mathbb{C})) \cap C(X, \mathbb{C})$ then there is a positive ϵ such that

$$\limsup_{n\to\infty} \omega(\phi_n, \epsilon) = \infty.$$

[*Hint:* Use the *Ascoli-Arzelà theorem*.]

Example 2.3.1.3. If the measure situation is $([0,1], 2^{[0,1]}, \mu)$ and

$$\mu(E) \stackrel{\text{def}}{=} \begin{cases} \#(E) & \text{if } E \text{ is finite} \\ \infty & \text{otherwise} \end{cases}$$

(μ is counting measure) then $\mathcal{X} \stackrel{\text{def}}{=} \{\chi_{\{x\}}\}_{x \in [0,1]}$ is an orthonormal set.

i. If $\{\chi_{\{x_n\}}\}_{n \in \mathbb{N}}$ is an infinite subsequence of $\mathcal{X}$ then

$$E \stackrel{\text{def}}{=} \left\{ x \ : \ \lim_{n \to \infty} \chi_{\{x_n\}} \stackrel{\text{def}}{=} \chi \text{ exists} \right\} = [0,1]$$

and $\chi = 0$ on E. This example highlights the absence of any hypothesis of σ-finiteness in THEOREM **2.3.1.1. 160**.

ii. Despite COROLLARY **2.3.1.1. 161** var $\left(\chi_{\{x\}}\right) = 1$, $x \in \mathbb{R}$. There is no contradiction because $([0,1], 2^{[0,1]}, \mu)$ is not $([0,1], \mathcal{L}, \lambda)$.

If $\{\mathbf{b}_n\}_{n \in \mathbb{N}}$ is a basis for a Banach space B then the biorthogonal set $\mathbf{B} \stackrel{\text{def}}{=} \{\mathbf{b}_m; \mathbf{b}_n^*\}_{m,n \in \mathbb{N}}$ is *maximal* in the sense that there is no biorthogonal set $\mathbf{B}'$ that properly contains $\mathbf{B}$.

Example 2.3.1.4. In $C([0,1], \mathbb{R})$ let M be the subspace for which there is no basis (cf. the discussion, pp. 157ff., above of the Davie-Enflo example). A direct generalization of the *Gram-Schmidt* orthonormalization process uses the *Hahn-Banach theorem* to produce for M a maximal biorthogonal set that cannot be a basis.

2.3.2. Dual spaces and reflexivity

For every Banach space B there is a *dual space* B^* consisting of the continuous linear functionals defined on B:

$$B^* \stackrel{\text{def}}{=} \left\{ \mathbf{x}^* \ : \ \mathbf{x}^* : B \ni \mathbf{x} \mapsto \mathbb{C}, \ \mathbf{x}^* \text{ is linear and continuous} \right\}.$$

The *Hahn-Banach theorem* shows that $B^* \neq \emptyset$. For B^* there is a *weakest* topology $\mathcal{T}(B^*, B)$ with respect to which each linear functional

$$\mathbf{x} : B^* \ni \mathbf{x}^* \mapsto \mathbf{x}^*(\mathbf{x}) \in \mathbb{C}$$

is continuous. With respect to $\mathcal{T}(B^*, B)$, the unit ball

$$\mathcal{B}_1 \stackrel{\text{def}}{=} \left\{ \mathbf{x}^* \ : \ \mathbf{x}^* \in B^*, \ \|\mathbf{x}^*\| \leq 1 \right\}$$

is compact. The *Krein-Milman theorem* [**Ber**] implies that the $\mathcal{T}(B^*, B)$-closure of the *convex hull* of the set of *extreme points* of $\mathcal{B}_1$ is $\mathcal{B}_1$ itself.

Example 2.3.2.1. Assume $B = C([0,1], \mathbb{R})$, the set of $\mathbb{R}$-valued functions continuous on $[0,1]$. The only extreme points of $\mathcal{B}_1$ are the constant functions $f_\pm \equiv \pm 1$. Indeed, if g is an extreme point of $\mathcal{B}_1$ assume there is in $[0,1]$ an x_0 such that $|g(x_0)| \overset{\text{def}}{=} 1 - 2a < 1$. For some neighborhood $U(x_0)$ of x_0, if $x \in U(x_0)$ then $-1 + a < g(x) < 1 - a$. Furthermore in B there is an h such that

$$0 \le h \le a$$
$$h(x_0) = a$$
$$h(x) = 0 \text{ if } x \notin U(x_0).$$

Thus $g \pm h \in \mathcal{B}_1$, $g \ne g \pm h$, and $g = \frac{1}{2}(g + h) + \frac{1}{2}(g - h)$, whence g is not an extreme point of $\mathcal{B}_1$. Thus the *closed convex hull* of the set of extreme points of $\mathcal{B}_1$ is the set of constant functions of norm not exceeding one, a proper subset of $\mathcal{B}_1$. The Krein-Milman theorem implies the following important conclusion.

The Banach space $C([0,1], \mathbb{R})$ is not the dual space of a Banach space.

Exercise 2.3.2.1. Use a modification of the argument in **Example 2.3.2.1** to show that $L^1([0,1], \mathbb{C})$ is not the dual space of a Banach space.

In $\mathbb{R}^\mathbb{N}$ let c_0 be the subset consisting of all sequences $\{a_n\}_{n \in \mathbb{N}}$ such that $\lim_{n \to \infty} a_n = 0$. For $\mathbf{a} \overset{\text{def}}{=} \{a_n\}_{n \in \mathbb{N}}$ in c_0 let $\|\mathbf{a}\|_\infty$ be $\sup_{n \in \mathbb{N}} |a_n|$. Then c_0 is a Banach space. Similarly c, consisting of all sequences $\{a_n\}_{n \in \mathbb{N}}$ such that $\lim_{n \to \infty} a_n \overset{\text{def}}{=} a_\infty$ exists and, endowed with the norm given to c_0, is a Banach space.

Exercise 2.3.2.2. Show that the in the unit ball of c_0 there is no extreme point. (Hence c_0 is also not the dual space of a Banach space.)

[**Remark 2.3.2.1:** The Banach space c_0 is in fact $C_0(\mathbb{N}, \mathbb{R})$, the set of $\mathbb{R}$-valued functions defined on $\mathbb{N}$, continuous in the *discrete* topology of $\mathbb{N}$, and *vanishing at infinity*. More generally, if X is a locally compact and noncompact Hausdorff space then the unit ball of $C_0(X, \mathbb{R})$ contains no extreme points (cf. [**Ge4**], Problem 83).]

Example 2.3.2.2. For $\mathbf{a} \overset{\text{def}}{=} \{a_n\}_{n \in \mathbb{N}}$ in c let $T(\mathbf{a})$ be $\{a_n - a_\infty\}_{n \in \mathbb{N}}$. Then $T(c) = c_0$, T is continuous and linear, and $\ker(T)$ is the (one-dimensional) subspace consisting of the constant sequences. The unit ball of c has just one extreme point, which gets "lost" under the action of T.

Exercise 2.3.2.3. Let C in $[0,1]$ be a Cantor set of positive measure: $\lambda(C) > 0$. Show that if $p \ge 1$ then $\chi_C \in L^p([0,1], \mathbb{C})$ and that there is no continuous function in the *equivalence class* of χ_C.

If B is a Banach space and B^* is *norm-separable* then B is also norm-separable [**Day**]. The converse is false.

Example 2.3.2.3. The Banach space l^1 is norm-separable and $(l^1)^*$ is **m** (denoted also l^∞), the Banach space of all bounded sequences:

$$\mathbf{m} \overset{\text{def}}{=} \left\{ \mathbf{x} \overset{\text{def}}{=} \{x_n\}_{n\in\mathbb{N}} \; : \; \|\mathbf{x}\| \overset{\text{def}}{=} \sup_n |x_n| < \infty \right\}.$$

Let S be $\left\{ \{\epsilon_n\}_{n\in\mathbb{N}} \; : \; \epsilon_n = \pm 1, \; n \in \mathbb{N} \right\}$, an uncountable subset of **m**. The uncountably many open sets $U(\mathbf{s}) \overset{\text{def}}{=} \{ \mathbf{t} \; : \; \|\mathbf{s} - \mathbf{t}\| < .25 \}$, $\mathbf{s} \in S$, are pairwise disjoint. Hence no countable set can meet every $U(\mathbf{s})$.

Each vector $\mathbf{x}$ in a Banach space B may be regarded as an element $\mathbf{f}$ of $(B^*)^* \overset{\text{def}}{=} B^{**}$:

$$\mathbf{x} : B^* \ni \mathbf{x}^* \mapsto \mathbf{x}^*(\mathbf{x}) \overset{\text{def}}{=} \mathbf{f}(\mathbf{x}^*) \in \mathbb{C}.$$

Hence there is a map $\theta : B \ni \mathbf{x} \mapsto \mathbf{f} \in B^{**}$ and $\|\theta(\mathbf{x})\| = \|\mathbf{x}\|$ (θ is a *linear isometry*). The Banach space B is *reflexive* iff $\theta(B) = B^{**}$.

All finite-dimensional Banach spaces and some infinite-dimensional Banach spaces are reflexive, e.g., if $p > 1$ then $L^p([0,1], \mathbb{C})$ is *reflexive*.

Example 2.3.2.4. In $\mathbf{c}_0$ let B_f be the Banach subspace consisting of the set S of all sequences $\mathbf{x} \overset{\text{def}}{=} \{x_n\}_{n\in\mathbb{N}}$ such that $\mathbf{x} \in \mathbf{c}_0$ and

$$\|\mathbf{x}\|_f \overset{\text{def}}{=} \sup_{n\in\mathbb{N}, \; p_1 < p_2 < \cdots < p_{n+1}} \left[\sum_{i=1}^{n} (x_{p_i} - x_{p_{i+1}})^2 + (x_{p_{n+1}} - x_{p_1})^2 \right]^{\frac{1}{2}} < \infty.$$

A direct calculation shows that if $\mathbf{x} \in B_f$ then $\|\mathbf{x}\|_s$ defined by

$$\|\mathbf{x}\|_s \overset{\text{def}}{=} \sup_{n\in\mathbb{N}, \; p_1 < p_2 < \cdots < p_{2n+1}} \left[\sum_{i=1}^{n} (x_{p_{2i-1}} - x_{p_{2i}})^2 + x_{p_{2n+1}}^2 \right]^{\frac{1}{2}}$$

is finite and, indeed, $\frac{1}{2}\| \; \|_s \leq \| \; \|_f \leq 3\| \; \|_s$. Since the norms $\| \; \|_f$ and $\| \; \|_s$ are *equivalent*, it follows that with respect to $\| \; \|_s$ the Banach space B_f is again a Banach space B_s. Since B_s is not reflexive [**Ja1**], B_f is not reflexive.

The set consisting, for all n in $\mathbb{N}$, of the vectors

$$\mathbf{z}_n \overset{\text{def}}{=} (0, \quad 0, \quad \ldots, \quad 1, \quad 0, \quad \ldots)$$
$$\uparrow$$
$$n\text{th component}$$

is a basis for B_f and B_f^{**} is the set of all sequences $\{F_n\}_{n \in \mathbb{N}}$ for which

$$\lim_{n \to \infty} \| \sum_{i=1}^{n} F_n \mathbf{z}_n \|_f < \infty.$$

The map

$$T : B_f \ni \mathbf{x} \mapsto (x_2 - x_1, \ldots, x_n - x_1, \ldots) \overset{\text{def}}{=} \mathbf{x}^{**}$$

is a linear isometric surjection of B_f onto B_f^{**}.

Thus there is a linear isometric surjection $T : B_f \mapsto B_f^{**}$ but the natural injection θ maps B_f into a subspace W of *codimension* 1 in B_f^{**}.

Thus B is a Banach space that is *not* reflexive and that is nevertheless linearly isometric to B^{**}. The construction is due to James. Details of the proof are to be found in [**Ja1, Ja2**].

2.3.3. Special subsets of Banach spaces

In the set of functions f defined on $\mathbb{R}$ and periodic with period 1 $(f(x) \equiv f(x + 1))$ let $C_p(\mathbb{R}, \mathbb{R})$ be the subset consisting of the $\mathbb{R}$-valued continuous periodic functions with period 1. Each function h in $C_p(\mathbb{R}, \mathbb{R})$ is bounded and with respect to the norm

$$\| \ \|_\infty : C_p(\mathbb{R}, \mathbb{R}) \ni h \mapsto \|h\|_\infty \overset{\text{def}}{=} \sup_{x \in \mathbb{R}} |h(x)|$$

$C_p(\mathbb{R}, \mathbb{R})$ is a Banach space. Moreover, if $\mathbb{T}$ is defined as $\mathbb{R}/\mathbb{Z}$ then $C_p(\mathbb{R}, \mathbb{R})$ may be identified with $C(\mathbb{T}, \mathbb{R})$, or with a closed subspace of $C([0, 1], \mathbb{R})$.

Example 2.3.3.1. The set

$$\mathcal{P}_T \overset{\text{def}}{=} \left\{ x \mapsto \sum_{k=0}^{n} (a_k \cos kx + b_k \sin kx) \ : \ n \in \mathbb{N} \cup \{0\}, a_k, b_k \in \mathbb{R} \right\}$$

of trigonometric polynomials is, owing to the *Stone-Weierstraß* or *Fejér* theorems, dense in $C(\mathbb{T}, \mathbb{R})$. Furthermore $\mathcal{P}_T$ consists of infinitely differentiable functions.

Example 2.3.3.2. The functions G_k defined in **Exercise 2.1.1.15. 50** are in $C_p(\mathbb{R}, \mathbb{R})$, all their difference quotients are large, i.e., everywhere more than k in absolute value, and $\|G_k\|_\infty \le 1$. Hence if U is an open set in $C(\mathbb{T}, \mathbb{R})$, there is in $\mathcal{P}_T$ a p, in $\mathbb{R}$ an α, and in $\mathbb{N}$ a K_k, depending on p and α, and such that $H_k \overset{\text{def}}{=} p + \alpha G_{K_k} \in U$ and everywhere H_k has difference quotients larger than k in absolute value.

Let $\mathcal{ND}$ be the set of nowhere differentiable functions in $C(\mathbb{T}, \mathbb{R})$ and let $\mathcal{A}_k$ be the set of functions having difference quotients everywhere larger than k in absolute value. Then, $H_k \in \mathcal{A}_k$: $\mathcal{A}_k$ is dense in $C(\mathbb{T}, \mathbb{R})$ and $\mathcal{A}_k \subset \mathcal{ND}$.

THEOREM **2.3.3.1.** IN $C(\mathbb{T}, \mathbb{R})$ THE SET $\mathcal{ND}$ OF NOWHERE DIFFER-
ENTIABLE CONTINUOUS FUNCTIONS IS A DENSE SET OF THE SECOND CAT-
EGORY.

PROOF. For each n in $\mathbb{N}$ let F_n in $C(\mathbb{T}, \mathbb{R})$ be the set of functions f
such that for some x in $\mathbb{T}$ and all positive $|h|$

$$\frac{|f(x+h) - f(x)|}{|h|} \leq n.$$

Owing to the compactness of $\mathbb{T}$, it follows that each F_n is a closed
subset of $C(\mathbb{T}, \mathbb{R})$. Since $\mathcal{A}_n$ is dense in $C(\mathbb{T}, \mathbb{R})$ it follows that the *interior*
of F_n is empty: $F_n^\circ = \emptyset$ and so each F_n is nowhere dense. However $\mathcal{SD}$, the
set of *somewhere differentiable* functions in $C(\mathbb{T}, \mathbb{R})$, is a subset of $\bigcup_{n \in \mathbb{N}} F_n$:

$$\mathcal{SD} \subset \bigcup_{n \in \mathbb{N}} F_n$$

whence $\mathcal{SD}$ is a set of the first category. Since $C(\mathbb{T}, \mathbb{R})$ is complete and
since $\mathcal{ND} = C(\mathbb{T}, \mathbb{R}) \setminus \mathcal{SD}$ it follows that, in $C(\mathbb{T}, \mathbb{R})$, $\mathcal{ND}$ is a set of the
second category.

$\square$

In $C(\mathbb{T}, \mathbb{R})$ the set $\mathcal{ND}$ of nowhere differentiable functions is a
dense subset of the second category and the set $\mathcal{SD}$ of somewhere
differentiable functions is a dense subset of the first category.

Exercise 2.3.3.1. Show that each F_n in the proof of THEOREM
2.3.3.1 is closed.

[*Hint:* Assume

$$\{f_n\}_{n \in \mathbb{N}} \subset F_n$$
$$\|f_n - f\|_\infty \to 0$$
$$\frac{|f_n(x_n + h) - f_n(x_n)|}{|h|} \leq n, \ |h| > 0, \ n \in \mathbb{N}.$$

Because $\mathbb{T}$ is compact it may be assumed that $\lim_{n \to \infty} x_n \overset{\text{def}}{=} x_\infty$
exists. Show that for each nonzero h and each positive ϵ

$$\frac{|(f(x_\infty + h) - f(x_\infty)|}{|h|} \leq n + \epsilon.]$$

THEOREM **2.3.3.2.** (TUKEY) IF A AND B ARE DISJOINT CONVEX SUBSETS OF A VECTOR SPACE V AND $\mathbf{x} \notin A \cup B$ THEN THERE EXIST DISJOINT CONVEX SETS C AND D SUCH THAT $A \subset C$, $B \subset D$ AND $\mathbf{x} \in C \cup D$ [**Tu2**].

PROOF. For three vectors $\mathbf{u}$, $\mathbf{v}$, $\mathbf{w}$ in V write $(\mathbf{uvw})$ iff $\mathbf{v}$ lies *between* $\mathbf{u}$ and $\mathbf{w}$, i.e., there is in $[0,1]$ a t such that $\mathbf{v} = t\mathbf{u} + (1-t)\mathbf{w}$. Note that if $\mathbf{a}_1, \mathbf{a}_2 \in A$ and $\mathbf{b}_1, \mathbf{b}_2 \in B$ then not both $(\mathbf{x}\mathbf{a}_1\mathbf{b}_1)$ and $(\mathbf{x}\mathbf{b}_2\mathbf{a}_2)$ can obtain. (Otherwise the (convex) *segments* $\overline{\mathbf{a}_1\mathbf{a}_2}$ and $\overline{\mathbf{b}_1\mathbf{b}_2}$ intersect in a point common to A and B.)

If $(\mathbf{x}\mathbf{a}_1\mathbf{b}_1)$ cannot obtain let D be the *convex hull* of $\mathbf{x}$ and B and let C be A. It follows that D and A, i.e., D and C are disjoint. If $(\mathbf{x}\mathbf{b}_2\mathbf{a}_2)$ cannot hold let C be the convex hull of $\mathbf{x}$ and A and let D be B.

$\square$

Exercise 2.3.3.2. Use some version of the *Axiom of Choice* to conclude from THEOREM **2.3.3.2** that if A and B are disjoint convex subsets of a vector space V there exist complementary (hence disjoint) convex sets C and D such that $C \supset A$ and $D \supset B$.

Example 2.3.3.3. Let B be a separable, infinite-dimensional, and normed vector space. The following construction in B leads to two sequences $Y \overset{\text{def}}{=} \{\mathbf{y}_n\}_{n \in \mathbb{N}}$ and $Z \overset{\text{def}}{=} \{\mathbf{z}_n\}_{n \in \mathbb{N}}$, each constituting a dense subset of B and such that $Y \cup Z$ is linearly independent.

Let $\{\mathbf{x}_n\}_{n \in \mathbb{N}}$ be a dense subset of B. For each n in $\mathbb{N}$ let U_n be $\left\{ \mathbf{x} : \|\mathbf{x} - \mathbf{x}_n\| < \frac{1}{n} \right\}$. Define $\mathbf{y}_1$ to be a nonzero vector in U_1. Because B is infinite-dimensional there is in $U_1 \setminus \text{span}(\mathbf{y}_1)$ a vector $\mathbf{z}_1$. Having defined linearly independent vectors $\mathbf{y}_1, \ldots, \mathbf{y}_n$, $\mathbf{z}_1, \ldots, \mathbf{z}_n$ such that

$$\text{span}(\{\mathbf{y}_1, \ldots, \mathbf{y}_n\}) \supset \text{span}(\{\mathbf{x}_1, \ldots, \mathbf{x}_n\})$$
$$\text{span}(\{\mathbf{z}_1, \ldots, \mathbf{z}_n\}) \supset \text{span}(\{\mathbf{x}_1, \ldots, \mathbf{x}_n\})$$

and such that as well, $\mathbf{y}_k, \mathbf{z}_k \in U_k$, $1 \le k \le n$, let M_n denote the span of $\{\mathbf{y}_k\}_{1 \le k \le n} \cup \{\mathbf{z}_k\}_{1 \le k \le n}$.

Every closed proper subspace M of B is nowhere dense. (Otherwise M contains a nonempty open ball $W \overset{\text{def}}{=} \{ \mathbf{x} : \|\mathbf{x} - \mathbf{a}\| < r \}$ and then M contains $U \overset{\text{def}}{=} -\mathbf{a} + W$, a nonempty open ball centered at $\mathbf{O}$. If $\mathbf{u} \in B$, then for a nonzero t, $t\mathbf{u} \in U$, whence $\mathbf{u} \in M$, i.e., M is not proper.) Thus, in particular, each M_n, as a closed proper subspace of B, is nowhere dense and hence $U_{n+1} \setminus M_n \ne \emptyset$. It follows that in U_{n+1} there are vectors $\mathbf{y}_{n+1}, \mathbf{z}_{n+1}$ such that $\mathbf{y}_1, \ldots, \mathbf{y}_{n+1}$, $\mathbf{z}_1, \ldots, \mathbf{z}_{n+1}$ are linearly independent.

Hence each of the sets $Y \overset{\text{def}}{=} \{\mathbf{y}_n\}_{n\in\mathbb{N}}$ and $Z \overset{\text{def}}{=} \{\mathbf{z}_n\}_{n\in\mathbb{N}}$ is linearly independent and dense and so each of the *convex hulls* Conv(Y) and Conv(Z) is dense in B. Since $Y \cup Z$ is linearly independent, Conv(Y)∩Conv(Z) = ∅.

Exercise 2.3.3.3. Use the results above to conclude that in any separable, normed, infinite-dimensional vector space B there are two complementary, dense, and convex subsets U and W.

Exercise 2.3.3.4. Let **u** resp. **w** be a point in U resp. W above. Show that in each of the sets

$$S_U \overset{\text{def}}{=} \{\, \mathbf{x} - \mathbf{u} \; : \; \mathbf{x} \in U \,\} \text{ and } S_W \overset{\text{def}}{=} \{\, \mathbf{y} - \mathbf{w} \; : \; \mathbf{y} \in W \,\}$$

there are sets Y_U and Z_U resp. Y_W and Z_W like Y resp. Z above and such that

$$\mathbf{u} + \operatorname{span}(Y_U) \cap U, \mathbf{u} + \operatorname{span}(Z_U) \cap U$$
$$\mathbf{w} + \operatorname{span}(Y_W) \cap W, \mathbf{w} + \operatorname{span}(Z_W) \cap W$$

are four pairwise disjoint, dense, and convex subsets of B. Show that for any n in $\mathbb{N}$ there are at least n pairwise disjoint, dense, and convex subsets the union of which is B. Show also that B itself is the boundary of each of the n sets:

In B there are arbitrarily large numbers of pairwise disjoint convex sets of which B is both the union and the common boundary.

[**Note 2.3.3.1:** Whereas there can be no pair of disjoint, dense, and convex subsets of $\mathbb{R}^n$, for every n and every k in $\mathbb{N}$ there are in $\mathbb{R}^n$ k disjoint regions, all with the same boundary (cf. **Subsection 3.1.2**).]

Exercise 2.3.3.5. Show that in the preceding results the hypothesis that B is separable is unnecessary.

[*Hint:* Use Zorn's lemma.]

Exercise 2.3.3.6. Let $\mathcal{B}$ be the category of Banach spaces and continuous homomorphisms and let $P(E)$ mean "E is separable." Show that P is a QL property: if B is a separable Banach subspace of the Banach space A and if $C \overset{\text{def}}{=} A/B$ is separable then A is separable.

2.3.4. Function spaces

The study of topological vector spaces began with the study of *linear function spaces*, e.g., $C([0,1], \mathbb{R})$. A function space B consisting of $\mathbb{R}$-valued functions may be *partially ordered* with respect to the natural *partial order*

$\succeq$: $f \succeq g$ iff for all x in the (common) domain of f and g, $f(x) \geq g(x)$. The same partially ordered function space B may be a *lattice*:

$$f, g \in B \Rightarrow \sup\{f(x), g(x)\} \overset{\text{def}}{=} f \vee g \in B$$

$$f, g \in B \Rightarrow \inf\{f(x), g(x)\} \overset{\text{def}}{=} f \wedge g \in B.$$

A function space B is an *algebra* iff

$$f, g \in B \Rightarrow f \cdot g \in B.$$

Of particular interest among the function spaces are, for a measure situation $(X, \mathcal{S}, \mu)$, the spaces $L^p(X, \mathbb{C})$, $1 \leq p$. For p in $[1, \infty)$

$$p' \overset{\text{def}}{=} \begin{cases} \frac{p}{p-1} & \text{if } 1 < p \\ \infty & \text{otherwise.} \end{cases}$$

The *Hölder inequality*, the theorems of *F. Riesz-Fischer*, and *M. Riesz-Thorin* imply:

i. if $X \in \mathcal{S}$ and $\mu(X) < \infty$, then $1 \leq p < q \Rightarrow L^q(X, \mathbb{C}) \subset L^p(X, \mathbb{C})$;

ii. if $\Phi \overset{\text{def}}{=} \{\phi_n\}_{n \in \mathbb{N}}$ is a *complete orthonormal set* in $\mathcal{H} \overset{\text{def}}{=} L^2(X, \mathbb{C})$ and $f \in \mathcal{H}$ then

$$\sum_{n=1}^{\infty} |(f, \phi_n)|^2 < \infty;$$

if $\sum_{n=1}^{\infty} |a_n|^2 < \infty$ then there is in $\mathcal{H}$ an f such that

$$(f, \phi_n) = a_n, \ n \in \mathbb{N};$$

iii. (Hausdorff-Young) if $X = [0, 1]$, if the functions in Φ are uniformly bounded, e.g., if Φ consists of the functions

$$x \mapsto 1, x \mapsto \frac{1}{\sqrt{2}} \cos 2\pi kx, \ x \mapsto \frac{1}{\sqrt{2}} \sin 2\pi kx, \ k \in \mathbb{N},$$

and if $1 < p \leq 2$ then

$$f \in L^p(X, \mathbb{C}) \Rightarrow \sum_{n=1}^{\infty} |(f, \phi_n)|^{p'} < \infty;$$

under the same hypotheses, if $\sum_{n=1}^{\infty} |a_n|^p < \infty$ there is in $L^{p'}(X, \mathbb{C})$ an f such that $(f, \phi_n) = a_n$, $n \in \mathbb{N}$.

However the following items indicate the degree of inflexibility of the hypotheses in i and iii.

Example 2.3.4.1. If the measure situation is $(\mathbb{R}, \mathcal{L}, \lambda)$ and $1 \le p < q$ then neither of $L^p(\mathbb{R}, \mathbb{C})$ and $L^q(\mathbb{R}, \mathbb{C})$ is a subset of the other. Indeed, since $\frac{1}{q} < \frac{1}{p}$ there is an a such that $0 < ap < 1 < aq$. Hence If

$$f(x) \stackrel{\text{def}}{=} \begin{cases} x^{-a} & \text{if } 0 < x \le 1 \\ 0 & \text{otherwise} \end{cases}$$

$$g(x) \stackrel{\text{def}}{=} \begin{cases} x^{-a} & \text{if } 1 \le x \\ 0 & \text{otherwise} \end{cases}$$

then $f \in L^p(\mathbb{R}, \mathbb{C}) \setminus L^q(\mathbb{R}, \mathbb{C})$ and $g \in L^q(\mathbb{R}, \mathbb{C}) \setminus L^p(\mathbb{R}, \mathbb{C})$. Thus the hypothesis $\mu(X) < \infty$ cannot be dropped from i.

The hypothesis $p \le 2$ cannot be dropped from iii. The details are somewhat complicated and can be found in [**Don, Zy**].

Exercise 2.3.4.1. Show that if $B \stackrel{\text{def}}{=} \left\{ f(x) \stackrel{\text{def}}{=} ax + b \; : \; a, b, x \in \mathbb{R} \right\}$ then B is a linear function space that is neither an algebra nor a lattice.

Exercise 2.3.4.2. Show that if $B \stackrel{\text{def}}{=} \{ p \; : \; p \text{ a polynomial over } \mathbb{R} \}$ then B is a linear function space that is an algebra but not a lattice.

Exercise 2.3.4.3. Show that if B is $L^1(\mathbb{R}, \mathbb{R})$ then B is a linear function space that is a lattice but not an algebra with respect to pointwise multiplication of functions: $f(x) \cdot g(x)$. Show also that B *is* an algebra with respect to multiplication as *convolution*: $f * g(x) \stackrel{\text{def}}{=} \int_{\mathbb{R}} f(x - t) g(t) \, dt$.

Let $\mathbb{I}_{\mathbb{R}}$ be the set of irrational real numbers, let $\mathbb{A}_{\mathbb{R}}$ be the set of algebraic real numbers, and let $\mathsf{Tr}_{\mathbb{R}}$ be the set of transcendental real numbers.

Exercise 2.3.4.4. Show that if

$$f \stackrel{\text{def}}{=} \chi_{\mathbb{I}_{\mathbb{R}}} - \chi_{\mathbb{Q}}$$

$$g \stackrel{\text{def}}{=} \chi_{\mathbb{A}_{\mathbb{R}}} - \chi_{\mathsf{Tr}_{\mathbb{R}}}$$

then f^2 and g^2 are Riemann integrable but $(f + g)^2$ and $f + g$ are not Riemann integrable. Thus the set of functions f such that f^2 is Riemann integrable is not a linear function space.

Exercise 2.3.4.5. Show that if E_1 resp. E_2 is a nonmeasurable subset of $[0, 1]$ resp. $[2, 3]$ and

$$f \stackrel{\text{def}}{=} \chi_{[0,1] \cup E_2} - \chi_{[2,3] \setminus E_2}$$

$$g \stackrel{\text{def}}{=} \chi_{[2,3] \cup E_1} - \chi_{[0,1] \setminus E_1}$$

then $f^2, g^2 \in L^1(\mathbb{R})$ but $(f + g)^2 \notin L^1(\mathbb{R})$. Thus, although $L^2(\mathbb{R}, \mathbb{R})$ is a linear function space, the set of functions f such that f^2 is Lebesgue integrable is not a linear function space.

Example 2.3.4.2. Assume that p is an integer and q is a nonzero integer and $(p, q) = 1$. If

$$F(x) \stackrel{\text{def}}{=} \begin{cases} \frac{4}{q} & \text{if } x = \frac{p}{q} \text{ and } q \text{ is odd} \\ -2 - \frac{4}{q} & \text{if } x = \frac{p}{q} \text{ and } q \text{ is even} \\ -2 & \text{if } x \in \mathbb{I}_{\mathbb{R}} \end{cases}$$

$$G(x) \stackrel{\text{def}}{=} \begin{cases} -1 - \frac{1}{q} & \text{if } x = \frac{p}{q} \text{ and } q \text{ is odd} \\ 1 + \frac{1}{q} & \text{if } x = \frac{p}{q} \text{ and } q \text{ is even} \\ -1 & \text{if } x \in \mathbb{I}_{\mathbb{R}} \end{cases}$$

$$H(x) \stackrel{\text{def}}{=} \begin{cases} -1 - \frac{1}{q} & \text{if } x = \frac{p}{q} \text{ and } q \text{ is odd} \\ 3 + \frac{1}{q} & \text{if } x = \frac{p}{q} \text{ and } q \text{ is even} \\ 3 & \text{if } x \in \mathbb{I}_{\mathbb{R}}. \end{cases}$$

Then each of F, G, and H is *semicontinuous* everywhere whereas

$$F(x) + G(x) + H(x) = \begin{cases} -2 + \frac{2}{q} & \text{if } x = \frac{p}{q} \text{ and } q \text{ is odd} \\ 2 - \frac{2}{q} & \text{if } x = \frac{p}{q} \text{ and if } q \text{ is even} \\ 0 & \text{if } x \in \mathbb{I}_{\mathbb{R}} \end{cases}$$

and $F + G + H$ is nowhere semicontinuous. Thus the set of semicontinuous functions in $\mathbb{R}^{\mathbb{R}}$ is not a linear function space.

Exercise 2.3.4.6. Show that if $\alpha \in \mathbb{I}_{\mathbb{R}}$ then both $x \mapsto \sin x$ and $x \mapsto \sin \alpha x$ are periodic but their sum is not. Hence the set of B of continuous periodic functions in $\mathbb{R}^{\mathbb{R}}$ is not a linear function space.

Example 2.3.4.3. In the function space $C([0,1], \mathbb{R})$ there are two possible norms: $\| \ \|_2$ resp. $\| \ \|_\infty$. Let P resp. Q be the corresponding *unit balls*:

$$P \stackrel{\text{def}}{=} \{ f \ : \ \|f\|_2 \leq 1 \}$$

$$Q \stackrel{\text{def}}{=} \{ f \ : \ \|f\|_\infty \leq 1 \}.$$

It follows that $Q \subset P$. Of greater interest is the fact that $P \backslash Q$ is $\| \ \|_2$-dense in P.

If $f \in P$ and $\|f\|_\infty > 1$ then $f \in P \backslash Q$. If $\|f\|_\infty \leq 1$ let x_0 be such that $|f(x_0)| = \|f\|_\infty$. It may be assumed that $f(x_0) \stackrel{\text{def}}{=} M > 0$. For a positive ϵ there is a neighborhood U of x_0 and in $C([0,1], \mathbb{R})$ a g such that:

i. for x in U, $f(x) > \frac{M}{2}$;
ii. $g(x_0) > 1 + M$, $g = 0$ off U, and $\|g\|_2^2 < \epsilon^2$;
iii. $\int_{[0,1] \backslash U} |f(x)|^2 \, dx + \|g\|_2^2 < 1$.

If

$$h(x) \stackrel{\text{def}}{=} \begin{cases} f(x) & \text{if } x \notin U \\ g(x) + f(x) & \text{if } x \in U \end{cases}$$

then $h \in P \backslash Q$ and $\|f - h\|_2 < \epsilon$.

2.4. Topological Algebras

2.4.1. Derivations

If B is a commutative *Banach algebra* there is a corresponding (possibly empty) set $\mathrm{Hom}(B, \mathbb{C}) \backslash \{\mathbf{O}\}$ of nonzero algebraic homomorphisms of B onto $\mathbb{C}$. The *radical* $\mathcal{R}$ of B is the set of *generalized nilpotents* or, equivalently, the intersection of the set of *kernels* of elements of $\mathrm{Hom}(B, \mathbb{C})$:

$$\mathcal{R} = \left\{ \mathbf{x} \ : \ \limsup_{n \to \infty} \|\mathbf{x}^n\|^{\frac{1}{n}} = 0 \right\}$$

$$= \bigcap_{h \in \mathrm{Hom}(B, \mathbb{C})} \ker(h).$$

In fact, $\lim_{n \to \infty} \|\mathbf{x}^n\|^{\frac{1}{n}}$ always exists and is the *spectral radius* of $\mathbf{x}$. The algebra B is *semisimple* iff $\mathcal{R} = \{\mathbf{O}\}$.

The set $C_b\,(\mathbb{R}, \mathbb{C})$ of $\mathbb{C}$-valued bounded continuous functions defined on $\mathbb{R}$ contains the subsets $C_b^{(k)}\,(\mathbb{R}, \mathbb{C})$ consisting of $\mathbb{C}$-valued functions with k continuous bounded derivatives. With respect to the norm

$$\| \ \|^{(k)} : C_b^{(k)}\,(\mathbb{R}, \mathbb{C}) \ni f \mapsto \sum_{j=0}^{k} \sup_{x \in \mathbb{R}} |f^{(j)}(x)| \overset{\mathrm{def}}{=} \|f\|^{(k)}$$

$C_b^{(k)}\,(\mathbb{R}, \mathbb{C})$ is a *Banach algebra*.

Example 2.4.1.1. The set

$$\mathcal{A} \overset{\mathrm{def}}{=} C_b^{(\infty)}\,(\mathbb{R}, \mathbb{C}) \overset{\mathrm{def}}{=} \bigcap_{k=0}^{\infty} C_b^{(k)}\,(\mathbb{R}, \mathbb{C})$$

in $\mathbb{R}^{\mathbb{R}}$ consisting of infinitely differentiable $\mathbb{C}$-valued functions for which each derivative is bounded on $\mathbb{R}$ is an *algebra* that is not a Banach algebra with respect to any norm. (Note that although no nonconstant polynomial belongs to $\mathcal{A}$, it is a rather rich algebra containing, e.g., the functions $f_n : \mathbb{R} \ni x \mapsto \exp(inx)$, $n \in \mathbb{Z}$ and all their finite linear combinations, the bridging functions (cf. **Example 2.1.2.4. 61**), etc.)

Assume the algebra $\mathcal{A}$ is a Banach algebra with respect to some norm $\| \ \|_b$. The Banach algebra B of continuous endomorphisms of $\mathcal{A}$ consists of *bounded* endomorphisms $T : \mathcal{A} \ni f \mapsto T(f) \in \mathcal{A}$ such that

$$\|T\| \overset{\mathrm{def}}{=} \sup \{ \|T(f)\|_b \ : \ \|f\|_b = 1 \} < \infty.$$

The maps $D : f \mapsto D(f) \overset{\text{def}}{=} f'$ and, for g in $\mathcal{A}$, $R_g : f \mapsto gf$ are in B and D is a *derivation*: $D(fg) = D(f)g + fD(g)$.

Since $DR_g - R_gD = R_{g'}$ and since $R_gR_{g'} = R_{g'}R_g$ it follows that R_g and $DR_g - R_gD$ commute: $R_g(DR_g - R_gD) = (DR_g - R_gD)R_g$.

Generally, if $T, S \in B$ and $TS - ST$ and S commute consider the endomorphism $\Delta : B \ni T \mapsto TS - ST$. Then Δ is a derivation on B. Since $TS - ST$ and S commute, $\Delta^2(T) = O$ whence by *Leibniz's rule*, $\Delta^k(T^l) = O$ if $k > l$. On the other hand, $\Delta^2(T^2) = 2(\Delta(T))^2$, whence by induction and Leibniz's rule it follows that $\Delta^n(T^n) = n!(\Delta(T))^n$. However if M is the norm of the operator Δ, then $\|\Delta^n(T^n)\| \le M^n\|T\|^n$ whence

$$\lim_{n \to \infty} \|(\Delta(T))^n\|^{\frac{1}{n}} \le \lim_{n \to \infty} \frac{M\|T\|}{n!^{\frac{1}{n}}} = 0,$$

i.e., $\Delta(T)$ is a generalized nilpotent in B.

If Δ is the map $D \mapsto DR_g - R_gD$ the conclusion above is that $\Delta(D)$ $(= R_{g'})$ is a generalized nilpotent in B and thus that g' is a generalized nilpotent in $\mathcal{A}$.

Fix t_0 in $\mathbb{R}$ and let ϕ_{t_0} be the functional that maps each f in $\mathcal{A}$ into $f'(t_0)$. Then let L_n be defined by

$$L_n : \mathcal{A} \ni f \mapsto \frac{f(t_0 + \frac{1}{n}) - f(t_0)}{\frac{1}{n}} \overset{\text{def}}{=} L_n(f).$$

Since $f \mapsto f(t_0)$ and $f \mapsto f(t_0 + \frac{1}{n})$ are algebraic homomorphisms of $\mathcal{A}$, they are $\|\ \|_b$-continuous whence each L_n is $\|\ \|_b$-continuous. Since, for all f in $\mathcal{A}$, $L_n(f) \to \phi_{t_0}(f)$ as $n \to \infty$, it follows that if $\mathcal{A}$ is a Banach algebra with respect to $\|\ \|_b$ then each ϕ_{t_0} is a $\|\ \|_b$-continuous linear functional. The *closed graph theorem* [**Rud**] implies that D is $\|\ \|_b$-continuous: $D \in B$.

However, if $\|\ \|_\infty$ designates the supremum norm in $\mathcal{A}$ then for all $\mathbf{x}$ in $\mathcal{A}$, $\|\mathbf{x}\|_\infty \le \|\mathbf{x}\|_b$. Indeed, if $t \in \mathbb{R}$ the evaluation map $E_t : \mathcal{A} \ni f \mapsto f(t) \in \mathbb{C}$ is an algebraic homomorphism and hence $|E_t(f)| = |f(t)| \le \|f\|_b$. Thus $\|f\|_\infty \le \|f\|_b$. Hence in the context above,

$$\|(g')^n\|_b^{\frac{1}{n}} \ge \|(g')^n\|_\infty^{\frac{1}{n}} = \|g'\|_\infty$$

whence $g' = 0$. Since g is an arbitrary element of $\mathcal{A}$ it follows that $D = 0$, a contradiction [**SiW**].

An alternative to the argument in the last paragraph stems from the characterization of the set $\mathcal{N}$ of generalized nilpotents of a commutative Banach algebra B as the *radical* $\mathcal{R}$ of B, i.e., $\mathcal{N}$ is the intersection of the kernels of all algebraic homomorphisms of B into $\mathbb{C}$: $\mathcal{N} = \mathcal{R} = \bigcap_{h \in \text{Hom}(B,\mathbb{C})} \ker(h)$ [**Ber, Loo**]. Among the elements of $\text{Hom}(\mathcal{A}, \mathbb{C})$ are the evaluation maps E_t described above. Hence the generalized nilpotent g' vanishes at each t in $\mathbb{R}$, i.e., $g' = 0$, and the contradiction achieved earlier is repeated.

[**Note 2.4.1.1:** Although there is no norm with respect to which the algebra $\mathcal{A}$ is a Banach algebra, there is a nontrivial *topology* $\mathcal{T}$ with respect to which $\mathcal{A}$ is a topological algebra, i.e., each of the maps

$$\mathcal{A} \times \mathcal{A} \ni (\mathbf{x}, \mathbf{y}) \mapsto \mathbf{x} + \mathbf{y} \in \mathcal{A}$$
$$\mathcal{A} \times \mathcal{A} \ni (\mathbf{x}, \mathbf{y}) \mapsto \mathbf{xy}$$
$$\mathbb{C} \times \mathcal{A} \ni (\alpha, \mathbf{x}) \mapsto \alpha\mathbf{x}$$

is continuous with respect to the product topologies in the *domain* and the topology $\mathcal{T}$ in the *range*.

The topology $\mathcal{T}$ is defined by the family

$$\left\{ p_n \ : \ p_n(f) \stackrel{\text{def}}{=} \sum_{k=0}^{n} \sup_{x \in \mathbb{R}} |f^{(k)}(x)|, \ n \in \mathbb{N} \right\}$$

of *seminorms*. A neighborhood of $\mathbf{O}$ is defined by an n in $\mathbb{N}$ and a positive number ϵ:

$$U_{n,\epsilon} \stackrel{\text{def}}{=} \{ f \ : \ f \in \mathcal{A}, \ p_k(f) < \epsilon, \ 0 \le k \le n \}.]$$

Exercise 2.4.1.1. Verify that $\mathcal{T}$ described above is a *Hausdorff topology* for $\mathcal{A}$ and that with respect to $\mathcal{T}$ the algebra $\mathcal{A}$ is a topological algebra.

2.4.2. Semisimplicity

Semisimple commutative Banach algebras are of particular interest in that they may be represented as algebras of continuous functions on locally compact Hausdorff spaces.

Example 2.4.2.1. Semisimplicity in the category of Banach algebras is a *QL* property, i.e., if I is a closed ideal in a Banach algebra B and if both I and B/I are semisimple then B is semisimple. [PROOF: For I a closed *ideal* in B and $B/I \stackrel{\text{def}}{=} C$, assume $P(I)$ and $P(C)$. The *quotient norm* $\| \ \|_Q$ in C is such that for $\mathbf{x}$ in B,

$$\|\mathbf{x}/I\|_Q \le \|\mathbf{x}\|.$$

Hence if $\mathbf{x}$ is a generalized nilpotent in B then $\mathbf{x}/I$ is a generalized nilpotent in C, whence $\mathbf{x}/I = \mathbf{O}$ and thus $\mathbf{x} \in I$. Since I is semisimple it follows that $\mathbf{x} = \mathbf{O}.$ $\square$]

In connection with the notion of semisimplicity assume that

$$0 \hookrightarrow B \hookrightarrow A \stackrel{\phi}{\mapsto} C \stackrel{\psi}{\mapsto} 0$$

is a short exact sequence in the category of commutative Banach algebras
and continuous algebraic homomorphisms.

Example 2.4.2.2. If $A = C^{(1)}([0,1], \mathbb{C})$, normed according to the
formula

$$\|f\| \overset{\text{def}}{=} \|f\|_\infty + \|f'\|_\infty,$$

if $0 < p < 1$, and $B = \{\, f \; : \; f \in A, \; f(p) = f'(p) = 0\,\}$ then B is a closed
ideal in A and the quotient $A/B \overset{\text{def}}{=} C$ is norm-equivalent to

$$\{\, x \mapsto a + (x - p)b \; : \; a, b \in \mathbb{C}\,\}.$$

(In C

$$\|a + (x - p)b\| \overset{\text{def}}{=} |a| + \max\{p, 1 - p\}|b|$$

and multiplication is performed "modulo $x \mapsto (x-p)^2$," i.e., $x \mapsto (x-p)^2 \overset{\text{def}}{=}$
O.) The radical $\mathcal{R}$ in C consists of all complex multiples of $x \mapsto x - p$ whence
although the original algebra A is semisimple, C is not.

Semisimplicity is not preserved under quotient mappings, i.e., A
and (hence) B can be semisimple while the quotient A/B is not.

[**Note 2.4.2.1:** Malliavin showed [**M**] that if G is a nondiscrete
locally compact abelian group, there is in the semisimple *group
algebra* $L^1(G)$ a closed ideal I such that $L^1(G)/I$ is not semisimple. Since $L^1(G)/I$ is semisimple iff I is the intersection of the
regular maximal ideals that contain I: $I = \text{kernel}(\text{hull}(I))$, his
argument is based on showing there is a closed ideal I such that
$I \subsetneq \text{kernel}(\text{hull}(I))$.]

For a commutative Banach algebra A there is the question of how
extensive $\text{Hom}(A, \mathbb{C})$ can be. Certainly the *zero homomorphism*

$$\mathbf{O} : A \ni \mathbf{x} \mapsto 0 \in \mathbb{C}$$

is always in $\text{Hom}(A, \mathbb{C})$. The next example shows that nothing more than
this conclusion is generally available.

Example 2.4.2.3. Let A be the set $H(D(0,1)^\circ)$ of functions f analytic in

$$D(0,1)^\circ \overset{\text{def}}{=} \{\, z \; : \; z \in \mathbb{C}, \; |z| < 1\,\}$$

and continuous on

$$D(0,1) \overset{\text{def}}{=} \overline{D(0,1)^\circ} \overset{\text{def}}{=} \{\, z \; : \; z \in \mathbb{C}, \; |z| \le 1\,\}.$$

With respect to *convolution* $f * g(z) \stackrel{\text{def}}{=} \int_0^z f(z-s)g(s)\, ds$ as multiplication and normed by the formula $\|f\| \stackrel{\text{def}}{=} \sup_{|z| \le 1} |f(z)|$, A is a Banach algebra. However if $f \in A$ then a direct calculation shows

$$\|f^{n*}\| \le \frac{\|f\|^n}{(n-1)!}, \quad n \in \mathbb{N} \tag{2.4.2.1}$$

whence $\|f^{n*}\|^{\frac{1}{n}} \le \|f\|(n-1)!^{-\frac{1}{n}}$ and *Stirling's formula* implies that each f in A is a generalized nilpotent. The radical $\mathcal{R}$ of A is A itself, i.e., A is a *radical algebra* and $\mathrm{Hom}(A, \mathbb{C}) = \{\mathbf{O}\}$: A is the antithesis of a semisimple Banach algebra.

There are nontrivial commutative Banach algebras A for which the only homomorphism of A into $\mathbb{C}$ is the zero homomorphism.

Exercise 2.4.2.1. Show the validity of the inequality (2.4.2.1) and apply Stirling's formula to draw the conclusions stated above.

[*Hint:* If $z = re^{i\theta}$, $0 \le r \le 1$, then $|f * f(z)| \le \|f\|^2 r$. Use mathematical induction.]

It is customary to confine one's attention to $\mathrm{Hom}\,(A, \mathbb{C}) \setminus \{\mathbf{O}\}$. The reasoning behind this choice is simple and revealing. By definition, an ideal I in an algebra A is a *proper* subalgebra such that $AI \cup IA \subset I$. The kernel of the zero homomorphism is A, which is *not* an ideal whereas the kernel of each element of $\mathrm{Hom}\,(A, \mathbb{C}) \setminus \{\mathbf{O}\}$ *is* an ideal in A.

For the commutative Banach algebra A above, $\mathrm{Hom}\,(A, \mathbb{C}) \setminus \{\mathbf{O}\}$ is empty and so the radical of A, as the intersection of the kernels of the maps in the empty set, is A!

A recurrent theme in the study of abstract structures is the replacement of the abstract by something more familiar, concrete, and amenable to study. For an abstract commutative Banach algebra A, the concrete paradigm is a set of continuous functions on a topological space. Each $\mathbf{a}$ in A is regarded as a function

$$\hat{\mathbf{a}} : \mathrm{Hom}\,(A, \mathbb{C}) \setminus \{\mathbf{O}\} \ni h \mapsto h(\mathbf{a}) \stackrel{\text{def}}{=} \hat{\mathbf{a}}(h) \in \mathbb{C}.$$

The set $\mathrm{Hom}\,(A, \mathbb{C}) \setminus \{\mathbf{O}\}$ is endowed with the weakest topology that makes each such function continuous. As explained in [**Ber, Loo**], this technique leads to useful developments, e.g., the spectral theorem for normal operators in Hilbert space.

If A is a radical algebra, $\mathrm{Hom}\,(A, \mathbb{C}) \setminus \{\mathbf{O}\}$ is empty and there is no possibility of this kind of analysis.

2.5. Differential Equations

2.5.1. Wronskians

THEOREM **2.5.1.1.** IF THE FUNCTIONS $a_0(x), \ldots, a_{n-1}(x)$ ARE CONTINUOUS ON $[a, b]$ AND IF $Y \overset{\text{def}}{=} \{y_1, \ldots, y_n\}$ IS A SET OF SOLUTIONS OF THE *homogeneous linear differential equation*

$$y^{(n)} + \sum_{k=n-1}^{0} a_k(x) y^{(k)} = 0 \qquad (2.5.1.1)$$

OF *order* n THEN Y IS LINEARLY INDEPENDENT ON $[a, b]$ IFF THE *Wronskian*

$$W(Y) \overset{\text{def}}{=} W(y_1, \ldots, y_n) \overset{\text{def}}{=} \det \begin{bmatrix} y_1 & \cdots & y_n \\ y_1' & \cdots & y_n' \\ \vdots & \ddots & \vdots \\ y_1^{(n-1)} & \cdots & y_n^{(n-1)} \end{bmatrix}$$

IS NOT 0 [**CodL**].

On the other hand, if Y is not the set of solutions of a differential equation like (2.5.1.1) $W(Y)$ may be 0 on $[a, b]$ while Y is nevertheless linearly independent (cf. [**Kr**] for developments).

Exercise 2.5.1.1. Show that if

$$y_1(x) \overset{\text{def}}{=} \begin{cases} x^2 & \text{if } -1 \leq x \leq 0 \\ 0 & \text{if } 0 < x \leq 1 \end{cases}$$

$$y_2(x) \overset{\text{def}}{=} \begin{cases} 0 & \text{if } -1 \leq x \leq 0 \\ x^2 & \text{if } 0 < x \leq 1 \end{cases}$$

then $W(y_1, y_2) = 0$ on $[-1, 1]$ but $Y \overset{\text{def}}{=} \{y_1, y_2\}$ is linearly independent on $[-1, 1]$. Show directly that Y is not the set of solutions of a second order homogeneous linear differential equation like (2.5.1.1).

2.5.2. Existence/uniqueness theorems

The *existence/uniqueness theorem* for a differential equation situation of the form

$$y' = f(x, y), \ y(x_0) = y_0$$

is frequently given as follows:

THEOREM **2.5.2.1.** IF

i. f IS CONTINUOUS IN A NEIGHBORHOOD OF (x_0, y_0),

ii. L IS LEBESGUE INTEGRABLE IN A NEIGHBORHOOD U OF x_0,

iii. FOR ALL y_1, y_2 IN A NEIGHBORHOOD V OF y_0 AND ALL x IN U

$$|f(x, y_1) - f(x, y_2)| \leq L(x)|y_1 - y_2|, \tag{2.5.2.1}$$

THEN THERE IS A UNIQUE SOLUTION y, VALID IN U, OF THE SYSTEM

$$y' = f(x, y), \ y(x_0) = y_0.$$

[**Remark 2.5.2.1:** The condition in (2.5.2.1) is a kind of generalized *Lipschitz condition* in which a Lebesgue integrable function L is substituted for the customary Lipschitz constant in the definition of the class *Lip* α.]

Exercise 2.5.2.1. Show that if $f(x, y) = 3y^{\frac{2}{3}}$, $|y| \leq 1$ then

$$y' = f(x, y), \ y(0) = 0$$

has two solutions: $f_1 : x \mapsto x^3$ and $f_2 : x \mapsto 0$, valid if $|x| < 1$.

A system of a more general form is

$$F\left(x, y, y', \ldots, y^{(k)}\right) = 0$$

$$y(x_0) = y_0, \ldots, y^{(k-1)}(x_0) = y_{k-1}. \tag{2.5.2.2}$$

The following **Example 2.5.2.1**, due to Rubel [**Rub**], shows that even if F is a polynomial function of its arguments the existence of unique solutions for (2.5.2.2) may be out of the question.

Example 2.5.2.1. The differential equation

$$3y'^4 y'' y''''^2 - 4y'^4 y'''^2 y'''' + 6y'^3 y''^2 y''''$$
$$+ 24y'^2 y''^4 y'''' - 12y'^3 y'' y'''^3$$
$$- 29y'^2 y''^3 y'''^2 + 12y''^7 = 0 \tag{2.5.2.3}$$

is such that if $\phi, \epsilon \in C\left((-\infty, \infty)\right)$ there is a solution y such that

$$|y(t) - \phi(t)| < \epsilon(t), \ t \in (-\infty, \infty).$$

In fact, if $g \in \mathbb{R}^{\mathbb{R}}$ and

$$g(t) \overset{\text{def}}{=} \begin{cases} \exp\left(-\frac{1}{(1-t^2)}\right) & \text{if } t \in (-1,1) \\ 0 & \text{otherwise} \end{cases}$$

and if $f' = g$ then for any constants A, B, α, β the function $Af(\alpha t + \beta) + B$ is a solution of (2.5.2.3).

If J is a closed interval $[a, b]$ let $F_J(t)$ be a bridging function (cf. **Exercise 2.1.2.6. 62**) of the form $af(\alpha t + \beta) + b$, and such that for a δ in $\left(0, \frac{(a+b)}{4}\right)$ and for given constants A, B, α, β

$$F_J(t) = \begin{cases} A & \text{if } t \in (-\infty, a + \delta] \\ \text{a monotone function} & \text{if } t \in [a + \delta, b - \delta] \\ B & \text{if } t \in [b - \delta, \infty). \end{cases}$$

If ϕ as described above is given then, on any compact interval K, ϕ may be approximated by a piecewise linear function ψ. If K is *partitioned* finely enough into subintervals and if bridging functions of the type F_J are pieced together to a *chain* that *interpolates* ψ at the endpoints of the partitioning intervals, the chain approximates ψ, hence ϕ, within $\epsilon(t)$ on K. The result of piecing together such chains over $\mathbb{R}$ is a solution of (2.5.2.3) that approximates ϕ within ϵ on $\mathbb{R}$.

Thus the solutions of (2.5.2.3) that pass through any point (c, d) are not only not unique but can be chosen to stay within any prescribed smooth open "$\epsilon(t)$-*channel*." This phenomenon is a form of *superbifurcation* in that if a solution y of (2.5.2.3) is known for all x in $(-\infty, a]$ the solution may be continued *as a solution of* (2.5.2.3) into any $\epsilon(t)$-channel containing the point with coordinates $(a, y(a))$.

Rubel's differential equation has too many solutions. Lewy's, described next, has no solutions at all.

Example 2.5.2.2. There is in $C^\infty\left(\mathbb{R}^3, \mathbb{C}\right)$ an f (a smooth function) such that the partial differential equation

$$-i\frac{\partial u}{\partial x} + \frac{\partial u}{\partial y} - 2(x + iy)\frac{\partial u}{\partial z} = f(x, y, z) \tag{2.5.2.4}$$

has not even a *weak* or *distribution* solution on any open subset of $\mathbb{R}^3$ [**Le**].

In [**Hö**] there is an extended study of the phenomenon above. Hörmander derives conditions that a *partial differential operator* P must satisfy if the equation $P(u) = f$ is to have a solution for every smooth function f.

On the other hand, the *Cauchy-Kowalewski* theorem implies that if the coefficients of a homogeneous partial differential equation are *analytic* then the equation has (local) analytic solutions.

Example 2.5.2.3. If P is partial differential operator and f is a smooth function such that $P(u) = f$ has no solution then $P(u) - fu = 0$ is a homogeneous partial differential equation admitting no solution. Indeed, if v is a solution of $P(u) - fu = 0$ then $\ln v \overset{\text{def}}{=} w$ is a solution of $P(u) = f$.

[**Note 2.5.2.1:**   The typical situation in which an ordinary differential equation, or a system of ordinary differential equations, or a partial differential equation, or a system of partial differential equations fails to have a solution is that in which the solution is required to satisfy boundary conditions, initial conditions, or smoothness conditions. The Lewy example does not involve such side conditions and yet there is no solution of the differential equation (2.5.2.4).]

2.6. Complex Variable Theory

2.6.1. Morera's theorem

The conventions

$$D(a, r) \overset{\text{def}}{=} \{ z \ : \ |z - a| \leq r \}$$

$$H(\Omega) \overset{\text{def}}{=} \{ f \ : \ f \text{ is holomorphic in the region } \Omega \}$$

are observed in this **Section**.

THEOREM 2.6.1.1. (MORERA) IF f IS CONTINUOUS IN A *region* Ω OF $\mathbb{C}$, IF $p \in \Omega$, AND IF $\int_{\partial\triangle} f(z)\, dz = 0$ FOR EVERY TRIANGLE $\triangle$ CONTAINED IN $\Omega \setminus \{p\}$ THEN $f \in H(\Omega)$ [**Rud**].

Exercise 2.6.1.1. Show that if $\Omega \overset{\text{def}}{=} D(0, 1)^\circ$ and

$$f(z) = \begin{cases} \frac{1}{z^2} & \text{if } z \neq 0 \\ 0 & \text{otherwise} \end{cases}$$

then for every triangle $\triangle$ contained in $\Omega \setminus \{0\}$, $\int_{\partial\triangle} f(z)\, dz = 0$. (However $f \notin H(\Omega)$ since f is not continuous at 0.)

2.6.2. Natural boundaries

The power series $\sum_{n=0}^{\infty} z^n \overset{\text{def}}{=} g(z)$ converges if $|z| < 1$ and diverges if $z = 1$. Hence its *radius of convergence* is 1. On the other hand, the function $f : \mathbb{C} \setminus \{1\} \ni z \mapsto (z - 1)^{-1}$ is holomorphic in its domain and $f = g$ in $\Omega \overset{\text{def}}{=} D(0, 1)^\circ$. Thus g has an *analytic continuation* beyond Ω.

Exercise 2.6.2.1. Show that if

$$f(z) = \sum_{n=0}^{\infty} z^{n!}, \ z \in \Omega \overset{\text{def}}{=} D(0,1)^{\circ} \qquad (2.6.2.1)$$

then $f \in H(\Omega)$ and there is no analytic continuation of f beyond Ω. (The boundary $\partial\Omega$ of Ω is a *natural boundary* for f.)

[*Hint:* For θ in $\pi\mathbb{Q} \cap [0, 2\pi)$ consider $f\left(\frac{n-1}{n} r e^{i\theta}\right)$, $0 \le r < 1$, $n \in \mathbb{N}$.]

The sequence $A \overset{\text{def}}{=} \{a_n\}_{n \in \mathbb{N}}$ of coefficients of the series (2.6.2.1) has *gaps* of increasing size, e.g., if $\{n_k\}_{k \in \mathbb{N}}$ is the sequence of indices for which $a_{n_k} \ne 0$ then

$$\frac{n_{k+1}}{n_k} = \frac{(k+1)!}{k!} = k+1.$$

Indeed, according to the *Hadamard gap theorem*, if $\lambda > 1$ and

$$c_k \ne 0, \ \frac{n_{k+1}}{n_k} \ge \lambda, \ k \in \mathbb{N},$$

the function represented by the series $\sum_{k=1}^{\infty} c_k z^{n_k}$ cannot be continued analytically beyond the circle of convergence of the series [**Rud**].

If the nonempty region Ω is not $\mathbb{C}$ there is in $H(\Omega)$ a function f such that $\partial\Omega$ is the natural boundary for f: each point of Ω is a singularity of f. Indeed, if $\partial\Omega$ has no limit points a *Mittag-Leffler expansion* provides a function having $\partial\Omega$ as its set of poles. Otherwise there is in Ω a countable set Z such that $Z' = \partial\Omega$. The *Weierstraß infinite product representation* leads to an f in $H(\Omega)$ and equal to zero precisely on Z. The Identity Theorem implies that $\partial\Omega$ is a natural boundary for f [**Hil, Rud**].

If the power series $\sum_{n=0}^{\infty} a_n z^n$ has a positive radius of convergence R then the series represents a function f in $H\left(D(0,R)^{\circ}\right)$. There is a considerable body of theorems dealing with the nature of the sequence $A \overset{\text{def}}{=} \{a_n\}_{n=0}^{\infty}$ and the nature of the set $S_R(f)$ of *singularities* of f on $\mathbb{T}_R \overset{\text{def}}{=} \{ z : |z| = R \}$, the boundary of the *circle of convergence* of the series. Since a limit point of singularities is a singularity, $S_R(f)$ is closed.

Example 2.6.2.1. Let F be a closed subset of $\mathbb{T}_R$.

i. If F is empty then for the function $f : z \mapsto z$, $S_R(f) = F$.

ii. If F is finite, say $F = \{z_1, \ldots, z_N\}$, then for the function

$$f : z \mapsto \sum_{n=1}^{N} \frac{1}{z - z_n},$$

$S_R(f) = F$.

iii. If F is infinite, for each n in $\mathbb{N}$ there are in F finitely many points $P_{n1}, \ldots, P_{nm_n}$ such that

$$F \subset \bigcup_{k=1}^{m_n} D\left(P_{nk}, \frac{R}{n+1}\right)^\circ.$$

In each $D(0, R)^\circ \cap D\left(P_{nk}, \frac{R}{n+1}\right)^\circ$ there is a z_{nk} such that

$$\inf_k |z_{nk}| > \sup_l |z_{n-1,l}|, \quad n = 2, 3, \ldots.$$

It follows that F is the set of limit points of the set

$$L \overset{\text{def}}{=} \{z_{nk}, \ n \in \mathbb{N}, \ 1 \le k \le m_n\},$$

i.e., $F = L'$. Via a Weierstraß infinite product representation there can be defined a function f holomorphic in $\mathbb{C} \setminus F$ and such that L is the set of zeros of f.

In i–ii above, f is not identically zero and f is representable by a power series $\sum_{n=0}^\infty a_n z^n$ valid in $D(0, R)$. In ii, iii the circle of convergence for that power series is $D(0, R)^\circ$. For iii, owing to the Identity Theorem for holomorphic functions, $S_R(f) = F$.

Example 2.6.2.2. The series $\sum_{n=0}^\infty \frac{1}{n^2} z^{n!}$ represents a function f in $H(D(0, 1)^\circ)$ and for which $\mathbb{T}$ is the natural boundary. Nevertheless the series converges uniformly in the closed disc $D(0, 1)$.

Example 2.6.2.3. If $0 < a < 1$ the series $\sum_{n=0}^\infty a^n z^{n^2}$ represents a function f in $H(D(0, 1)^\circ)$. The Hadamard gap theorem implies that $\mathbb{T}$ is the natural boundary for f. The series converges uniformly in the closed disc $D(0, 1)$. Furthermore:

i.

$$\left\{\phi_n : \theta \mapsto f\left(\frac{n-1}{n} e^{i\theta}\right)\right\}_{n \in \mathbb{N}}$$

is a sequence of infinitely differentiable functions on $[0, 2\pi]$;

ii. for $k \in \mathbb{N}$, $\{\phi_n^{(k)}\}_{n \in \mathbb{N}}$ converges uniformly on $[0, 2\pi]$.

An application of THEOREM **2.1.2.1. 53** to the sequence $\{\phi_n\}_{n \in \mathbb{N}}$ shows that $h(\theta) \overset{\text{def}}{=} f(e^{i\theta})$ is an infinitely differentiable function of θ. Yet $e^{i\theta}$ is, for each real θ, a singular point of f.

Exercise 2.6.2.2. Show that if, for k in $\mathbb{N}$,

$$\limsup_{n \to \infty} |n(n-1)\cdots(n-k+1)a_n|^{\frac{1}{n}} = 1$$

then $\sum_{n=0}^{\infty} a_{n^2} z^{n^2}$ represents a function f in $H\left(D(0,1)^{\circ}\right)$, $S_1(f) = \mathbb{T}$, $h(\theta) \overset{\text{def}}{=} f\left(e^{i\theta}\right)$ exists for all θ in $\mathbb{R}$ and is infinitely differentiable.

2.6.3. Square roots

If Ω is a region, then Ω is *simply connected* iff any one of the following obtains [**Rud**]:

   i. the region Ω is *conformally equivalent* to $D(0,1)^{\circ}$;

  ii. for every $f \in H(\Omega)$, if $f \neq 0$ in Ω then there is in $H(\Omega)$ a function h such that $f = e^h$ (h may be regarded as "$\ln f$");

 iii. for every $f \in H(\Omega)$, if $f \neq 0$ in Ω then there is in $H(\Omega)$ a function g is such that $f = g^2$ (g may be regarded as "$\sqrt{f}$").

(Note the elementary implication: $ii \Rightarrow iii$ since $e^{\frac{h}{2}}$ serves for g.)

Example 2.6.3.1. If $\Omega \overset{\text{def}}{=} D(0,1)^{\circ}$ then $f : z \mapsto z^2$ is holomorphic in Ω. Although $f(0) = 0$ yet $g : z \mapsto z$ is holomorphic in Ω and $f = g^2$. Correspondingly, although $\Omega \setminus \{0\}$ is *not* simply connected, nevertheless $g \in H\left(\Omega \setminus \{0\}\right)$ and $f = g^2$.

2.6.4. Uniform approximation

The *Weierstraß approximation theorem* is valid in the set of $\mathbb{R}$-valued continuous functions defined on a fixed compact interval or on a compact subset of $\mathbb{R}^n$. Indeed the *Stone-Weierstraß theorem* is valid in the set $C(X, \mathbb{R})$ of continuous $\mathbb{R}$-valued functions defined on a compact Hausdorff space X. The situation is quite different for $C(X, \mathbb{C})$, i.e., when $\mathbb{R}$ is replaced by $\mathbb{C}$.

Example 2.6.4.1. If $r > 0$ the set

$$\mathcal{P} \overset{\text{def}}{=} \left\{ z \mapsto \sum_{k=0}^{n} a_k z^k \ : \ a_k, \ z \in \mathbb{C}, \ n \in \mathbb{N} \right\}$$

is *not* dense (with respect to the $\| \ \|_{\infty}$-induced topology of uniform convergence) in $C\left(D(0,1), \mathbb{C}\right)$. Otherwise the special function $f : z \mapsto \bar{z}$ would be the uniform limit of a sequence of polynomials in $\mathcal{P}$. Since f is not differentiable it is not *holomorphic* in $D(0,1)^{\circ}$ and so f cannot be the uniform limit of a sequence of polynomials, since every polynomial is *entire* and the uniform limit of a sequence of holomorphic functions is holomorphic.

Exercise 2.6.4.1. Show that if $f \in H\left(D(0,1)^{\circ}\right) \cap C\left(D(0,1), \mathbb{C}\right)$ then there is a sequence $\{p_n\}_{n \in \mathbb{N}}$ of polynomials such that $p_n \overset{u}{\to} f$ on $D(0,1)$.

[*Hint:* Use Fejér's theorem and the *maximum modulus theorem*.]

2.6.5. Rouché's theorem

The statement of *Rouché's theorem* is an instance in which the replacement of the symbol $<$ by the symbol $\leq$ changes a valid theorem into one that is, in the vein of Landau humor, completely invalid.

Example 2.6.5.1. The functions $f : \mathbb{C} \ni z \mapsto z^2$ and $g : \mathbb{C} \ni z \mapsto 1$ are such that $|g(z)| \leq |f(z)|$ and $|f(z)| \leq |g(z)|$ on $C \stackrel{\text{def}}{=} \{ z : |z| = 1 \}$. Yet, Z_{h,γ^*} denoting the number of zeros of the function h inside the *rectifiable Jordan contour* γ^*,

$$2 = Z_{f,C} \neq Z_{f+g,C} = 0$$

although

$$0 = Z_{g,C} = Z_{g+f,C} = 0.$$

[**Remark 2.6.5.1:**   One proof of Rouché's theorem uses the $\mathbb{Z}$-valued integral

$$\frac{1}{2\pi i} \int_\gamma \frac{f'(z) + tg'(z)}{f(z) + tg(z)} \, dz,$$

which, if the strict inequality $|g(z)| < |f(z)|$ obtains on γ^*, exists and is a continuous, hence constant, function of t on $[0,1]$. If the (Rouché) condition $|g(z)| < |f(z)|$ on γ^* is replaced by $|g(z)| \leq |f(z)|$ the integral above might fail to exist when $t = 1$.]

2.6.6. Bieberbach's conjecture

Experimentation and some theoretical calculations led Bieberbach in 1916 to conjecture the next result about *univalent* (injective) holomorphic functions [**Bi**].

THEOREM (BIEBERBACH). IF f IS HOLOMORPHIC AND UNIVALENT (INJECTIVE) IN $D(0,1)^\circ$ AND IF, FOR $z \in D(0,1)^\circ$,

$$f(z) \stackrel{\text{def}}{=} \sum_{n=1}^{\infty} a_n z^n$$

THEN FOR ALL n IN $\mathbb{N}$,

$$|a_n| \leq n|a_1|.$$

The record of progress, before the decisive result of de Branges in 1985, in the proof of the Bieberbach conjecture is in the following list, where "19xy, Name(s), $n = k$" signifies that the result was confirmed in 19xy by Name(s) for the case in which $n = k$:

1916, L. Bieberbach, $n = 2$
1923, K. Löwner, $n = 3$;
1955, P. R. Garabedian and M. Schiffer, $n = 4$;
1968, R. N. Pederson and, independently, M. Ozawa, $n = 6$;
1972, R. N. Pederson and M. Schiffer, $n = 5$.

De Branges showed the truth of a stronger result, the Milin conjecture described below, that implies the validity of the *Bieberbach conjecture*. In [Br] the proof of the Bieberbach conjecture itself is given and references to proofs of the stronger results are provided. The THEOREM is sharp since if $\beta \in \mathbb{R}$ and f is given by

$$z \mapsto f(z) \overset{\text{def}}{=} \frac{z}{(1 + e^{i\beta}z)^2} \overset{\text{def}}{=} \sum_{n=1}^{\infty} a_n z^n$$

then f is holomorphic and univalent in $D(0,1)^\circ$ and for all n in $\mathbb{N}$, $|a_n| = n$.

The validity of Bieberbach's conjecture is implied by the validity of the *Robertson conjecture* [Rob] put forth in 1936.

THEOREM (ROBERTSON). IF f IS HOLOMORPHIC AND UNIVALENT IN $D(0,1)^\circ$ AND

$$f(z) = \sum_{n=1}^{\infty} b_n z^{2n-1}, \ |z| < 1$$

THEN

$$\sum_{k=1}^{n} |b_k|^2 \le n|b_1|^2.$$

In turn, the validity of Robertson's conjecture is implied by the validity of the *Milin conjecture* [Mi] announced in 1971.

THEOREM (MILIN). IF f IS HOLOMORPHIC AND UNIVALENT IN $D(0,1)^\circ$ THERE IS A POWER SERIES

$$\sum_{n=1}^{\infty} \gamma_n z^n$$

CONVERGENT IN $D(0,1)^\circ$ AND SUCH THAT

$$f(z) = zf'(0) \exp\left(\sum_{n=1}^{\infty} \gamma_n z^n\right).$$

FURTHERMORE

$$\sum_{n=1}^{r}(r + 1 - n)n|\gamma_n| \le \sum_{n=1}^{r}(r + 1 - n)\frac{1}{n}.$$

On the other hand, if the hypothesis of univalency is dropped, the conclusion in the Bieberbach conjecture cannot be drawn.

Example 2.6.6.1. If f is $z \mapsto z + 3z^2$ then f is holomorphic but not univalent in $D(0,1)^\circ$ and $|a_2| = 3 > 1 = |a_1|$.

3. Geometry/Topology

3.1. Euclidean Geometry

3.1.1. Axioms of Euclidean geometry

Hilbert [**Hi2**] reformulated Euclid's axioms for plane (and solid) geometry. Not unexpectedly, Hilbert's contribution was decisive in the subsequent study of Euclidean geometry both in the schools and in research. His axioms are grouped as follows.

   i. axioms relating points, lines, and planes, e.g., two points determine exactly one line, two lines determine at most one point, there exist three noncollinear points, there exist four noncoplanar points, etc.;
   ii. axioms about *order* or "betweenness" of points on a line;
   iii. axioms about *congruent* ("≅"):
      a. line segments;
      b. angles;
      c. triangles ($\triangle ABC \cong \triangle A'B'C'$ if $AB \cong A'B'$, $AC \cong A'C'$, and $\angle BAC \cong \angle B'A'C'$, the "SAS" criterion);
   iv. the axiom about parallel lines: if L is a line and if P is a point not on L then, in the plane determined by L and P, there is precisely one line L' through P and not meeting L (Euclid's "fifth postulate").
   v. the axiom of continuity and completeness (versions of the Archimedean ordering and completeness of $\mathbb{R}$).

Among the topics of research interest are those dealing with *logical independence* and *logical consistency* of axioms and theorems. Hilbert treated these problems with great thoroughness. The interested reader is urged to consult [**Hi2**] for all the details.

Even before Hilbert's work, many questions about the axioms of geometry, in particular the parallel axiom, were resolved by Riemann's example of spherical geometry.

Example 3.1.1.1. Let S_1 be the surface of the unit ball in $\mathbb{R}^3$:

$$S_1 \overset{\text{def}}{=} \left\{ (x, y, z) \; : \; x^2 + y^2 + z^2 = 1 \right\}.$$

If "line" is taken to mean "great circle" then most of the axioms of plane Euclidean geometry are *not* satisfied and, e.g., if L and L' are two distinct lines then they must meet (twice!): there are no parallel lines.

On the other hand, Lobachevski offered a model in which all axioms of *plane* geometry save the parallel axiom are satisfied but in which for a line L and point P not on L more than one line passes through P and does not meet L. In **Example 3.1.1.2** there is a description of Poincarè's alternative model with similar properties.

Example 3.1.1.2. Let Π be the interior of the unit disc in $\mathbb{R}^2$:

$$\Pi \overset{\text{def}}{=} \left\{ (x, y) \; : \; x^2 + y^2 < 1 \right\} \; (= D(0, 1)^\circ).$$

In Π let a "line" be either a diameter of Π or the intersection of Π and a circle orthogonal to the circumference of Π. Then it is possible to define the terms of Hilbert's system so that his axioms in *i, ii, iii, v* are satisfied. However if a "line" L is not a diameter of Π then through the center **O** of Π there are infinitely many diameters, i.e., "lines", not meeting L.

A more subtle question arose in the study of *Desargue's theorem* illustrated in **Figure 3.1.1.1** and stated next.

THEOREM 3.1.1.1. (DESARGUE) WHEN CORRESPONDING SIDES OF TWO TRIANGLES IN A PLANE ARE PARALLEL, THE LINES JOINING CORRESPONDING VERTICES ARE PARALLEL OR HAVE A POINT IN COMMON (ARE "COAXIAL") [**Hi2**].

Despite the fact that Desargue's theorem is about triangles in a plane and refers not at all to congruence, many proofs of it depend on constructions involving the use of points outside the plane of the triangles in question and other proofs depend on the "SAS" criterion for the congruence of triangles. Moulton [**Mou**] showed that the proof cannot be given unless resort is made either to the axiom asserting the existence of four points that are not coplanar, i.e., to the use of solid geometry, or to the congruence axiom for triangles.

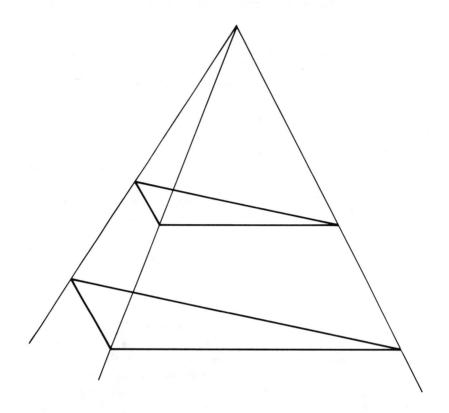

Figure 3.1.1.1. Desargue's theorem.

Example 3.1.1.3. As in **Figure 3.1.1.2** below, in $\mathbb{R}^2$ let "line" mean
any of the following:

 i. a horizontal line;
 ii. a vertical line;
 iii. a line with negative slope;
 iv. the union of the sides L and U of an angle having its vertex on the
    horizontal axis, L lying in the lower half-plane, U lying in the upper
    half-plane, the slopes of L and U positive, and

$$\frac{\text{slope of } L}{\text{slope of } U} = 2.$$

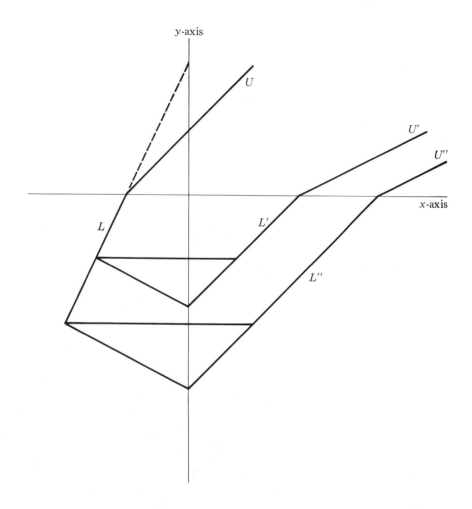

Figure 3.1.1.2. Moulton's plane.

In the resulting model of the "plane" all the axioms save the congruence axiom for triangles are satisfied. Nevertheless the two "Desarguesian" triangles in **Figure 3.1.1.2** are such that the "lines" joining corresponding vertices are neither parallel nor coaxial.

3.1.2. Topology of the Euclidean plane

Example 3.1.2.1. In the square having vertices at $(\pm 1, \pm 1)$ in the plane let C_1 and C_2 be defined as follows:

$$C_1 \stackrel{\text{def}}{=} \left\{ (-1 + t, -1 + \frac{7}{8}t) \; : \; t \in [0,1] \right\}$$

$$\cup \left\{ (t, \frac{1}{2}\sin\left(\frac{\pi}{2t}\right) + \frac{1}{4}) \; : \; t \in (0,1) \right\}$$

$$\cup \left\{ (1, \frac{3}{4} + \frac{1}{4}t) \; : \; t \in [0,1] \right\}$$

$$C_2 \stackrel{\text{def}}{=} \left\{ (-1 + t, 1 - \frac{7}{8}t) \; : \; t \in [0,1] \right\}$$

$$\cup \left\{ (t, \frac{1}{2}\sin\left(\frac{\pi}{2t}\right) - \frac{1}{4}) \; : \; t \in (0,1) \right\}$$

$$\cup \left\{ (1, -1 + \frac{5}{4}t) \; : \; t \in [0,1] \right\}.$$

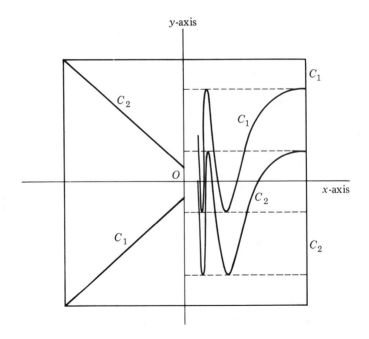

Figure 3.1.2.1.

Then C_1 and C_2 are disjoint connected sets, each of which is the union of two closed arc-images and one open arc-image. Furthermore

$$\{(-1,-1)\} \cup \{(1,1)\} \subset C_1 \text{ and } \{(-1,1)\} \cup \{(1,-1)\} \subset C_2,$$

i.e., C_1 and C_2 are two disjoint connected sets contained in a square and connecting diagonally opposed vertices, cf. **Figure 3.1.2.1.**

Exercise 3.1.2.1. Show that a simple arc-image or a simple open arc-image is nowhere dense in the plane.

  [*Hint:* The removal of a single point from a connected open subset of the plane does not disconnect the set.]

Since an arc-image, which is a compact connected set, can be a square it is nevertheless true that there are compact connected sets that are not arc-images.

Example 3.1.2.2. Let S be the union of the graph of $y = \sin(\frac{1}{x})$, $0 < x \le 1$ and the interval $\{0\} \times [-1, 1]$. Then S is compact and connected. On the other hand, regarded as a space topologized by heredity from $\mathbb{R}^2$, S is not *locally connected*, e.g., the neighborhood N that is the intersection of S and the open disc centered at the origin and of radius $\frac{1}{2}$ contains no connected neighborhood. Since every arc-image is locally connected **[Ne]**, S is a compact connected set that is not an arc-image.

Exercise 3.1.2.2. Show that:

i. the simple arc S_1 defined by the parametric equations

$$x = t$$
$$y = \begin{cases} t\sin\frac{1}{t} & \text{if } t \ne 0 \\ 0 & \text{if } t = 0 \end{cases}, \ 0 \le t \le 1$$

  is nonrectifiable;

ii. by contrast, the simple arc S_2 defined by the parametric equations

$$x = t$$
$$y = \begin{cases} t^2\sin\frac{1}{t} & \text{if } t \ne 0 \\ 0 & \text{if } t = 0 \end{cases}, \ 0 \le t \le 1$$

  is rectifiable;

iii. the simple arc S_3 defined by the parametric equations

$$x = t$$
$$y = \begin{cases} t^2\sin\frac{1}{t^2} & \text{if } t \ne 0 \\ 0 & \text{if } t = 0 \end{cases}, \ 0 \le t \le 1$$

  is nonrectifiable but that y' exists on $[0, 1]$.

  [*Hint:* For S_1 and S_3 inscribe a polygon with vertices at the origin, at a finite number of maxima, and at a finite number of minima.

For S_2 show that if $0 < \epsilon \leq 1$ then $\int_\epsilon^1 \sqrt{1 + y'(x)^2}\, dx$ is bounded by a number independent of ϵ. For S_3 the corresponding integral is not bounded.]

Example 3.1.2.3. Let g be the nowhere differentiable function of **Exercise 2.1.1.15. 50**. Let C be the simple arc defined by the parametric equations $x = t$, $y = g(t)$, $t \in [0, 1]$. Since g is nowhere differentiable, g is not of *bounded variation* on any interval $[a, b]$. Hence if $0 \leq a < b \leq 1$ then the simple arc C_{ab} corresponding to the parameter domain $[a, b]$ is nonrectifiable.

Example 3.1.2.4. Without recourse to LEMMA **2.2.1.1. 113** one can construct "by hand" an arc-image that fills the square $S \overset{\text{def}}{=} [0, 1] \times [0, 1]$.

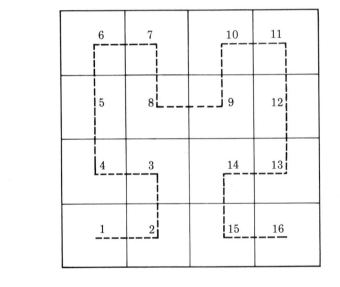

Figure 3.1.2.2.

The construction parallels the procedure in **Example 2.2.1.8. 115**. For each partition of the interval $I \overset{\text{def}}{=} [0, 1]$ into 4^n subintervals of equal length, partition S into 4^n pairwise congruent squares. Order the squares linearly so that they correspond to the natural linear ordering of the partition intervals in the manner indicated in **Figure 3.1.2.2**. Each point t of I is the intersection of a sequence of nested partition intervals. The intersection of the sequence of corresponding squares is a point $P(t)$ of S. The map $\gamma : [0, 1] \ni t \mapsto P(t) \stackrel{\text{def}}{=} (x(t), y(t)) \in S$ is continuous because as nested partition intervals descend on a t in $[0, 1]$ their corresponding squares descend on $P(t)$ in S. The image γ^* of the arc $\gamma : t \mapsto P(t)$ is "space-filling." The idea is due to Hilbert. The first construction of this kind is due to Peano, after whom such arc-images are named.

Exercise 3.1.2.3. Show that for the arc γ above, if $0 \leq a < b \leq 1$ the corresponding arc-image γ^*_{ab} is nonrectifiable.

[*Hint:* The arc-image in $\mathbb{R}^2$ of a rectifiable arc is a two-dimensional null set (λ_2).]

Example 3.1.2.5. The following construction produces in $\mathbb{R}^{\mathbb{R}}$ a dense set A that meets every vertical and every horizontal line of $\mathbb{R}^2$ in exactly one point, i.e., A is the dense graph of a bijection $f : \mathbb{R} \mapsto \mathbb{R}$.

The function f is defined on a countable dense set D by induction and then on $\mathbb{R}$ according to the formula

$$f(x) = \begin{cases} f(x) & \text{if } x \in D \\ x & \text{otherwise.} \end{cases}$$

Let $S \stackrel{\text{def}}{=} \{(x_n, y_n)\}_{n \in \mathbb{N}}$ be an enumeration of $\mathbb{Q}^2$, let B_1 be the square $[-1, 1)^2$ in $\mathbb{R}^2$, and let (x_{11}, y_{11}) be the first term in $S \cap B_1$. Then $f(x_{11}) \stackrel{\text{def}}{=} y_{11}$. Divide $B_2 \stackrel{\text{def}}{=} [-2, 2)^2$ into 2^2 congruent subsquares $B_{21}, \ldots, B_{22^2}$. Let (x_{21}, y_{21}) be the first term in $S \cap B_{21}$ and such that $x_{21} \neq x_{11}, y_{21} \neq y_{11}$. Then $f(x_{21}) \stackrel{\text{def}}{=} y_{21}$. $\ldots$

Let (x_{22^2}, y_{22^2}) be the first term in $S \cap B_{22^2}$ and such that x_{22^2} is none of the x_{ij} and y_{22^2} is none of the y_{ij} previously chosen for defining values of f. Then $f(x_{22^2}) \stackrel{\text{def}}{=} y_{22^2}$.

Divide $B_n \stackrel{\text{def}}{=} [-n, n)^2$ into 2^n congruent subsquares $B_{n1}, \ldots, B_{n2^n}$. Choose (x_{n1}, y_{n1}) as the first term of $S \cap B_{n1}$ and such that x_{n1} is none of the x_{ij} and y_{n1} is none of the y_{ij} previously chosen for defining values of f. Then $f(x_{n1}) \stackrel{\text{def}}{=} y_{n1}$. $\ldots$ Choose (x_{n2^n}, y_{n2^n}) as the first term of $S \cap B_{n2^n}$ and such that x_{n2^n} is none of the x_{ij} and y_{n2^n} is none of the y_{ij} previously chosen for defining values of f. Then $f(x_{n2^n}) \stackrel{\text{def}}{=} y_{n2^n}$. $\ldots$

Exercise 3.1.2.4. Show that the procedure described above yields: a) a set D dense in $\mathbb{R}$; b) a set $\{(x, f(x)) : x \in D\}$ dense in $\mathbb{R}^2$; c) a bijective function f for which the graph $\{(x, f(x)) : x \in \mathbb{R}\}$ is dense in $\mathbb{R}^2$.

Example 3.1.2.6. Knaster and Kuratowski constructed in $\mathbb{R}^2$ a connected set S containing a point P such that $S \setminus \{P\}$ is, like the Cantor set C_0, *totally disconnected* **[KnKu]**.

In C_0 there is the set X of endpoints of intervals deleted in the construction of C_0 and there is the complementary set $Y \overset{\text{def}}{=} C_0 \setminus X$. Let P be the point $(0.5, 0.5)$ in $\mathbb{R}^2$ and for each point c in C_0 let $L(c)$ be the line segment joining P to c. For each x in X let R_x in $L(x)$ be the set of points having rational ordinates. For each y in Y let R_y in $L(y)$ be the set of points having irrational ordinates. Then, as the following argument reveals, the properties described in the preceding paragraph obtain for $S \overset{\text{def}}{=} \bigcup_{x \in X} R_x \cup \bigcup_{y \in Y} R_y$. (Note that $Y \cap S = \emptyset$.)

To show that $S \setminus \{P\}$ is totally disconnected note that for every c in C_0, $S \cap L(c)$ is (totally) disconnected. Hence if Z is a nontrivial connected subset of $S \setminus \{P\}$ then Z meets two different line segments $L(c_1)$ and $L(c_2)$. Thus there is through P a line L that meets $[0,1] \setminus C_0$ between c_1 and c_2. If Z is connected then Z meets L, hence only at P, which is impossible since $P \notin Z$. Thus $S \setminus \{P\}$ is totally disconnected.

The proof, due to Knaster and Kuratowski, that S is connected is offered in the format of a sequence of pairs **Statement**(*Reason*).

Assume S is the union of two sets A and B such that $P \in A$ and $(\overline{A} \cap B) \cup (A \cap \overline{B}) = \emptyset$. It is shown below that $B = \emptyset$, which proves that S is connected. For each c in C_0 let $\ell(c)$ be the lower endpoint of the longest line segment Λ_c contained in $L(c)$, containing P, and not meeting B.

1. For all c in C_0, $\ell(c) \in \overline{A}$ and $\ell(c) \notin B$. (*All the points on $\Lambda_c \cap R_c$ are in A and $\overline{A} \cap B = \emptyset$.*)

2. If $\ell(c) \notin C_0$ then $\ell(c) \notin S$. (*If $\ell(c) \neq c$ then $\ell(c) \in \overline{B}$ whence $\ell(c) \notin A$. See **1**.*)

3. The ordinate of each $\ell(y)$ is rational. (*See **2**. and the definition of $L(y)$.*)

Let E be the set of all $\ell(y)$, $y \in Y$ and let $\{r_n\}_{n \in \mathbb{N}}$ be an enumeration of $(0, 0.5] \cap \mathbb{Q}$. Then, E_0 denoting $E \cap (x\text{-axis})$ and E_n denoting the intersection of E with the horizontal line through $(0, r_n)$, $E = E_0 \cup \bigcup_{n=0}^{\infty} E_n$.

4. Each of the following relations is valid.

$$E_0 \subset Y$$
$$E_n \subset \overline{A} \cap \overline{B}, \ n \in \mathbb{N}$$
$$\overline{E_n} \subset \overline{A} \cap \overline{B}, \ n \in \mathbb{N}$$
$$\overline{E_n} \cap S = \emptyset, \ n \in \mathbb{N}.$$

(*See the definitions and **1**.*)

5. If $n \in \mathbb{N}$ and $x \in X$ then $\overline{E_n} \cap L(x) = \emptyset$. (*If $Z \in \overline{E_n} \cap L(x)$ then the ordinate of Z is rational whence $Z \in S$. Since the ordinate of Z is not zero, $Z \notin S$. See **2**.*)

Let Y_n be $\{c : c \in C_0, \overline{E_n} \cap L(c) \neq \emptyset\}$.

6. For n in $\mathbb{N}$, $Y_n \subset Y \subset E_0 \cup \bigcup_{n \in \mathbb{N}} Y_n$. (*See **5**.*)

7. $C_0 = X \cup E_0 \cup \bigcup_{n \in \mathbb{N}} Y_n$. (*See **4**.*)

8. Each Y_n is compact. (*Each Y_n is the central projection from* $\left(\frac{1}{2}, \frac{1}{2}\right)$ *of the compact set* $\overline{E_n}$ *onto the x-axis.*)

9. Each set Y_n is nowhere dense in C_0 (*The interior (relative to C_0) of Y_n is empty since every point of Y is a limit point of X and $Y_n \subset Y$, see* **6**. *Furthermore Y_n is closed (**8**) and so $Y_n^\circ = (\overline{Y_n})^\circ$.*)

10. The set E_0 is dense in C_0. (*The set $Z \overset{\text{def}}{=} X \cup \bigcup_{n \in \mathbb{N}} Y_n$ is of the first category in C_0. The set C_0 is a complete metric space, and $E_0 = C_0 \setminus Z$* [Kur].)

11. If $e \in E_0$ and $z \in S \cap L(e)$ then $z \in A$. (*There is in Y a y such that $e = \ell(y)\ (= \ell(e))$, see* **1**.)

12. The closure of A contains S: $\overline{A} \supset S$, i.e., $\overline{A} \supset A \cup B$, whence $\overline{A} \cap B\ (= \emptyset) = B$. (*Because E_0 is dense in C_0 it follows that $S \cap \bigcup_{e \in E_0} L(e)$ is dense in S.*)

$\square$

The work of Brouwer led to a method for defining in $\mathbb{R}^2$ three regions R_1, R_2, R_3 having a common *boundary*: $\partial R_i = \partial R_j$, $1 \le i, j \le 3$, [Bro]. An appealing description of his construction can be given in the form of a algorithm for revising the topography of a rectangular and sea-bound island in which there are two square lakes, one consisting of warm fresh water and the other consisting of cold fresh water. For each n in $\mathbb{N}$ three canals are dug successively so that:

i. first, a canal brings water from the sea within a distance of $\frac{1}{n}$ from every point of dry land on the island;

ii. second, a canal brings water from the lake containing warm fresh water within a distance of $\frac{1}{n}$ from every point of the dry land remaining on the island after the construction of the canal for sea water;

iii. third, a canal brings water from the lake containing cold fresh water within a distance of $\frac{1}{n}$ from every point of dry land remaining on the island after the construction of canal for sea water and the canal for warm fresh water.

What is left of the dry land on the island when the cycle *i–iii* has been carried out for all n in $\mathbb{N}$, e.g., in a "one-year plan" in which cycle 1 is completed in half a year, cycle 2 in a quarter of a year, ..., cycle n in the 2^{-n}th part of a year, is the common boundary of the newly created regions R_1 resp. R_2 resp. R_3 consisting of of sea water resp. warm fresh water resp. cold fresh water.

The description above is vivid but does not deal with, e.g., the problem of building the canal for the sea water so that the lakes are thereby not separated from each other. In what follows, a systematic formulation of a more general construction is offered.

In **Figure 3.1.2.3** there are square regions $R_1, \ldots, R_n$ inside a fixed closed rectangular region K.

$$K$$

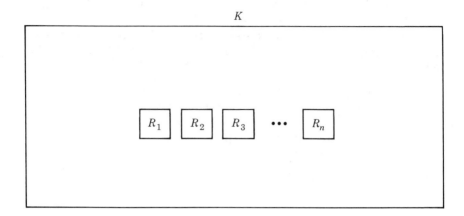

Figure 3.1.2.3.

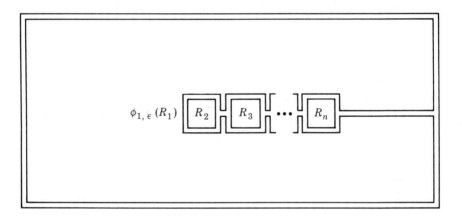

Figure 3.1.2.4.

In **Figure 3.1.2.4** the region R_1 is "exploded" by an *auteomorphism* $\phi_{1,\epsilon} : K \mapsto K$ so that the image $\phi_{1,\epsilon}(R_1)$ of R_1 embraces, in the manner indicated, R_2, R_3, ..., R_n.

Similarly R_2 is exploded by an auteomorphism $\phi_{2,\epsilon}$ so that the image $\phi_{2,\epsilon}(R_2)$ of R_2 embraces R_1, R_3, ...; then R_3 is exploded to embrace R_1, R_2, R_4, ...; ... .

As a result,

i. $\phi_{k,\epsilon}(K) = K$ (by definition of *auteomorphism*);

ii. for $1 \leq i \leq n$, $i \neq k$, and a positive ϵ, $R_{i,\epsilon}$ denoting the compact set that is the union of R_i together with a compact ϵ-*pad* covering ∂R_i, $i \neq k$,

$$\phi_{k,\epsilon}(R_k) \subset K \setminus \bigcup_{i \neq k} R_{i,\epsilon}$$

$$\phi_{k,\epsilon}(R_{i,\epsilon}) = R_{i,\epsilon}, \ i \neq k,$$

whence every point in the compact set $K_{k,\epsilon} \overset{\text{def}}{=} K \setminus \bigcup_{i \neq k} R_i$ is not farther then ϵ from $\phi_{k,\epsilon}(R_k)$;

iii. $K \setminus \phi_{k,\epsilon}(R_k)$ is connected.

[**Remark 3.1.2.1:** The image of R_k under the map $\phi_{k,\epsilon}$ occupies most of the complement in K of the union of other sets $R_{i,\epsilon}$, i.e., for i different from k:

$$\phi_{k,\epsilon}(R_k) \subset K \setminus \left(\bigcup_{i \neq k} R_i \right).]$$

Let Φ_ϵ be the homeomorphism that is the composition of the maps $\phi_{i,\epsilon}$, $1 \leq i \leq n$: $\Phi_\epsilon \overset{\text{def}}{=} \phi_{n,\epsilon} \circ \cdots \circ \phi_{1,\epsilon}$. Then, according to Brouwer's *invariance of domain* theorem, each $\Phi_\epsilon(R_i)$ is a region and thus every point of the *compact* set

$$K_\epsilon \overset{\text{def}}{=} K \setminus \Phi_\epsilon \left[\bigcup_{i=1}^{n} R_i \right]$$

is not farther than ϵ from *each* of the regions $\Phi_\epsilon(R_i)$, $1 \leq i \leq n$. Furthermore, if $\epsilon' < \epsilon$ then

$$K_{\epsilon'} \subset K_\epsilon$$

$$\Phi_{\epsilon'} \left[\bigcup_{i=1}^{n} R_i \right] \supset \Phi_\epsilon \left[\bigcup_{i=1}^{n} R_i \right].$$

Hence, if

$$F \overset{\text{def}}{=} \bigcap_{k \in \mathbb{N}} K_{\frac{1}{k}}$$

$$\mathcal{R}_i \overset{\text{def}}{=} \bigcup_{k \in \mathbb{N}} \Phi_{\frac{1}{k}}(R_i), \ 1 \leq i \leq n,$$

then each point of the nonempty compact set F is at distance zero from each region $\mathcal{R}_i$, i.e., F is the common boundary of the n disjoint and intertwined regions $\mathcal{R}_i$, $1 \leq i \leq n$.

For n in $\mathbb{N}$, there are in $\mathbb{R}^2$ n pairwise disjoint regions

$$\mathcal{R}_1, \ldots, \mathcal{R}_n$$

having a compact set F as their common boundary, i.e., if $P \in F$ then every neighborhood of P meets each region and its complement.

[**Note 3.1.2.1:**   Appel, Haken, and Koch [**ApH1–ApH4, ApHK**] positively resolved the *four-color problem*. When $n > 4$ the construction above negatively resolves the four-color problem if it is stated loosely, viz.: Can any map in $\mathbb{R}^2$ be colored with four or fewer colors?]

Example 3.1.2.7. Each of the regions $\mathcal{R}_k$ is a *non-Jordan region* since the complement of the union of a Jordan region and its boundary is precisely one region. Nevertheless each $\mathcal{R}_k$ is the interior of the closure $\overline{\mathcal{R}_k}$ of $\mathcal{R}_k$: $\mathcal{R}_k = \left(\overline{\mathcal{R}_k}\right)^{\circ}$. Only the inclusion $\mathcal{R}_k \supset \left(\overline{\mathcal{R}_k}\right)^{\circ}$ needs proof. However, $\overline{\mathcal{R}_k} = \mathcal{R}_k \cup \partial\mathcal{R}_k$ and $K \setminus \left(\overline{\mathcal{R}_k}\right)$ is the union of the other $\mathcal{R}_i$. Hence, if $P \in \partial\mathcal{R}_k$ and P is in $\left(\overline{\mathcal{R}_k}\right)^{\circ}$ then some neighborhood of P fails to meet each of the other $\mathcal{R}_i$, in contradiction of the fact that $\partial\mathcal{R}_k$ is also $\partial\mathcal{R}_i$, i.e., $\mathcal{R}_k \supset \left(\overline{\mathcal{R}_k}\right)^{\circ}$.

Example 3.1.2.8. In contrast to the non-Jordan regions $\mathcal{R}_k$ above, the non-Jordan region

$$R \stackrel{\text{def}}{=} \left\{ (x,y) \ : \ x^2 + y^2 < 1 \right\} \setminus \left\{ (x,0) \ : \ 0 \le x < 1 \right\},$$

consisting of a circular region from which a slit is deleted, is not the interior of its closure:

$$\left(\overline{R}\right)^{\circ} = \left\{ (x,y) \ : \ x^2 + y^2 < 1 \right\} \ne R.$$

In [**KnKu**] there is a wealth of other results about pathologies of the Euclidean plane.

3.2. Topological Spaces

3.2.1. Metric spaces

Exercise 3.2.1.1. Show that if $\mathbb{R}$ is metrized according to the formula:

$$d(x,y) \stackrel{\text{def}}{=} \frac{|x - y|}{1 + |x - y|}$$

then for n in $\mathbb{N}$, each set $F_n \stackrel{\text{def}}{=} [n,\infty)$ is closed and bounded and yet $\bigcap_{n \in \mathbb{N}} F_n = \emptyset$.

[**Note 3.2.1.1:**   Although each set F_n is bounded and closed, it is *not* compact in the d-induced topology for $\mathbb{R}$.]

Exercise 3.2.1.2. Show that if $\mathbb{N}$ is metrized according to the formula

$$d(m,n) \stackrel{\text{def}}{=} \frac{|m-n|}{mn}$$

then $\{n\}_{n\in\mathbb{N}}$ is a nonconvergent Cauchy sequence.

[**Note 3.2.1.2:**   Metrized in the manner just described $\mathbb{N}$ is not Cauchy complete while in the metric inherited from $\mathbb{R}$, $\mathbb{N}$ *is* Cauchy complete. In each metric $\mathbb{N}$ is homeomorphic to itself in the other metric. Hence Cauchy completeness is *not* a topological invariant.]

In a complete metric space a decreasing sequence

$$B_1 \supset B_2 \supset \cdots$$

of balls with diameters tending to 0 has a nonempty intersection:

$$\bigcap_{n\in\mathbb{N}} B_n \neq \emptyset.$$

If the sequence of balls merely decreases, i.e., if $B_n \subset B_{n+1}$, $n \in \mathbb{N}$, but the diameters do not tend to 0, the intersection can be empty:

$$\bigcap_{n\in\mathbb{N}} B_n = \emptyset.$$

Exercise 3.2.1.3. Show that if $\mathbb{N}$ is metrized according to the formula

$$d(m,n) \stackrel{\text{def}}{=} \begin{cases} 1 + \frac{1}{m+n} & \text{if } m \neq n \\ 0 & \text{if } m = n \end{cases}$$

then:

   i. $\mathbb{N}$ is a complete metric space;
  ii. each set $B_n \stackrel{\text{def}}{=} \{ m \ : \ d(m,n) \leq 1 + \left(\frac{1}{2n}\right) \}$, $n \in \mathbb{N}$, is a *closed ball*;
 iii. $B_n \supset B_{n+1}$, $n \in \mathbb{N}$;
 iv. $\bigcap_{n\in\mathbb{N}} B_n = \emptyset$.

In a *normed vector space* the closure $\overline{U}$ of an *open ball*:

$$U \stackrel{\text{def}}{=} \{ x \ : \ \|x - a\| < r \}$$

is the corresponding closed ball:

$$B \stackrel{\text{def}}{=} \{ x \ : \ \|x - a\| \leq r \}.$$

Furthermore, if a vector space contains closed balls B_1 resp. B_2 of radii r_1 resp. r_2 and if $B_1 \subset B_2$ then $r_1 < r_2$. In a metric space where the metric is not derived from a norm such relations can be different.

Exercise 3.2.1.4. Assume X is a set and $\#(X) \geq 2$. Metrize X according to the formula

$$d(x,y) \overset{\text{def}}{=} \begin{cases} 1 & \text{if } x \neq y \\ 0 & \text{otherwise.} \end{cases}$$

For a fixed x in X let U resp. B be the open resp. closed ball centered at x and of radius 1. Then $\{x\} = U \subset B = X$ and $\overline{U} = U \neq B$.

Exercise 3.2.1.5. Let X be the closed disk $\{ (x,y) : x^2 + y^2 \leq 9 \}$ in the metric inherited from $\mathbb{R}^2$. Show that if $B_2 \overset{\text{def}}{=} X$ and

$$B_1 \overset{\text{def}}{=} B_2 \cap \{ (x,y) : (x-2)^2 + y^2 \leq 16 \}$$

then $B_1 \subset B_2$ while their respective radii are 4 and 3.

3.2.2. General topological spaces

In metric spaces, a) sequences have unique limits, b) *derived sets* are closed, c) the *Bolzano-Weierstraß theorem* obtains, etc. In the looser domain of *nonmetrizable* spaces there are correspondingly less intuitive phenomena. The next discussion provides only a very small sample of the richness of *topopathology*. In [SS] there is a far more extensive treatment of the subject.

In what follows, O denotes the set of open sets of the topology of a set X. For the *trivial topology* is $O = \{\emptyset, X\}$ and for the *discrete topology* D, O is $\mathcal{P}(X) \overset{\text{def}}{=} 2^X$, the *power set* consisting of all subsets of X. When Y is a subset of a topological space X, the derived set (the set of *limit points* of Y) is denoted Y'.

Let $\mathcal{R}^n$ denote the usual topology for $\mathbb{R}^n$: a countable base for $\mathcal{R}^2$ is the set of all open disks $\{ (x,y) : (x-a)^2 + (y-b)^2 < r^2, \ a,b,r \in \mathbb{Q} \}$.

Example 3.2.2.1. Let X be a space with the trivial topology.

i. If $\#(X) \geq 2$, $y \in X$, and $Y \overset{\text{def}}{=} \{y\}$ then $Y' = X \setminus \{y\}$ and hence Y is not closed.

ii. If $\#(X) \geq 2$ and $\mathcal{N} \overset{\text{def}}{=} \{x_\lambda\}_{\lambda \in \Lambda}$ is a net (in particular if $\mathcal{N}$ is a sequence) then every point in X is a limit of $\mathcal{N}$, even if the net is a constant, i.e., if there is in X a point y such that for all λ, $x_\lambda = y$.

Exercise 3.2.2.1. Let X be $\mathbb{R}$ in which, by abuse of language, $O = \mathcal{R}^1 \cap \mathbb{Q}$. Show:

    i. $\mathbb{Q}$ is a countable dense subset of X;

   ii. $X \setminus \mathbb{Q} \overset{\text{def}}{=} \mathbb{I}_{\mathbb{R}}$ (the set of irrational real numbers) inherits from O the discrete topology;

  iii. there is in $\mathbb{I}_{\mathbb{R}}$ no countable dense subset.

Exercise 3.2.2.2. Let X be the *closed upper half-plane*

$$\{\,(x,y) \ : \ x,y \in \mathbb{R}, \ y > 0\,\} \cup \{\,(x,y) \ : \ x \in \mathbb{R}, y = 0\,\} \overset{\text{def}}{=} A \cup B.$$

Let a set U be in the base for the topology O of X iff U is an open subset of A or U is of the form $\{\,(x,y) \ : \ (x-a)^2 + (y-b)^2 < b^2, \ b > 0\,\} \cup \{(a,0)\}$. Show that the (countable) set of all points with rational coordinates is dense in X but that $B \ (= \mathbb{R})$ inherits from X the discrete topology and thus contains no countable dense subset.

Exercise 3.2.2.3. Show that the spaces X in **Exercises 3.2.2.1, 3.2.2.2** are not *separable*, i.e., that neither contains a countable base for its topology.

    The topology of a space can be specified by the set of all convergent nets. On the other hand, the set of all convergent sequences can fail to determine the topology of a nonmetrizable space.

Exercise 3.2.2.4. Assume $\#(X) > \#(\mathbb{N})$. Let O consist of $\emptyset$ and the complements of all sets S such that $\#(S) \leq \#(\mathbb{N})$. Show that:

    i. the sequence $\{x_n\}_{n\in\mathbb{N}}$ converges iff x_n is *ultimately constant*, i.e., iff there is in X an x and there is in $\mathbb{N}$ an m such that $x_n = x$ if $n > m$;

   ii. O is strictly *weaker* than the discrete topology D and, in D, a net $\mathcal{N}$ converges iff it is ultimately constant;

  iii. if A is an uncountable proper subset of X and $y \in X \setminus A$ then y is a limit point of A and yet no subsequence of A converges to y;

  iv. if A is a proper subset of X and $y \in X \setminus A$ there is a net $\{a_\lambda\}_{\lambda\in\Lambda}$ contained in A and converging to y.

    [*Hint:* For *iv* let Λ be the set of all neighborhoods of y and partially order Λ by inclusion: $\lambda \succ \lambda'$ iff $\lambda \subset \lambda'$. For each λ in Λ let a_λ be a point in $A \cap \lambda$.]

    If $1 \leq p < \infty$, for l^p there are the norm-induced topology N derived from the metric $d(\mathbf{a},\mathbf{b}) \overset{\text{def}}{=} \|\mathbf{a} - \mathbf{b}\|_p$ and the weak or $\sigma\left(l^p, (l^p)^*\right)$ topology W for which a typical neighborhood of O is

$$U_{\mathbf{x}_1^*,\dots,\mathbf{x}_K^*;\epsilon} \overset{\text{def}}{=} \{\,\mathbf{x} \ : \ |\mathbf{x}_k^*(\mathbf{x})| < \epsilon, \ \mathbf{x}_k^* \in (l^p)^*, \ 1 \leq k \leq K\,\}.$$

Exercise 3.2.2.5. Show that in l^p every weak neighborhood of $\mathbf{O}$ contains a norm-induced neighborhood of $\mathbf{O}$, but that every weak neighborhood of $\mathbf{O}$ is norm-unbounded. (Hence N is strictly stronger than W and every N-convergent sequence is W-convergent.)

Exercise 3.2.2.6. Show that in l^1 every weakly convergent sequence is norm-convergent. (Hence, although W is strictly weaker than N the sets of convergent sequences for the two topologies are the same.)

[*Hint:* Assume that for some positive δ, all the terms of a sequence S converging weakly to $\mathbf{O}$ have norms not less than δ. Let $\mathbf{x}^{(n)} \stackrel{\text{def}}{=} \left(x_1^{(n)}, \ldots, x_m^{(n)}, \ldots \right)$ be the nth term of S. Then, $\mathbf{d}_m$ denoting the sequence $\{\delta_{mn}\}_{n\in\mathbb{N}}$ (an element of l^{1^*}), it follows that for m in $\mathbb{N}$, $\mathbf{d}_m\left(\mathbf{x}^{(n)}\right) = x_m^{(n)} \to 0$ as $n \to \infty$. Let n_1 be 1 and let m_1 be such that $\sum_{m=m_1+1}^{\infty} \left| x_m^{(n_1)} \right| < \frac{\delta}{2}$. There is an n_2 such that $\sum_{m=1}^{m_1} \left| x_m^{(n_2)} \right| < \frac{\delta}{2^2}$. Let m_2 be such that $\sum_{m=m_2+1}^{\infty} \left| x_m^{(n_2)} \right| < \frac{\delta}{2^2}$, whence $\sum_{m=m_1+1}^{m_2} \left| x_m^{(n_2)} \right| > \frac{\delta}{2}$. In this manner there are definable strictly increasing sequences $\{n_1, n_2, \ldots\}$ and $\{m_1, m_2, \ldots\}$ such that

$$\sum_{m=m_k+1}^{m_{k+1}} \left| x_m^{(n_{k+1})} \right| > \frac{\delta}{2}, \ k \in \mathbb{N}.$$

If

$$a_m \stackrel{\text{def}}{=} \begin{cases} 0 & \text{if } 1 \le m \le m_2 \\ \text{sgn}\left(x_m^{(n_{k+1})}\right) & \text{if } m_k + 1 \le m \le m_{k+1}, \ 2 \le k < \infty, \end{cases}$$

then $\mathbf{a} \stackrel{\text{def}}{=} \{a_m\}_{m\in\mathbb{N}}$ is in $l^\infty = l^{1^*}$. For n in the sequence $\{n_2, n_3, \ldots\}$, $\mathbf{a}(\mathbf{x}_n) \ge \frac{\delta}{2}$ and thus an infinite subsequence of S lies outside the weak neighborhood $U_{\mathbf{a}, \frac{\delta}{2}}$, a contradiction.]

[**Remark 3.2.2.1:** The construction above uses the "moving hump" technique, the quantities

$$\sum_{m=m_k+1}^{m_{k+1}} \left| x_m^{(n_{k+1})} \right| \ \left(> \frac{\delta}{2} \right), k \in \mathbb{N},$$

serving as "humps."]

Exercise 3.2.2.7. Show that if

$$\mathbf{x}^{(n)} \stackrel{\text{def}}{=} \{\delta_{nm}\}_{m\in\mathbb{N}}, \ \mathbf{y}^{(m,n)} \stackrel{\text{def}}{=} \mathbf{x}^{(m)} + m\mathbf{x}^{(n)} \in l^2, \ m, n \in \mathbb{N}$$

then $\mathbf{O}$ is a weak limit point of $S \stackrel{\text{def}}{=} \{\mathbf{y}^{(n,m)}\}_{m,n\in\mathbb{N}}$ and that no subsequence of S converges weakly to $\mathbf{O}$.

[**Remark 3.2.2.2:** Despite the topopathologies exhibited above, the following should be observed. If E is a Banach space then B_1^*, the unit ball of the dual space E^* is compact in the weak* or $\sigma(E^*, E)$ topology for E^*. (A typical weak* neighborhood of $\mathbf{O}$ in E^* is, for n in $\mathbb{N}$ and $\mathbf{x}_1, \ldots, \mathbf{x}_n$ in E, a set of the form

$$U_{\mathbf{x}_1,\ldots,\mathbf{x}_n;\epsilon} \stackrel{\text{def}}{=} \{\mathbf{x}^* : |\mathbf{x}^*(\mathbf{x}_i)| < \epsilon,\ 1 \le i \le n\}.)$$

If E is separable then B_1^* in the weak* topology satisfies the first axiom of countability.]

A map $f : X \mapsto Y$ between topological spaces is *continuous, open,* or *closed* iff, correspondingly, f^{-1}(open set) is open, f(open set) is open, or f(closed set) is closed.

Exercise 3.2.2.8. Show that $f : \mathbb{R} \ni x \mapsto e^x \cos x$ is continuous but neither open nor closed.

Exercise 3.2.2.9. Show that if $X \stackrel{\text{def}}{=} \{(\cos\theta, \sin\theta) : 0 \le \theta < 2\pi\}$ then:

i. $X \ni (\cos\theta, \sin\theta) \mapsto \theta \in [0, 2\pi)$ is both open and closed but is not continuous (at $(1, 0)$);

ii.

$$g : X \ni (\cos\theta, \sin\theta) \stackrel{\text{def}}{=} x \mapsto \begin{cases} 0 & \text{if } 0 \le \theta \le \pi \\ \theta - \pi & \text{if } \pi < \theta < 2\pi \end{cases}$$

is closed but neither continuous nor open.

[*Hint:* For *ii* note that $\text{Discont}(g) = \{(1,0)\} \stackrel{\text{def}}{=} \{P\}$. Reduce the problem to: $x_n \to P$ and $g(x_n) \not\to g(P)$ and a contradiction.]

Exercise 3.2.2.10. Show that

$$\mathbb{R}^2 \ni (x, y) \mapsto x \in \mathbb{R}$$

is continuous and open but not closed.

Exercise 3.2.2.11. Show that H in **Example 2.1.1.3. 51** is open but neither closed nor continuous.

[*Hint:* If $n < x_n < n + 1$ and $H(x_n) \in \left(\frac{1}{n+1}, \frac{1}{n}\right)$ for n in $\mathbb{N}$ then $X \stackrel{\text{def}}{=} \{x_n\}_{n\in\mathbb{N}}$ is closed and $H(X)$ is not.]

Exercise 3.2.2.12. Show that

$$[0, 2] \ni x \mapsto \begin{cases} 0 & \text{if } x \in [0, 1] \\ x - 1 & \text{if } x \in (1, 2] \end{cases}$$

is continuous and closed but not open.

Example 3.2.2.2. Assume

$$X \stackrel{\text{def}}{=} \bigcup_{n=0}^{\infty} ((3n, 3n+1) \cup \{3n+2\}), \ Y \stackrel{\text{def}}{=} (X \setminus \{2\}) \cup \{1\}$$

$$f : X \ni x \mapsto \begin{cases} x & \text{if } x \neq 2 \\ 1 & \text{if } x = 2 \end{cases} \in Y$$

$$g : Y \ni y \mapsto \begin{cases} \frac{y}{2} & \text{if } y \leq 1 \\ \frac{y}{2} - 1 & \text{if } 3 < y < 4 \in X \\ y - 3 & \text{if } y \geq 5. \end{cases}$$

Then both f and g are continuous bijections but neither is a homeomorphism since in any homeomorphism $h : Y \mapsto X$, $h((0,1))$ must be some interval $(3n, 3n+1)$ and then $h(1)$ cannot be defined.

In **Figure 3.2.2.1** there are depicted two subsets X and Y of $\mathbb{R}^2$. Each is a continuous bijective image of the other, although X and Y are not homeomorphic. Of course, although the maps S and T described below are continuous bijections, one from X onto Y and the other from Y onto X, they are *not* inverses of each other.

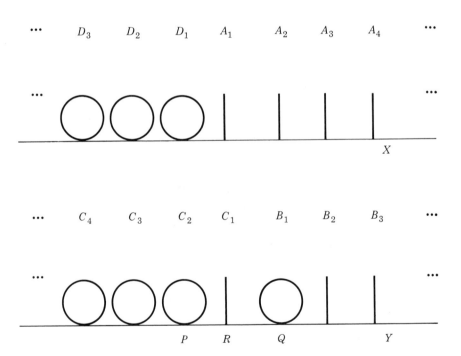

Figure 3.2.2.1.

The vertical segments are of length 2 and are shorn of their upper endpoints. The circles are of diameter 2.

The continuous bijection $S : X \mapsto Y$ is defined as follows:

$$S : \text{horizontal line} \mapsto \text{horizontal line}$$
$$S : A_1 \mapsto C_1$$
$$S : A_2 \mapsto B_1, \ (S : [0,2) \ni t \mapsto (\cos \pi t, \sin \pi t))$$
$$S : A_{n+2} \mapsto B_{n+1}, \ n \in \mathbb{N}$$
$$S : D_n \mapsto C_{n+1}, n \in \mathbb{N}.$$

The continuous bijection $T : Y \mapsto X$ is defined as follows:

$$T : \text{horizontal line} \mapsto \text{horizontal line}$$
$$T : B_1 \mapsto D_1$$
$$T : B_{n+1} \mapsto A_n, \ n \in \mathbb{N}$$
$$T : C_1 \mapsto D_2 \ (T : [0,2) \ni t \mapsto (\cos \pi t, \sin \pi t))$$
$$T : C_{n+1} \mapsto D_{n+2}, \ n \in \mathbb{N}.$$

If $H : Y \mapsto X$ is a homeomorphism, $H(B_1)$ and $H(C_2)$ must be circles D_{i_1} and D_{i_1+1} and the segment $\overline{PQ}$ must map into a corresponding segment $\overline{P'Q'}$ in X. Removal of R from Y leaves a set with three components. Removal of $H(R)$ from X leaves a set with two components. Hence X and Y are not homeomorphic.

In $\mathbb{R}^3$ there are many homeomorphic images of $S_1 \stackrel{\text{def}}{=} \{ \mathbf{x} : \|\mathbf{x}\| = 1 \}$, the surface of the unit ball centered at the origin. For example, the octahedron or double pyramid

$$P_2 \stackrel{\text{def}}{=} \{ (x, y, z) : |x| + |y| + |z| = 1 \},$$

is homeomorphic (via central projection) to S_1 and is regarded as *tamely embedded* or *tame*, because the complement of P_2 is simply connected: every simple closed curve $\gamma : [0,1] \ni t \mapsto \mathbb{R}^3 \setminus P_2$ in $\mathbb{R}^3$ is homotopic (in $\mathbb{R}^3 \setminus P_2$) to a point.

By contrast, the complement of the surface of a torus T swept out in $\mathbb{R}^3$ by a unit circle orthogonal to the xy-plane and centered on the circle

$$\{ (x, y, z) : x^2 + y^2 = 4, \ z = 0 \}$$

is such that the circle

$$C_1 \stackrel{\text{def}}{=} \{ (x, y, z) : (y - 2)^2 + z^2 = 5, \ x = 0 \}$$

is not homotopic to a point in $\mathbb{R}^3 \setminus T$. The complement of T is not simply connected.

Alexander **[Al]** produced a surface Σ homeomorphic to S_1 and yet such that $\mathbb{R}^3 \setminus \Sigma$ is not simply connected. The surface Σ is a *wildly embedded* or *wild* sphere. The surface depicted in **Figure 3.2.2.2** is a version of Σ, *Alexander's horned sphere*, on which the set of "horn-tips," a homeomorphic image of the Cantor set, constitutes a set of "obstacles to homotopy."

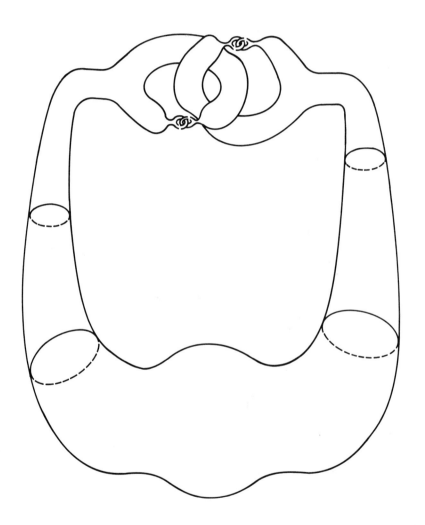

Figure 3.2.2.2. Alexander's horned sphere.

Subsequently Artin and Fox **[FoAr]** developed the theory of wild *em-*

beddings. They gave an example of a simple arc $\gamma : [0,1] \ni t \mapsto \mathbb{R}^3$ such that γ^* is snarled and (wild) cf. **Figure 3.2.2.3**. The endpoints $\gamma(0)$ and $\gamma(1)$ are obstacles to homotopy because there is a closed curve $\eta : [0,1] \mapsto \mathbb{R}^3 \setminus \gamma^*$ that is not homotopic to a point in $\mathbb{R}^3 \setminus \gamma^*$.

The arc-image γ^* can be used to form a wild sphere as follows. Let a long hose be tapered to a point at each end and then snarled so that the axis of the hose lies along γ^*. The *surface* Σ of the resulting snarled hose is homeomorphic to the surface S_1. Nevertheless Σ is a wild sphere since the points that are the two pinched ends of the hose are obstacles to homotopy in the unbounded component of $\mathbb{R}^3 \setminus \Sigma$.

There exist wild spheres in $\mathbb{R}^3$.

Figure 3.2.2.3. The Artin-Fox arc-image γ^*.

In **Figure 3.2.2.4** there is an attempt to indicate the early stages of the construction of *Antoine's necklace* $\mathcal{N}$. For n in $\mathbb{N}$, at the nth stage there are 2^n linked solid tori, the union of which is a compact set K_n. The sequence of the K_n is nested: $K_{n+1} \subset K_n$, and $\bigcap_{n \in \mathbb{N}} K_n = \mathcal{N}$.

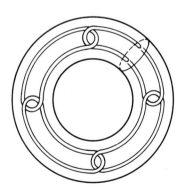

Figure 3.2.2.4. Antoine's necklace.

Antoine's necklace is homeomorphic to the Cantor set C_0, i.e., $\mathcal{N}$ is a nowhere dense, perfect, and totally disconnected set, and yet $\mathbb{R}^3 \setminus \mathcal{N}$ is not simply connected, e.g., the circle lying on the surface of the first torus is not homotopic to a point in $\mathbb{R}^3 \setminus \mathcal{N}$.

3.3. Exotica in Differential Topology

For homeomorphic topological spaces X and Y on each of which there is a *differential geometric structure* there arises the delicate question:

Are X and Y *diffeomorphic*, i.e., is there a C^∞ homeomorphism, a *diffeomorphism*, $f : X \mapsto Y$?

Such questions aroused great interest when it was discovered by Brieskorn, Hirzebruch, and Milnor [**Bri, Miln1**] that the sphere S^7 is susceptible of several differential geometric structures no two of which are diffeomorphic. The surfaces are rather easily described, although the absence of diffeomorphisms between them requires more discussion than can be given here.

Example 3.3.1. When $\mathbb{N} \ni k \geq 3$, $\mathbb{N} \ni a_j \geq 2$ there are defined:

$$X_k(a_0, \ldots, a_k) \overset{\text{def}}{=} \left\{ (z_0, \ldots, z_k) \ : \ z_j \in \mathbb{C}, \ 0 \leq j \leq k, \ \sum_{j=0}^{k} z_j^{a_j} = 0 \right\}$$

$$S^{2k+1} \overset{\text{def}}{=} \left\{ (z_0, \ldots, z_k) \ : \ z_j \in \mathbb{C}, \ 0 \leq j \leq k, \ \sum_{j=0}^{k} z_j \overline{z_j} = 1 \right\}$$

$$W^{2k-1}(a_0, \ldots, a_k) \overset{\text{def}}{=} X_k(a_0, \ldots, a_k) \cap S^{2k+1}.$$

Thus X_k is a set of *complex dimension* k, S^{2k+1} is a set of complex dimension k, and $W^{2k-1}(a_0, \ldots, a_k)$ is a set of complex dimension $k - 1$ in $\mathbb{C}^{k+1}$. Viewed in $\mathbb{R}^{2k+2}$ the sets in question have real dimensions $2k + 1$, $2k + 1$, and $2k - 1$. Then $W^7(3, 6r - 1, 2, 2, 2)$, $1 \leq r \leq 28$, are all homeomorphic to S^7 and are pairwise nondiffeomorphic. Furthermore, any oriented (7-dimensional) differentiable manifold homeomorphic to S^7 is diffeomorphic, for some natural number r in $[0, 28]$, to $W^7(3, 6r - 1, 2, 2, 2)$.

In [**Law**] there is an exposition of and a large bibliography about *exotica*, i.e., homeomorphic but nondiffeomorphic differential geometric structures. In particular, there is an extensive discussion of the following phenomenon, among many others.

There are (uncountably many) nondiffeomorphic differential geometric structures for $\mathbb{R}^4$. On the other hand, if $4 \neq k \in \mathbb{N}$ then any two differential geometric structures for $\mathbb{R}^k$ *are* diffeomorphic.

Related to the developments just described is the almost complete resolution of the *Poincaré conjecture*, relating homotopy and homeomorphism of differential geometric structures.

Two arcs $\gamma_i : [0,1] \ni t \mapsto X$, $i = 1, 2$, are homotopic iff there is a continuous map $h : [0,1]^2 \ni (s,t) \mapsto X$ such that

$$h(0,t) = \gamma_1(t) \text{ and } h(1,t) = \gamma_2(t).$$

Arcs are special instances of continuous maps (between the topological space $[0,1]$ and a topological space X). There is a more general notion of homotopy of arbitrary maps $f_i : X \mapsto Y$, $i = 1, 2$, between arbitrary topological spaces X and Y. Two such maps are homotopic iff there is a continuous map $h : [0,1] \times X \mapsto Y$ such that for all x in X,

$$h(0,x) = f_1(x) \text{ and } h(1,x) = f_2(x)$$

in which case one writes $f_1 \cong f_2$. Two topological spaces X and Y are *homotopically equivalent* iff there are continuous maps

$$f : X \mapsto Y \text{ and } g : Y \mapsto X$$

such that the composition maps

$$f \circ g : Y \mapsto Y \text{ resp. } g \circ f : X \mapsto X$$

are homotopic to the identity maps

$$\mathbf{1}_Y : Y \ni y \mapsto y \in Y \text{ resp. } \mathbf{1}_X : X \ni x \mapsto x \in X.$$

Poincaré's conjecture in modern terms asserts that an *n-dimensional manifold* homotopically equivalent to the sphere

$$S^n \overset{\text{def}}{=} \{\mathbf{x} \ : \ \mathbf{x} \in \mathbb{R}^n, \|\mathbf{x}\| = 1\}$$

is actually homeomorphic to S^n. In the early part of the twentieth century the conjecture was known to be valid if $n = 1$ or $n = 2$. In 1961 Smale [**Sm1**] showed the conjecture is valid if $n > 4$ and in 1982 Freedman [**Fr**] showed it is valid if $n = 4$. At this writing, the resolution of the conjecture for $n = 3$ has not been announced.

[**Remark 3.3.1:** In [**Fr**] the existence, noted earlier, of nondiffeomorphic and yet homeomorphic images of $\mathbb{R}^4$, falls out as a by-product of the general thrust of the work.]

4. Probability Theory

4.1. Independence

The theory of probability deals with a *probabilistic measure situation* $(X, \mathcal{S}, \mathrm{P})$, i.e, a measure situation specialized by the assumptions a) $X \in \mathcal{S}$ and b) $P(X) = 1$. Elements of $\mathcal{S}$ are *events* and measurable functions in $\mathbb{R}^X$ are called *random variables*. If f is an integrable random variable then its *expected value* or *expectation* is $E(f) \overset{\text{def}}{=} \int_X f(x)\, dP$.

According to Kolmogorov, the founder of the modern theory of probability, the distinguishing features of the subject are the notions of *independence* of events and independence of random variables [**Ko1, Ko2**].

A set $\mathcal{E} \overset{\text{def}}{=} \{A_\lambda\}_{\lambda \in \Lambda}$ of events A_λ is said to be independent iff for every subset $\{ A_n \ : \ A_i \neq A_j \text{ if } i \neq j, \ 1 \leq n \leq N \}$ of $\mathcal{E}$,

$$P\left(\bigcap_{n=1}^{N} A_n\right) = \prod_{n=1}^{N} P(A_n).$$

A set $\mathcal{F} \overset{\text{def}}{=} \{f_\lambda\}_{\lambda \in \Lambda}$ of random variables is said to be independent iff for every finite subset $\{ f_n \ : \ f_i \neq f_j, \text{ if } i \neq j, \ 1 \leq n \leq N \}$ of $\mathcal{F}$ and every finite set $\{B_1, \ldots, B_N\}$ of Borel subsets of $\mathbb{R}$, $\{f_1^{-1}(B_1), \ldots, f_N^{-1}(B_N)\}$ is independent. One says that the events belonging to a set $\mathcal{E}$ are themselves independent if $\mathcal{E}$ is; similarly one says that the random variables belonging to a set $\mathcal{F}$ are themselves independent if $\mathcal{F}$ is.

210

Exercise 4.1.1.

i. Show that if E and F are independent events then

$$\int_X \chi_E(x)\chi_F(x)\,dP = \int_X \chi_E(x)\,dP \int_X \chi_F(x)\,dP.$$

ii. Show that if $f_1,\ldots,f_n$ are independent random variables, if $\{p_i\}_{i=1}^n \subset \mathbb{N}$, and if $f_i^{p_i}$, is integrable, $1 \le i \le n$, then

$$\int_X \prod_{i=1}^n f_i^{p_i}(x)\,dP = \prod_{i=1}^n \int_X f_i^{p_i}(x)\,dP. \tag{4.1.1}$$

iii. Show that if, in the definition of independent random variables, the set $\{f_1^{-1}(B_1),\ldots,f_N^{-1}(B_N)\}$ is independent whenever the Borel sets B_i, $1 \le i \le N$, are open intervals then the random variables are independent.

For any event A, $\{\emptyset, X, A\}$ is independent and for any random variable f and any constant function c, $\{c, f\}$ is independent. These are the *trivial* instances of independence. Note also that if $\mathcal{E}$ is an independent set of events and if some events in $\mathcal{E}$ are replaced by their complements to produce a new set $\widetilde{\mathcal{E}}$, then $\widetilde{\mathcal{E}}$ is also independent.

A discussion motivating the definitions of independence of events and of random variables is given in [**Halm**].

A significant fraction of classical probability theory concerns itself with theorems about sets of independent random variables [**Loe**]. In the remainder of this **Section** there is an attempt to reveal the "reluctance" of sets of events resp. sets of random variables to be independent. (Salomon Bochner said that since (4.1.1) is "repugnant" to a set of functions, he was not surprised at the singularity and hence the importance of independence as a phenomenon.)

Exercise 4.1.2. Let P in the measure situation $(\mathbb{N}, 2^{\mathbb{N}}, P)$ be the discrete measure:

$$P(n) \overset{\text{def}}{=} \begin{cases} 1 - \sum_{n=2}^\infty 2^{-n!} & \text{if } n = 1 \\ 2^{-n!} & \text{if } n \ge 2. \end{cases}$$

Show:

i. if $1 < m, n \in \mathbb{N}$ then there is in $\mathbb{N}$ no k such that $k! = m! + n!$;

ii. trivial instances aside, there are no independent sets of events in $2^{\mathbb{N}}$;

iii. trivial instances aside, there are no independent sets of random variables defined on $\mathbb{N}$ endowed with the σ-algebra $2^{\mathbb{N}}$ and the measure P.

[*Hint:* If $\mathcal{E}$ is an independent set of events then any two-element subset of $\mathcal{E}$ is independent.]

Example 4.1.1. If

$$X \stackrel{\text{def}}{=} \{a, b, c, d\}, \ \mathcal{S} \stackrel{\text{def}}{=} 2^X, \ P(a) = \cdots = P(d) = \frac{1}{4}$$

then any two of the events in $\mathcal{E} \stackrel{\text{def}}{=} \{\{a, b\}, \{a, c\}, \{b, c\}\}$ are independent but $\mathcal{E}$ is not independent. Furthermore if $P(a) = 0.1, P(b) = 0.2, P(c) = 0.3, P(d) = 0.4$ then two events are independent iff one of them is X or $\emptyset$: only trivial instances of independence are at hand.

Events as such are not independent. The independence of events is completely determined only with respect to the probability measure P.

A large class of probabilistic measure situations consists of those that are measure-theoretically equivalent to

$$\mathcal{S} \stackrel{\text{def}}{=} ([0, 1], \mathcal{L}, \lambda)$$

[**Halm**]. Hence, counterexamples for $\mathcal{S}$ have a large range of relevance. In the following discussion it is assumed that the probabilistic measure situation is $\mathcal{S}$. Two Borel sets A_1 and A_2 are regarded as equal if their *symmetric difference*

$$A_1 \Delta A_2 \stackrel{\text{def}}{=} (A_1 \setminus A_2) \cup (A_2 \setminus A_1)$$

is a null set. Two random variables f and g are regarded as equal if $f - g = 0$ a.e. This kind of "equality" is *equality modulo null sets*.

For a set S of random variables let $\text{Ind}(S)$ denote the set of random variables f such that $S \cup \{f\}$ is independent. In particular let Const denote the set of constant functions. Hence for any set S of random variables, $\text{Const} \subset \text{Ind}(S)$.

Exercise 4.1.3. Show that if $\{f_{ij}\}_{1 \leq i \leq N, \ 1 \leq j \leq J_i}$ is an independent set of random variables, if ϕ_i is a Borel measurable function in $\mathbb{R}^{\mathbb{R}^{J_i}}$, $1 \leq i \leq N$, then

$$\left\{ g_1 \stackrel{\text{def}}{=} \phi_1 \left(f_{11}, \ldots, f_{1J_1} \right), \ldots, g_N \stackrel{\text{def}}{=} \phi_N \left(f_{N1}, \ldots, f_{NJ_N} \right) \right\}$$

is independent.

[*Hint:* Use the result in **Exercise 4.1.1***iii*. **211**.]

Example 4.1.2. Let f and g be the functions for which the graphs are given in **Figure 4.1.1** ($f = g$ on $(-\infty, d]$).

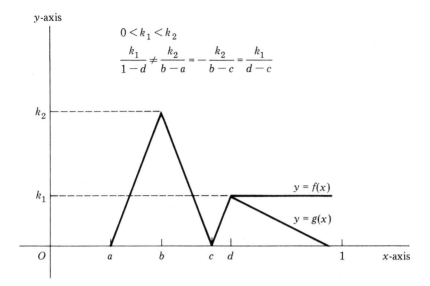

y-axis

$0 < k_1 < k_2$

$$\frac{k_1}{1-d} \neq \frac{k_2}{b-a} = -\frac{k_2}{b-c} = \frac{k_1}{d-c}$$

Figure 4.1.1.

Then $\mathrm{Ind}\,(\{f\}) = \mathrm{Ind}\,(\{g\}) = \mathrm{Const}$. The proof is based on the *metric density theorem* [**Rud**]:

THEOREM. IF $E \in \mathcal{L}$ THEN THERE IS IN E A NULL SET N SUCH THAT IF $x \in E \setminus N$ THEN

$$\lim_{\epsilon \downarrow 0} \frac{\lambda\left(E \cap (x - \epsilon, x + \epsilon)\right)}{2\epsilon} = 1.$$

Assume B is a Borel set in $[0,1]$ and that $\{B, J \overset{\text{def}}{=} f^{-1}\left([\alpha, \beta]\right)\}$ is independent. If $0 < \alpha < \beta < k_1$, then J is the union of three equally long and disjoint intervals I_1 resp. I_2 resp. I_3 that are subintervals of $[a, b]$ resp. $[b, c]$ resp. $[c, d]$ and $P(J) = 3P\left(I_j\right)$, $1 \le i \le 3$. Thus

$$P(B)P(J) = P(B \cap J) = P\left(B \cap I_1\right) + P\left(B \cap I_2\right) + P\left(B \cap I_2\right).$$

If $|\beta - \alpha| \to 0$, $0 < \alpha < \beta < k_1$, then metric density arguments show that the only possible values for $P(B)$ are $0, \frac{1}{3}, \frac{2}{3}, 1$. On the other hand, if $k_1 < \alpha < \beta < k_2$ and $|\beta - \alpha| \to 0$ then similar arguments show that the only possible values for $P(B)$ are $0, \frac{1}{2}, 1$. Hence, *modulo null sets*, B must be $[0, 1]$ or $\emptyset$. Hence $\mathrm{Ind}(f) = \mathrm{Const}$. Arguments of the same kind show that $\mathrm{Ind}(g) = \mathrm{Const}$.

[**Note 4.1.1:** If f and g are functions for which the conclusions above are valid, validity persists if f and g are replaced by homothetic translates of infinitely differentiable functions $\widetilde{f}$ and $\widetilde{g}$, constructed by smoothly rounding the corners of the graphs of f and g. Since k_1 and k_2 are arbitrary it follows that if p is a polynomial over $\mathbb{R}$ then an arbitrarily small part of the graph of p can be excised and replaced by the graph of a translate of a smoothed homothetic version $\widetilde{g}$ of g so that the result $\widetilde{p}$ is a function again in $C^\infty([0,1],\mathbb{R})$ and $\mathrm{Ind}(\widetilde{p})$ is again Const.

Hence in each of the familiar function spaces built on $[0,1]$, e.g., $C([0,1],\mathbb{R})$, $L^p([0,1],\mathbb{R})$ et al., there is a dense subset D consisting entirely of functions f such that $\mathrm{Ind}(f) = \mathrm{Const}$.

Thus if $S \overset{\text{def}}{=} \{f_n\}_{n\in\mathbb{N}}$ is an independent sequence of random variables and $f_n \to f$ in any of a number of different modes of convergence, e.g., uniform, in L^p, pointwise, etc., then each f_n may be replaced by a g_n such that $g_n \to f$ in the same mode and yet for each n, $\mathrm{Ind}(g_n) = \mathrm{Const}$. In other words, although the independence of a sequence of random variables is frequently an important part of the basis for a conclusion about the convergence of the sequence and the nature of the limit f, e.g., in the *central limit theorem*, f may well be the limit (in the same mode of convergence) of a sequence that is as far from independence as it is possible to be.]

The set of trigonometric functions spans a dense subset of each of the familiar function spaces. Yet, as the next result shows, neither this set nor any other orthogonal set that spans a dense subset of one of these spaces can be independent.

THEOREM 4.1.1. LET $(X, \mathcal{S}, P)$ BE A PROBABILISTIC MEASURE SITUATION SUCH THAT $L^2(X, \mathbb{R})$ IS A Hilbert SPACE $\mathcal{H}$ OF DIMENSION NOT LESS THAN 3. IF $S \overset{\text{def}}{=} \{f_\lambda\}_{\lambda \in \Lambda}$ IS INDEPENDENT, ITS CLOSED SPAN M IS SUCH THAT THE *orthogonal complement* $M^\perp$ IS NOT $\{\mathbf{O}\}$. IF $\mathcal{H}$ IS INFINITE-DIMENSIONAL THEN $M^\perp$ IS ALSO INFINITE-DIMENSIONAL.

PROOF. Since each f_λ is integrable, mathematical induction shows that

$$\{g_\lambda \overset{\text{def}}{=} f_\lambda - \int_X f_\lambda(x)\,dP\}_{\lambda\in\Lambda}$$

is independent and $\int_X g_\lambda(x)\,dP = 0$, $\lambda \in \Lambda$. Thus it may be assumed that for all λ, $\int_X f_\lambda\,dP = 0$. Finally it may be assumed that no f_λ is constant and the question is whether $\Sigma \overset{\text{def}}{=} S \cup \{1\}$ can span $\mathcal{H}$.

From (4.1.1) page 211 it follows that each product $\phi \overset{\text{def}}{=} f_{\lambda_1} \cdots f_{\lambda_n}$ of pairwise different members of S is a nonzero element of $\mathcal{H}$ and is orthogonal to any f_λ different from each of the factors. On the other hand, if f is one of the factors, say f_{λ_i}, of ϕ and if $n \geq 2$ then (4.1.1) page 211 implies $(f_{\lambda_i}, \phi) = 0$. In other words, M denoting the closed linear span of Σ, if $n \geq 2$ then $\phi \in M^{\perp}$.

$\square$

[**Remark 4.1.1:** If X consists of two elements x_1, x_2 and if $f_1 \equiv 1$, $f_2(x_1) = -f_2(x_2) = 1$, $P(\{x_1\}) = P(\{x_2\}) = 0.5$ then $\{f_1, f_2\}$ is an independent set consisting of orthonormal functions that span $\mathcal{H}$. Hence the number 3 in the statement of THEOREM **4.1.1** is optimal.]

On $[0, 1]$ the *Rademacher functions* are defined as follows:

$$r_n(x) \stackrel{\text{def}}{=} \begin{cases} 1 & \text{if } n = 0 \\ \operatorname{sgn}(\sin 2\pi 2^n x) & \text{if } n \in \mathbb{N} \end{cases}.$$

Exercise 4.1.4. Show that the set of Rademacher functions (random variables) is a maximal independent set for the measure situation $([0, 1], \mathcal{L}, \lambda)$, i.e., that if a nonconstant random variable f is adjoined to the set of Rademacher functions the resulting set is not independent.

[*Hint:* To prove the maximality of the set it may be assumed that $A \stackrel{\text{def}}{=} f^{-1}(\{1\}) \neq \emptyset$. Then if $n > 1$,

$$\lambda\left(r_n^{-1}(\{1\}) \cap A\right) = \frac{1}{2}\lambda(A).$$

The form of $r_n^{-1}(\{1\})$ and the metric density theorem imply that $\lambda(A)$ is unbounded.]

The construction in the proof of THEOREM **4.1.1** is related to the construction of the *Walsh functions* from the Rademacher functions [**Zy**]. Since the Walsh functions constitute a complete orthonormal set in

$$L^2([0, 1], \mathbb{C})$$

there arises the question: Does the general construction above always yield a complete orthonormal set, at least if the set Σ is a *maximal independent set*?

Example 4.1.3. Let f be a random variable f such that $\operatorname{Ind}(f) = $ Const. It follows that $\Sigma_1 \stackrel{\text{def}}{=} \{f\} \cup \operatorname{Const}$ is a maximal independent set. The general construction used for Σ and applied to Σ_1 leads to a an orthonormal set containing no more than three elements and thus, if the associated Hilbert space is of dimension at least four, the orthonormal set that emerges is not complete.

In [**Ge6**] there is a more extensive discussion of independence phenomena.

4.2. Stochastic Processes

For a probabilistic measure situation $(X, \mathcal{S}, P)$ and a set $\{f_\lambda\}_{\lambda \in \Lambda}$ of random variables in $\mathbb{R}^X$ there is the set

$$\mathcal{F} \overset{\text{def}}{=} \{ F_{\lambda_1,\dots,\lambda_n} \; : \; \lambda_1, \dots, \lambda_n, \; n \in \mathbb{N} \}$$

of associated *distribution functions*. Each $F_{\lambda_1,\dots,\lambda_n}$ is in $\mathbb{R}^{\mathbb{R}^n}$ and

$$F_{\lambda_1,\dots,\lambda_n}(x_1,\dots,x_n) = P(\{ x \; : \; x \in X, f_{\lambda_i}(x) \le x_i, \; 1 \le i \le n \}).$$

It follows that the functions in $\mathcal{F}$ satisfy the five *Kolmogorov criteria* [**Ko2**]:

i. if $1 \le i \le n$ then $\lim_{x_i \downarrow -\infty} F_{\lambda_1,\dots,\lambda_n}(x_1,\dots,x_n) = 0$;

ii. $\lim_{x_1 \uparrow \infty,\dots,x_n \uparrow \infty} F_{\lambda_1,\dots,\lambda_n}(x_1,\dots,x_n) = 1$;

iii. for each i,

$$x_i \le x_i' \Rightarrow$$
$$F_{\lambda_1,\dots,\lambda_n}(x_1,\dots,x_i,\dots,x_n) \le F_{\lambda_1,\dots,\lambda_n}(x_1,\dots,x_i',\dots,x_n);$$

iv. if $\{i_1, i_2, \dots, i_n\}$ is a permutation of $\{1, 2, \dots, n\}$ then

$$F_{\lambda_{i_1},\dots,\lambda_{i_n}}(x_{i_1},\dots,x_{i_n}) = F_{\lambda_1,\dots,\lambda_n}(x_1,\dots,x_n);$$

v. if $k < n$ then

$$F_{\lambda_1,\dots,\lambda_k}(x_1,\dots,x_k) = F_{\lambda_1,\dots,\lambda_n}(x_1,\dots,x_k,\infty,\dots,\infty).$$

Kolmogorov showed that conversely, if a set $\mathcal{F}$ satisfying his criteria is given then there is definable on a σ-algebra $\mathcal{Z}$ in $\mathbb{R}^\Lambda$ a probability measure P, i.e., a measure situation $\left(\mathbb{R}^\Lambda, \mathcal{Z}, P\right)$, and random variables $f_\mu : \mathbb{R}^\Lambda \ni (x_\lambda)_{\lambda \in \Lambda} \mapsto x_\mu$ for which $\mathcal{F}$ constitutes the set of distribution functions. The σ-algebra $\mathcal{Z}$ is generated by the set of all *cylinder sets* based on Borel sets in $\mathbb{R}^n, n \in \mathbb{N}$: a typical cylinder set has the following form:

$$Z_{\lambda_1,\dots,\lambda_n}(A) \overset{\text{def}}{=} \left\{ (x_\lambda)_{\lambda \in \Lambda} \; : \; (x_\lambda)_{\lambda \in \Lambda} \in \mathbb{R}^\Lambda, \; (x_{\lambda_1},\dots,x_{\lambda_n}) \in A \right\},$$

A a Borel set in $\mathbb{R}^n$.

When A is a *corner* of the form $\{ (a_1, \dots, a_n) \; : \; a_i \le x_i, \; 1 \le i \le n \}$ the probability $P[Z_{\lambda_1,\dots,\lambda_n}(A)]$ of the corresponding cylinder set is

$$F_{\lambda_1,\dots,\lambda_n}(x_1,\dots,x_n) \; \left(= P\left[\{ (x_\lambda)_{\lambda \in \Lambda} \; : \; f_{\lambda_i}((x_\lambda)_{\lambda \in \Lambda}) \le x_i, \; 1 \le i \le n \}\right] \right).$$

Thus the value at $(x_1, \ldots, x_n)$ of the joint distribution function of the random variables f_{λ_i}, $1 \leq i \leq n$, is $F_{\lambda_1, \ldots, \lambda_n}(x_1, \ldots, x_n)$ as required. The extension of P to $\mathcal{Z}$ follows readily [**Ko2**].

It should be noted that $\mathsf{S} \stackrel{\text{def}}{=} \mathbb{R}^\Lambda$ is a vector space of functions. The evaluation map: $\phi \mapsto \phi(\lambda_i) \in \mathbb{R}$ taking the function ϕ in S to its value at λ_i is a special kind of linear functional on S. The natural extension of this observation leads to the next construction when a) Λ is itself an ingredient of a measure situation $(\Lambda, \mathcal{T}, \mu)$ or b) when Λ is a topological space or c) when Λ is an n-dimensional manifold.

For a) there are considerations of spaces $L^p(\Lambda, \mathbb{R})$, $1 \leq p \leq \infty$; for b) there are considerations of $C(\Lambda, \mathbb{R})$; for c) there are considerations of vector spaces of functions satisfying differentiability requirements. In each instance there is the question of whether P induces a probability measure on the subspace to be studied.

Stripped to essentials, the context is the following.

i. Let V be a topological vector space, e.g., a Banach space, let $(\Omega, \mathcal{S}, \Pi)$ be a probability measure situation, and let T be a linear map of the dual space V^* of V into the set $\mathcal{R}$ of random variables on Ω. For each finite set $\{\mathbf{x}_1^*, \ldots, \mathbf{x}_n^*\}$ in V^* and each Borel set A in $\mathbb{R}^n$ let

$$Z_{\mathbf{x}_1^*, \ldots, \mathbf{x}_n^*; A} \stackrel{\text{def}}{=} \{\mathbf{x} \ : \ \mathbf{x} \in V, \ (\mathbf{x}_1^*(\mathbf{x}), \ldots, \mathbf{x}_n^*(\mathbf{x})) \in A\}$$

be the *cylinder set* based on $\{\mathbf{x}_1^*, \ldots, \mathbf{x}_n^*; A\}$.

ii. The set Z of all cylinder sets is an algebra on which one can define the finitely additive set function

$$\mu : Z \ni Z_{\mathbf{x}_1^*, \ldots, \mathbf{x}_n^*; A} \mapsto \Pi(\{\omega \ : \ \omega \in \Omega, \ (T(\mathbf{x}_1^*)(\omega), \ldots, T(\mathbf{x}_n^*)(\omega)) \in A\}).$$

There arises the question:

In what circumstances can μ be extended to a countably additive measure on $\mathcal{Z}$, the σ-algebra generated by Z?

The answer is far from simple since it depends on V, on $(\Omega, \mathcal{S}, \Pi)$, and on T. However, among the choices for A is the corner

$$\{(s_1, \ldots, s_n) \ : \ s_i \leq t_i, \ 1 \leq i \leq n\}$$

in which case the measure of the corresponding cylinder set is the value

$$F_{\mathbf{x}_1^*, \ldots, \mathbf{x}_n^*}(t_1, \ldots, t_n)$$

of a finite-dimensional distribution function $F_{\mathbf{x}_1^*, \ldots, \mathbf{x}_n^*}$. The set of these is parametrized by the set of all finite subsets of V^* and constitutes a family $\mathcal{F}$ satisfying Kolmogorov's criteria. Consequently there is defined on the σ-algebra generated by the cylinder sets in $\mathbb{R}^{V^*}$ a probability measure P.

Since V may be viewed as a subset of $\mathbb{R}^{V^*}$ the question above is reduced to whether P, restricted to V, yields a probability measure on the σ-algebra generated by the set of intersections of cylinder sets in $\mathbb{R}^{V^*}$ with V. A not very helpful answer is the near-tautology: iff V is a *thick* subset of $\mathbb{R}^{V^*}$ [**Halm**].

In a more concrete fashion, Hemasinha [**Hem**] showed that if D is a bounded region in $\mathbb{C}$, if Π is Lebesgue measure normalized on D so that $\Pi(D) = 1$, if V is the Hilbert space of functions f harmonic in D and such that

$$\int_D |f(z)|^2 \, d\Pi < \infty,$$

and if T is any bounded endomorphism of V then the corresponding measure μ is countably additive. (For convenience and generality, Hemasinha worked with holomorphic functions defined on D.)

On the other hand, if T is required to operate on a Hilbert space $\mathcal{H} \stackrel{\text{def}}{=} L^2_{\mathbb{R}}(\Omega, \Pi)$ of $\mathbb{R}$-valued functions, which is its own dual, and, additionally, the endomorphism T is to map orthogonal pairs of functions into pairs of independent functions, the set function μ is *not* countably additive.

The following sequence of results yields the conclusion above. All functions considered are assumed to be in $L^2_{\mathbb{R}}(\Omega, \Pi)$.

LEMMA **4.2.1.**   IF ϕ, γ IS A PAIR OF ORTHONORMAL $\mathbb{R}$-VALUED FUNCTIONS THEN $\phi \pm \gamma$ IS A PAIR OF ORTHOGONAL FUNCTIONS.

PROOF.

$$(\phi + \gamma, \phi - \gamma) = \|\phi\|^2 - \|\gamma\|^2 - (\phi, \gamma) + (\gamma, \phi) = 0.$$

$\square$

LEMMA **4.2.2.**   IF BOTH $\{f, g\}$ AND $\{f \pm g\}$ ARE SETS OF INDE-PENDENT FUNCTIONS AND AT LEAST ONE OF THE FUNCTIONS IS NOT A CONSTANT, THEN EACH OF $f, g, f + g$ AND $f - g$ IS *normally distributed*, i.e., THE FORM OF THE DISTRIBUTION FUNCTION FOR EACH IS

$$\frac{1}{\sqrt{2\pi}\sigma} \int_{-\infty}^x \exp(-\frac{t^2}{2\sigma^2}) \, dt$$

[**Ge8**].

PROOF. It may be assumed that

$$\int_\Omega f(\omega) \, d\Pi = \int_\Omega g(\omega) \, d\Pi = 0$$
$$\int_\Omega f(\omega)^2 \, d\Pi = \int_\Omega g(\omega)^2 \, d\Pi = 1.$$

Let $(f \pm g)/\sqrt{2}$ be $h_\pm$. Because f and g are linear combinations of $h_\pm$ the result in **Exercise 4.1.3. 212** implies that if

$$\phi(t) \overset{\text{def}}{=} \int_\Omega \exp(itf(\omega))\, d\Pi$$

$$\gamma(t) \overset{\text{def}}{=} \int_\Omega \exp(itg(\omega))\, d\Pi$$

then

$$\xi(t) \overset{\text{def}}{=} \int_\Omega \exp\left(ith_+(\omega)\right)\, d\Pi = \phi\left(\frac{t}{\sqrt{2}}\right)\gamma\left(\frac{t}{\sqrt{2}}\right)$$

$$\eta(t) \overset{\text{def}}{=} \int_\Omega \exp\left(ith_-(\omega)\right)\, d\Pi = \phi\left(\frac{t}{\sqrt{2}}\right)\overline{\gamma\left(\frac{t}{\sqrt{2}}\right)}$$

$$\phi(t) = \xi\left(\frac{t}{\sqrt{2}}\right)\eta\left(\frac{t}{\sqrt{2}}\right)$$

$$\gamma(t) = \xi\left(\frac{t}{\sqrt{2}}\right)\overline{\eta\left(\frac{t}{\sqrt{2}}\right)}.$$

Mathematical induction applied to the process of substituting the right members of the first two equations for the appearances of $\xi(t)$ and $\eta(t)$ in the last two equations leads to the following formula:

$$\phi(t) = \left[\frac{\phi\left(\frac{t}{2^{2n-1}}\right)}{\left|\phi\left(\frac{t}{2^{2n-1}}\right)\right|}\right]^{2^{2n-1}} \cdot \left|\phi\left(\frac{t}{2^{2n-1}}\right)\right|^{2^{2n-1}} \cdot \left|\gamma\left(\frac{t}{2^{2n-1}}\right)\right|^{2^{2n-1}}, \quad n \in \mathbb{N},$$

and a similar formula for $\gamma(t)$.

Since, as $t \to 0$,

$$\phi(t) = \left(1 - \frac{t^2}{2} + o\left(t^2\right)\right)$$

$$\gamma(t) = \left(1 - \frac{t^2}{2} + o\left(t^2\right)\right)$$

it follows that

$$\phi(t) = \gamma(t) = \exp\left(-\frac{t^2}{2}\right).$$

The *inversion formula* for Fourier transforms shows that f and g are normally distributed. Since f and g are independent, the distribution functions for $f \pm g$ are convolutions of the distribution functions for f and $\pm g$. Since convolutions of normal distribution functions are themselves normal, the result follows.

$$\square$$

Let T be an endomorphism of $\mathcal{H} \overset{\text{def}}{=} L^2_\mathbb{R}(\Omega, \Pi)$. Then T is said to induce a Gaußian measure μ on the algebra Z of cylinder sets in $\mathcal{H}$ if the joint distribution function for the set $\{T(\mathbf{x}_1), \ldots, T(\mathbf{x}_n)\}$ of independent random variables is of the form

$$(2\pi)^{-\frac{n}{2}} \prod_{i=1}^n \left(\int_{-\infty}^{t_i} \exp\left(-\frac{s_i^2}{2}\right) ds_i \right), \tag{4.2.1}$$

whence $\{T(\mathbf{x}_1), \ldots, T(\mathbf{x}_n)\}$ is independent.

LEMMA **4.2.3.** IF THE ENDOMORPHISM T OF THE INFINITE-DIMEN-SIONAL HILBERT SPACE $\mathcal{H} \overset{\text{def}}{=} L^2_\mathbb{R}(\Omega, \Pi)$ MAPS ORTHOGONAL FUNCTIONS INTO INDEPENDENT FUNCTIONS THEN T INDUCES A GAUSSIAN MEASURE ON THE ALGEBRA Z OF CYLINDER SETS IN $\mathcal{H}$. THIS GAUSSIAN MEASURE CANNOT BE EXTENDED TO A COUNTABLY ADDITIVE MEASURE ON THE σ-ALGEBRA $\mathcal{Z}$ GENERATED BY Z.

PROOF. Since T maps orthogonal functions into independent functions it follows from LEMMA **4.2.2. 218** that if $\{\mathbf{x}_n\}_{n \in \mathbb{N}}$ is complete orthonormal set in $\mathcal{H}$ then the finite-dimensional distribution functions for the random variables $\{T(\mathbf{x}_n)\}_{n \in \mathbb{N}}$ take the form given in (4.2.1).

If the Gaußian measure μ may be extended from Z to a countably additive measure on $\mathcal{Z}$ there emerge the following contradictory relations [**Kur**]:

i. if $\mathbb{N} \ni k_n \uparrow \infty$ then

$$\mathcal{H} = \bigcup_{n=1}^\infty \{ \mathbf{x} \ : \ \mathbf{x} \in \mathcal{H}, |(\mathbf{x}, \mathbf{x}_i)| \leq n, \ 1 \leq i \leq k_n \} \overset{\text{def}}{=} \bigcup_{n=1}^\infty H_n;$$

ii. for any M in $(0, 1)$ the numbers k_n can be chosen so that

$$\mu(H_n) = \left(\int_{-n}^n \exp\left(\frac{s^2}{2}\right) ds \right)^{k_n} \leq M^n;$$

iii. $\mu(\mathcal{H}) \leq \sum_{n=1}^\infty M^n = \frac{M}{1-M}$, whence $\mu(\mathcal{H}) = 0$;

iv. $\mathcal{H} = \{ \mathbf{x} \ : \ (\mathbf{x}, \mathbf{x}_1) \in \mathbb{R} \}$ and so $\mu(\mathcal{H}) = (2\pi)^{-\frac{1}{2}} \int_\mathbb{R} \exp\left(-\frac{s^2}{2}\right) ds = 1$.

$\square$

The reconciliation between the result above and Hemasinha's work stems from the fact that the functions in his model of Hilbert space are holomorphic, and, trivialities aside, sets of holomorphic functions can*not* be independent [**Ge6**], i.e., in Hemasinha's model, there is *no* endomorphism T satisfying the hypotheses of LEMMA **4.2.3.**

Example 4.2.1. In an infinite-dimensional Hilbert space $\mathcal{H}$ there can be no nontrivial Borel measure that is translation-invariant or unitarily invariant. Indeed if μ is a nontrivial Borel measure, let $\{\phi_n\}_{n\in\mathbb{N}}$ be an orthonormal set and let B_n be a ball centered at $\frac{1}{2}\phi_n$ and of radius 0.1. If μ is translation-invariant or unitarily invariant, $\mu(B_n) > 0$. Since $i \neq j \Rightarrow$ $B_i \cap B_j = \emptyset$ it follows that if $B \overset{\text{def}}{=} \{\, \mathbf{x} \ : \ \mathbf{x} \in \mathcal{H}, \ \|\mathbf{x}\| \leq 1 \,\}$ then

$$B \supset \bigcup_{n\in\mathbb{N}} B_n \text{ and } \mu\left(\bigcup_{n\in\mathbb{N}} B_n\right) = \infty.$$

Hence the unit ball centered at $\mathbf{O}$ has infinite measure and, by a similar argument, every ball of positive radius and centered anywhere, has infinite measure and so, for every Borel set A, $\mu(A) = 0$ or $\mu(A) = \infty$, i.e., μ is trivial.

4.3. Transition Matrices

A *transition matrix* $P \overset{\text{def}}{=} (p_{ij})_{i,j=1}^{n,n}$ is characterized by the conditions

$$\sum_{j=1}^{n} p_{ij} = 1, \ 1 \leq i \leq n$$

$$p_{ij} \geq 0, \ 1 \leq i,j \leq n.$$

The number p_{ij} is interpreted as the probability that a system in "state" i will change into "state" j. For many transition matrices P it can be shown that

$$\lim_{n\to\infty} P^n \overset{\text{def}}{=} P_\infty$$

exists. For example, if for some k in $\mathbb{N}$, all entries in P^k are positive, then P_∞ exists [**Ge9**].

The matrix

$$A \overset{\text{def}}{=} \begin{pmatrix} 0 & 1 \\ 1 & 0 \end{pmatrix}$$

is a transition matrix whereas

$$A^k = \begin{cases} \begin{pmatrix} 0 & 1 \\ 1 & 0 \end{pmatrix} & \text{if } k \text{ is odd} \\ \begin{pmatrix} 1 & 0 \\ 0 & 1 \end{pmatrix} & \text{if } k \text{ is even,} \end{cases}$$

whence A_∞ does not exist. A clue to this behavior is found in an examination of the eigenvalues, ± 1 of A. The Jordan normal form of A is

$$\begin{pmatrix} 1 & 0 \\ 0 & -1 \end{pmatrix}$$

which immediately reveals why A_∞ does not exist.

For any transition matrix P, the vector $(1, 1, \ldots, 1)^t$ is an eigenvector corresponding to the eigenvalue 1, and for every eigenvalue λ, $|\lambda| \leq 1$.

THEOREM **4.3.1.** IF $P \overset{\text{def}}{=} (p_{ij})_{i,j=1}^{n,n}$ IS A TRANSITION MATRIX AND IF

$$
J_m \overset{\text{def}}{=} \begin{pmatrix}
\lambda_m & * & & & & \\
& \lambda_m & * & & & \\
& & \ddots & \ddots & & \\
& & & \lambda_m & * & \\
& & & & \lambda_m &
\end{pmatrix},
$$

$$* = 1 \text{ OR } * = 0, \ 1 \leq m \leq M$$

ARE THE *Jordan blocks* OF P THEN P_∞ EXISTS IFF:

  i. $|\lambda_m| < 1$ WHENEVER $* = 1$;
  ii. $\lambda_m = 1$ WHENEVER $|\lambda_m| = 1$.

PROOF. If $|\lambda_m| = 1$ then $\lim_{k\to\infty} \lambda_m^k$ exists iff $\lambda_m = 1$, cf. [**Ge9**]. □

Exercise 4.3.1. Regard each $n \times n$ transition matrix P as a vector in $\mathbb{R}^{n^2}$. Show that the set $\mathcal{P}$ of $n \times n$ transition matrices is the intersection of the *nonnegative orthant* $\mathbb{R}^{(n^2,+)}$ and n *hyperplanes*.

Exercise 4.3.2. View $\mathcal{P}$ as a "flat" part of $n^2 - n$-dimensional Euclidean space and thus as endowed with the inherited Euclidean topology and Lebesgue measure $\lambda_{n^2-n} \overset{\text{def}}{=} \mu$. Let $\mathcal{P}_\infty$ be the subset consisting of transition matrices P for which P_∞ exists. Show that

$$\mu(\mathcal{P}_\infty) = 0$$

and that $\mathcal{P} \setminus \mathcal{P}_\infty$ is a dense open subset of $\mathcal{P}$, cf. COROLLARY **1.3.1.1. 26**.

5. Foundations

5.1. Logic

From early times human language has been a source of counterexamples to the belief that normal discourse is consistent. The sentence, "This statement is false," can be neither true nor false. The phrase "not self-descriptive" is neither self-descriptive nor not self-descriptive. Can an omnipotent being overpower itself? In [**BarE**] there is an extensive discussion of those aspects of language that deal with grammatically accurate but logically daunting statements.

Mathematical versions of such paradoxes, antinomies, explicitly or implicitly self-referential words and sentences, etc., eventually led to the search for a formal system of logic in which the perils of inconsistency are absent or at least so remote that humankind need have no fear of their obtrusion into scientific discourse.

The next few paragraphs, summarizing the presentation in [**Me**], deal with the fundamental concepts of a *formal system of logic F*.

[**Note 5.1.1:**   However rigorous, however formal, however restrictive the formal systems themselves, the proving of theorems about these same systems inescapably leads to reliance upon the use of human language whence the problems first emerged. Thus, it appears, that in the drive to achieve consistency and to avoid paradox, the logicians resort to harshly restricted modes of reasoning that are no more formal than the modes that lead to the

223

paradoxes, the antinomies, the self-referential sentences, etc. The hope that success will crown the effort rests on the "finitism" of the approach.

The next paragraph, introducing the formalization of logic, adverts almost immediately to a "countable set" without defining a countable set. Presumably a countable set is a set (not defined) that can be put in bijective correspondence (not defined) with $\mathbb{N}$ (also not defined). Later developments of formal logic and set theory lead to an axiomatic formalization of $\mathbb{N}$ and its consequent structures, $\mathbb{Z}$, $\mathbb{Q}$, $\mathbb{R}$, $\mathbb{C}$, $\mathbb{H}$, et al. Is there no circularity in the procedure? For a profound discussion of these matters the reader is urged to consult [**HiB**].]

There is a countable set $\mathcal{S}$ of symbols, finite sequences of which are *expressions*. Some of the symbols are *logical connectives* such as $\vee$ ("or"), $\wedge$ ("and"), $\rightarrow$ ("implies"), and $\neg$ ("negation"). Others are *quantifiers* $\forall$ ("for all"), $\exists$ ("there exists"), *function letters* $f, g, \ldots$, *predicate letters* $P, Q, \ldots$, *variables* $x, y, \ldots$, and *constants* $a, b, \ldots$.

A predicate P or a function f always appears in association with a nonempty set consisting of finitely many predicates, constants and variables ("arguments"), e.g., $P(a)$, $f(x, y, P)$. A quantifier always appears in association with variables and predicates, e.g., $\forall(x)P(P, Q, x, y, a, b, c)$. A large part of formal logic, in particular the part discussed below, is devoted to the study of *first-order theories* in which the arguments of predicates may *not* be predicates or functions and in which the argument of a quantifier must be a variable. Thus in first-order theories forms such as $\exists(P)(P \rightarrow Q)$ are not included.

Within the set of expressions there is a subset WF consisting of *well-formed formulae* (*wfs*) and a subsubset A consisting of those wfs that are the *axioms*. There is a finite set R of *rules of inference* that permit the chaining together of axioms to lead to *consequences* and the chaining together of axioms and/or consequences to produce *proofs*. The last link in a proof-chain is a *theorem*, (which might be an axiom).

The objects above constitute a framework in terms of which specific mathematical entities, e.g., groups, $\mathbb{N}$, etc., can be discussed by adding to the logical symbols and axioms other symbols and axioms. For groups the symbols and axioms in **Subsection 1.1.1** are the added objects. For formal number theory, i.e., for the treatment of $\mathbb{Z}$, the symbols and axioms added are some carefully tailored version of those given originally by Dedekind but known more popularly as Peano's axioms.

Closely associated with a formal system $\mathcal{F}$ are *interpretations* and *models* for it. An interpretation is a "concrete" nonempty set D and assignments:

i. of each n-variable predicate to a relation in D, i.e., to a subset of D^n;

ii. of each n-variable function to a function $D^n \mapsto D$;

iii. of each constant to a fixed element of D.

The symbols $\neg$, $\rightarrow$, $\forall$, and $\exists$ are given their "usual" meanings.

There are systematic definitions (due to Tarski) of the notions of *satisfiability* and *truth* of wfs.

Informally, a wf $\mathcal{A}$ is *satisfiable for some interpretation* I, if $\mathcal{A}$ obtains for *some* substitution in $\mathcal{A}$. For example, in group theory, if the interpretation I is the set of nonzero real numbers regarded as an abelian group with respect to multiplication, then the wf $\mathcal{A} \overset{\text{def}}{=} \{x^2 = 1\}$ is satisfiable iff one substitutes for x the number 1 or the number -1. On the other hand, the same wf $\mathcal{A}$ (written additively $\{2x = 1\}$) is not satisfiable in $\mathbb{Z}$ regarded as the abelian additive group of integers. A wf is *satisfiable* iff it is satisfiable in some interpretation.

Again informally, a wf $\mathcal{A}$ is *true* in an interpretation I if $\mathcal{A}$ obtains for *every* substitution. For example the wf $\mathcal{A} \overset{\text{def}}{=} x + x = e$ obtains in $\mathbb{Z}_2$ for all (both) substitutions $x \mapsto 0$ and $x \mapsto 1$. A wf $\mathcal{A}$ is *logically valid* iff $\mathcal{A}$ is true in every interpretation.

There are natural (informal) definitions of *contradictory* wfs, of the phrase $\mathcal{A}$ *implies* $\mathcal{B}$, and of the phrase $\mathcal{A}$ *is equivalent to* $\mathcal{B}$.

An interpretation I is a *model* $\mathsf{M}(\mathsf{I})$ for a set of wfs iff each wf is true for I.

In the language and context of the outline above, Gödel, who was soon to become the pre-eminent logician among his contemporaries, proved the formal equivalence of the notions of *theorem* and *logical validity*.

GÖDEL'S COMPLETENESS THEOREM. IN A FORMAL SYSTEM $\mathcal{F}$ A WF $\mathcal{A}$ IS A THEOREM IFF $\mathcal{A}$ IS LOGICALLY VALID [**Gö1**].

In [**Gö1**] Gödel proved a more striking result:

GÖDEL'S COUNTABILITY THEOREM. EVERY CONSISTENT FIRST-ORDER SYSTEM HAS A COUNTABLE MODEL.

A consequence of Gödel's countability theorem is a result proved earlier by Löwenheim [**Löw**] and Skolem [**Sk**].

LÖWENHEIM-SKOLEM THEOREM. IF A FIRST-ORDER THEORY HAS A MODEL IT HAS A COUNTABLE MODEL.

[**Note 5.1.2:** Do the Gödel-Löwenheim-Skolem results imply that, despite what every mathematician knows, $\mathbb{R}$ is countable? A simple answer is "No!" The reason lies in the subtlety of the notion of model. In the countable model of the formal system for analysis the "uncountability" of $\mathbb{R}$ is the assertion that for the D of the interpretation there is no map $f : D \mapsto D$ such that *in the model* $f(\mathbb{N}) = \mathbb{R}$.]

The mechanism above having been established, its founders planned to produce a formal system $\mathcal{F}$ adequate to deal at least with number theory, i.e., to cope with theorems about $\mathbb{Z}$. In this system each wf $\mathcal{A}$ or its negation

$\neg \mathcal{A}$ was to be a theorem and not both $\mathcal{A}$ and $\neg \mathcal{A}$ were to be theorems (the latter desideratum was for *consistency*).

Gödel and Rosser [**Gö2, Ross**] proved that any consistent formal logical system $\mathcal{F}$ that deals with $\mathbb{N}$ contains *undecidable* wfs. No formal proof exists for each nor for its negation: $\mathcal{F}$ is *incomplete*. One among those undecidable wfs, has a striking self-referential interpretation:

"The system $\mathcal{F}$ is consistent."

[**Note 5.1.3:**   Since the wf $\mathcal{A}$ interpreted above is undecidable it may be adjoined to $\mathcal{F}$ to form a new system $\mathcal{F}'$ which is as consistent as $\mathcal{F}$. But then there is in $\mathcal{F}'$ an undecidable wf $\mathcal{A}'$ interpretable, like $\mathcal{A}$, as asserting that $\mathcal{F}'$ is consistent. In $\mathcal{F}'$ the wf $\mathcal{A}$ is an axiom, hence is a theorem, and thus is decidable.]

One view of Gödel's *incompleteness theorem* is the following.

If one can consistently axiomatize logic so that there are mechanical rules whereby one passes, step-by-step, from axioms to theorems then one can imagine a machine that systematically lists all proofs, e.g., proofs involving one step, proofs involving two steps, etc. In theory the machine creates a countably infinite list of *all* theorems, each preceded by its proof. Then if a wf $\mathcal{T}$ is given, the list can be consulted to determine whether $\mathcal{T}$ or $\neg \mathcal{T}$ appears in the list of theorems. To determine whether $\mathcal{T}$ or $\neg \mathcal{T}$ is in the list, the machine is programmed in some way, e.g., to compare $\mathcal{T}$ and then $\neg \mathcal{T}$ with each listed theorem. The original hope of the axiomatizers was that there is a program that, given a wf $\mathcal{T}$, checks $\mathcal{T}$ and then $\neg \mathcal{T}$ against each of the listed theorems and, in *finitely many steps*, finds either $\mathcal{T}$ or $\neg \mathcal{T}$.

There arises the question of whether the machine, however programmed to carry out the task, will, for a given wf, ever stop. Gödel's result says in effect that if the axiomatized system $\mathcal{F}$ is consistent and deals with theorems about $\mathbb{N}$ then there is a wf for which the machine will never stop. Neither the wf nor its negation will appear on the list of derivable theorems.

There is a wf $\mathcal{S}$ and its negation $\neg \mathcal{S}$. For any N in $\mathbb{N}$, the machine, having compared both $\mathcal{S}$ and $\neg \mathcal{S}$ with each of the first N theorems in the list, will have encountered neither $\mathcal{S}$ nor $\neg \mathcal{S}$. Hence at no stage of the process will there be a decision that $\mathcal{S}$ is a theorem or that $\neg \mathcal{S}$ is a theorem: $\mathcal{S}$ is an undecidable formula: $\mathcal{F}$ is incomplete.

There are various ways for coding or numbering wfs, proofs, theorems, etc. There are various ways for coding or numbering programs for machines. Each such coding method assigns to each wf, proof, theorem, or program a natural number. Such a coding can be prepared so that each natural number is the code for some wf and each natural number is also the code for some program.

Gödel's conclusion, says that there is a wf, say numbered n, such that for any checking program, say numbered m, the machine, using program m to check wf n (and the negation of wf n) against the list of theorems, will never halt.

The flavor of his argument can be conveyed in the following way by considering an analogous problem in computer operation.

Every computer program is ultimately a finite sequence of zeros and ones. Similarly, every data-set is also a finite sequence of zeros and ones. Since there are countably infinitely many programs and countably infinitely many data-sets, the programs may be numbered $1, 2, \ldots$, and the data-sets may be numbered $1, 2, \ldots$. Some programs applied to some data-sets stop after performing finitely many steps, others never stop. For example, the simplex method applied to some PLPP cycles endlessly. It is conceivable that, for a given pair (m, n) representing a program numbered m and a data-set numbered n, one can determine, say via some TESTPROGRAM *whether* program m, applied to data-set n, halts or fails to halt. In other words:

> Confronted with any pair (m, n), TESTPROGRAM processes the pair and reports EITHER that program m applied to data-set n stops after finitely many steps OR that program m applied to data-set n never stops.

The next discussion shows that no such TESTPROGRAM exists.

If TESTPROGRAM exists one may assume that TESTPROGRAM calculates the value of a function $f : \mathbb{N} \times \mathbb{N} \ni (m, n) \mapsto f(m, n)$ such that:

  i. $f(m, n) = 0$ if program numbered m applied to data-set numbered n stops;
  ii. $f(m, n) = 1$ if program numbered m applied to data-set numbered n never stops.

In the list of all programs there is one, STOPGO, numbered, say m_S, and operating as follows.

 iii. Given the number n, first STOPGO calculates $f(m_S, n)$.
  iv. If $f(m_S, n) = 1$, then STOPGO prints the number 2 and stops.
   v. If $f(m_S, n) = 0$ STOPGO engages in the task of printing the sequence of markers in the binary representation of π.

Thus if TESTPROGRAM reports that STOPGO (program m_S) applied to data-set n never stops, i.e., if $f(m_S, n) = 1$, then STOPGO applied to data-set n stops. If TESTPROGRAM reports that STOPGO (program m_S again) applied to data-set n stops, i.e., if $f(m_S, n) = 0$, then STOPGO never stops.

It follows that there is no program like TESTPROGRAM that can accurately decide about *all* pairs (m, n) whether program m applied to data-set n stops.

The conclusion reached is interpreted as follows:

> There is no algorithmic, systematic technique, defined a priori, that can be used to determine for each wf $\mathcal{T}$ whether $\mathcal{T}$ or $\neg \mathcal{T}$ is derivable as a theorem.

The technicalities of rigorously formalizing the discussion above are lengthy but straightforward. Excellent sources for the details are [**Me, Rog**]. There is an illuminating discussion of these matters in [**Jo**].

Gödel's work gave rise to the study of recursion, the definition of such terms as *algorithm, effectively computable, Turing machine*, ..., and a host of related topics and concepts. Theorems of varying degrees of strength and impressiveness emerged. It is the opinion of many that the result about the nonexistence of a TESTPROGRAM typifies the field. It is viewed as the unsolvability of the *halting problem*.

The work of Church, Gödel, Herbrand, Kleene, Post, and Turing all drove to the same conclusion that consistent formal logical systems rich enough to deal with $\mathbb{N}$ are perforce incomplete in that they contain meaningful and yet undecidable wfs. Following their work many others showed the undecidability of many "natural" wfs in mathematics, e.g., the wf corresponding to the word problem in finitely presented groups (cf. **Note 1.1.5.2. 11**).

In [**Bar, BarE, Chai1, Chai2, Davi1, Davi2, Kln, Lam, Me, Ross, T, TaMR, Tur**] there is more information on the topics discussed above.

It should be noted that once undecidability surfaced, all sorts of questions were attacked. An example is Hilbert's tenth problem.

A $\mathbb{Z}$-polynomial P is, for some n in $\mathbb{N}$ and a set

$$\{a_{i_1,\ldots,i_n}\}_{i_1,\ldots,i_n \leq M}$$

consisting of n-tuples of integers, the map

$$P : \mathbb{Z}^n \ni (x_1,\ldots,x_n) \mapsto \sum_{i_1,\ldots,i_n} a_{i_1,\ldots,i_n} x_1^{i_1} \cdots x_n^{i_n} \in \mathbb{Z}.$$

Let $\mathcal{D}$ be the set of all $\mathbb{Z}$-polynomials. Is there an algorithm such that for each P in $\mathcal{D}$ the algorithm determines in finitely many steps (the number of steps depending on the polynomial P) whether the *Diophantine* equation

$$P(x_1,\ldots,x_n) = 0$$

has a solution $\mathbf{s} \stackrel{\text{def}}{=} (s_1,\ldots,s_n)$ in $\mathbb{Z}^n$?

Matijasevič [**Mat1, Mat 2**] in 1970 and 1971 showed that no such algorithm exists.

Of somewhat independent interest was a fortuitous discovery by von Neumann. In the course of writing [**N3**] on operator theory, he could have used a rather general proposition about the measurability of images of analytic sets [**Kur**]. He showed that the particular image of the analytic set under consideration was measurable but he noted that the general proposition regarding the measurability of all such sets is undecidable.

One may speculate that, e.g., Fermat's theorem T is undecidable in the axiomatic framework for $\mathbb{N}$. If that is the case, then for all practical purposes, T is true, since any counterexample to the statement of T would constitute a proof of the $\neg T$ and thereby demonstrate that T *is* decidable. If T is undecidable never will there be found nonzero integers x, y, z and a natural number n greater than 2 such that $x^n + y^n = z^n$.

[**Note 5.1.4:** As recently as 1989, a new and apparently shorter proof of Gödel's undecidability theorem was offered by Boolos [**Bo**]. After examining that proof, Professor Richard Vesley at the State University of New York at Buffalo made the following observations [**V**]:

> *i.* Call an algorithm *correct* if it never lists a false theorem. A *truth omitted* by an algorithm is a true wf not listed by the algorithm.
>
> *ii.* Boolos's argument shows that if M is a correct algorithm then there is a truth omitted by M.
>
> *iii.* Gödel's original work produced an algorithm M_1 that, applied to any correct algorithm M of a restricted class of algorithms, yields a truth omitted by M. Subsequently there was produced an algorithm M_2 that, applied to *any* correct algorithm M yields a truth omitted by M.

The thrust of Vesley's comments is that Boolos's proof is *existential* and *nonconstructive*. On the other hand, Gödel's proof is *constructive* in the sense that it describes the algorithm M_1 that can be applied to any correct algorithm M and thereby exhibit a truth omitted by M. (What would M_1 applied to M_1 yield?)]

5.2. Set Theory

Closely related to the problem of formalizing logic is the problem of axiomatizing set theory. Current thinking has settled on the *Zermelo-Fraenkel (ZF)* formulation of the basic axioms for a theory of sets [**Me**]. These axioms, related to a more general system *NBG* proposed by von Neumann and modified in stages by Bernays and Gödel, involve objects called *classes* and only one predicate, symbolized by $\in$, intended to suggest "membership." Among the classes there are *sets* distinguished as follows: a class X is a set iff there is a class Y such that $X \in Y$. Customarily, sets are denoted by lower case letters, classes by capital letters.

Every set is a class but not every class is necessarily a set. In terms of $\in$ there are defined relations $\subset$ (*inclusion*), $\subsetneq$ (*proper inclusion*), and $=$ (*equality*) among sets. The axioms provide for an empty set $\emptyset$, for subsets

y of set x, for the power class 2^X of any class X, for the Cartesian product $X \times Y$ of two classes X and Y, etc. Ordinal numbers are defined without reference to the Axiom of Choice or any of its logical equivalents, the Axiom of Zermelo, i.e., the Well-ordering Principle, etc.

[**Note 5.2.1:**   The Gödel-Löwenheim-Skolem theorem as it bears on the uncountability of $\mathbb{R}$ may be viewed as follows.

To say that $\mathbb{R}$ is uncountable is to say that there is no surjection $f : \mathbb{N} \mapsto \mathbb{R}$. A surjection, like any map, may be regarded as the graph of the map in $D \times D$. To say that f does not exist is to say that the graph of f does not exist, i.e., that there is in $D \times D$ no *set*, as distinguished from a *class*, that can serve as the graph of the surjection in question. In the countable model for analysis, the countable set $\mathbb{R}$ is not the surjective image of $\mathbb{N}$, hence in the countable model for analysis, the countable set representing $\mathbb{R}$ is not countable in the language of the model.]

Once these axioms are accepted, extensions are considered, so that, e.g., the *Axiom of Choice* C, the *Continuum Hypothesis* CH, the *Generalized Continuum Hypothesis* GCH may be added to the axioms of ZF. The corresponding axiom systems are ZFC, ZFCCH, ZFCGCH, etc.

In 1940 Gödel **[Gö3]** showed that if ZF is consistent then the three extensions cited are also consistent.

Finally, in 1963 Cohen **[Coh1, Coh2, Coh3]** showed that C, CH, and GCH are independent of ZF, in other words, C, CH, and GCH are undecidable propositions in ZF, i.e., ZF with any one or more of C, CH, and GCH adjoined is just as consistent as ZF with any one or more of ¬C, ¬CH, and ¬GCH adjoined.

Cohen invented a new technique, *forcing*, whereby, starting with a consistent model of ZF, he replaced the model (via forcing) by larger consistent models in which various consistent combinations of C, ¬C, CH, ¬CH, GCH, and ¬GCH obtain. Excellent references for this topic are **[Bar, Coh3, Je]**.

Following upon Cohen's accomplishment, Solovay **[Sol]** showed that ZF may be extended in another consistent way by adding the following axiom:

AXIOM OF SOLOVAY. EVERY FUNCTION $f : \mathbb{R}^n \mapsto \mathbb{R}$ IS LEBESGUE MEASURABLE.

Since the Axiom of Choice implies the negation of the Axiom of Solovay it follows that, although both ZFC and ZFS are consistent, they are mutually incompatible axiom systems.

If ZFS replaces ZFC as a basis for set theory there arises the following situation.

Let the topology of a topological vector space V be defined by a *separating* and *filtering* set $P \stackrel{\text{def}}{=} \{p_\lambda\}_{\lambda \in \Lambda}$ of *seminorms*, i.e., V is a *separated locally convex vector space*, LCV. Define such a topological vector space V

as *good* if every seminorm π defined on V is continuous in the sense that there is a constant C_π such that for each p_λ,

$$\mathbf{x} \in V \Rightarrow \pi(\mathbf{x}) \leq C_\pi p_\lambda(\mathbf{x}).$$

Most of the familiar locally convex (topological) vector spaces are good.

Garnir showed that if ZFS is used instead of ZFC then every linear map $T : V \mapsto W$ of a good space V into a locally convex vector space W is continuous [**Gar**].

BIBLIOGRAPHY

Ad — Adian, S. I., *The Burnside problem and identities in groups*, Ergebnisse der Mathematik und ihre Grenzgebiete, **95**, Springer-Verlag, New York, 1979.

Al — Alexander, J. W., *An example of a simply connected surface bounding a region which is not simply connected*, Proc. Nat. Acad. Sci., **10**, (1924), 8–10.

AH — Alexandrov, P., and Hopf, H., *Topologie*, Springer, Berlin, 1935.

ApH1 — Appel, K. and Haken, W., *Every planar map is four-colorable*, Bull. Amer. Math. Soc., **82**, (1976), 711–2.

ApH2 — ————, *Every planar map is four-colorable, I. Discharging*, Illinois J. of Math., **21**, (1977), 429–90.

ApH3 — ————, *Supplement to: Every map is four colorable, I. Discharging; II. Reducibility*, Illinois J. of Math., **21**, (1977), 1–251.

ApH4 — ————, *Every planar map is four-colorable*, Contemporary Mathematics, **98**, Amer. Math. Soc., Providence, 1989.

ApHK — Appel, K., Haken, W., and Koch, J., *Every planar map is four-colorable, II. Reducibility*, Illinois J. of Math., **21**, (1977), 491–567.

Arn — Arnol'd, V. I., *On functions of three variables*, Dokl. Akad. Nauk SSSR, **114**, (1957), 953–6.

Art — Artin, E., *Geometric algebra*, Interscience Publishers, New York, 1957.

Ban — Banach, S., *Théorie des opérations linéaires*, Monografje Matematyczne, Tom I, Warszawa, 1932.

BanT — Banach, S. and Tarski, A., *Sur la décomposition des ensembles de points en parties respectivement congruentes*, Fund. Math., **6**, (1924), 244–77.

Bar — Barwise, J. (editor), *Handbook of mathematical logic*, North Holland Publishing Company, New York, 1977.

BarE — Barwise, J. and Etchemendy, J., *The Liar: An essay on truth and circularity*, Oxford University Press, New York, 1987.

BBN — Baumslag, G., Boone, W. W., and Neumann, B. H., *Some unsolvable problems about elements and subgroups of groups*, Mathematica Scandinavica, **7**, (1959), 191–201.

Bea — Beale, E. M. L., *Cycling in the dual simplex algorithm*, Naval Research Logistics Quarterly, **2**, (1955), 269–76.

Ber — Berberian, S. K., *Lectures in functional analysis and operator theory*, Springer-Verlag, New York, 1974.

Bes1 — Besicovitch, A. S., *On Kakeya's problem and a similar one*, Mathematische Zeitschrift, **27**, (1928), 312–20.

Bes2 — ———, *On the definition and value of the area of surface,* Quarterly Journal of Mathematics, **16**, (1945), 86–102.

Bes3 — ———, *The Kakeya problem,* American Mathematical Monthly, **70**, 7, (August-September, 1963), 697–706.

Bi — Bieberbach, L., *Über die Koeffizienten derjenigen Potenzreihen, welche eine schlichte Abbildung des Einheitkreises vermitteln,* Sitzungberichte Preussische Akademie der Wissenschaften, (1989), 940–55.

Bl — Bland, R. G., *New finite pivoting rules for the simplex method,* Mathematics of Operations Research, **2**, (May,1977), 103–7.

Bo — Boolos, G., *A new proof of the Gödel incompleteness theorem,* Notices of the American Mathematical Society, **36**, (April, 1989), 388–90.

Boo — Boone, W. W., *The word problem,* Ann. Math., **70**, (1959), 207–65.

Bor — Borgwardt, K. H., *The average number of pivot steps required by the simplex method is polynomial,* Zeitschrift für Operations Research, Serie A-B, **26**, (1982) no. 5, A157–A177.

Bou — Bourbaki, N., *General topology,* Addison-Wesley Publishing Company, Reading, Massachusetts, 1966.

Br — Branges, L. de, *A proof of the Bieberbach conjecture,* Acta Math., **154**, (1985), 137–52.

Bri — Brieskorn, E. V., *Beispiele zur Differentialtopologie von Singularitäten,* Inventiones Mathematicae, **2**, (1926), 1–14.

Brit — Britton, J. L., *The word problem,* Ann. Math., **77**, (1963), 16–32.

Bro — Brouwer, L. E. J., *Zur Analysis Situs,* Math. Ann., **68**, (1910), 422–34.

Ca — Carleson, L., *On convergence and growth of partial sums of Fourier series,* Acta Mathematica, **116**, (1966), 135–57.

Chai1 — Chaitin, G. J., *Algorithmic information theory,* Cambridge University Press, Cambridge, 1987.

Chai2 — ———, *Information[,] randomness[,] & incompleteness,* World Scientific, Singapore, 1987.

Char — Charnes, A., *Optimality and degeneracy in linear programming,* Econometrica, **20**, (1952), 160–70.

CodL — Coddington, E. A. and Levinson, L., *Theory of ordinary differential equations,* McGraw-Hill Book Company, Inc., New York, 1955.

Coh1 — Cohen, P. J., *The independence of the continuum hypothesis,* Proc. Nat. Acad. Sci., **50**, (1963), 1143–8.

Coh2 — ———, *The independence of the continuum hypothesis,* ibid., **51**, (1964), 105–10.

Coh3 — ———, *Set theory and the continuum hypothesis,* The Benjamin Cummings Publishing Company, Inc., Reading, 1966.

Cou — Coury, J. E., *On the measure of zeros of coordinate functions,* Proc. Amer. Math. Soc., **25**, (1970), 16–20.

Cs — Császár, Á., *General topology*, Akadémiai Kiadó, Budapest, 1978.

Dan — Dantzig, G. B., *Maximization of linear functions of variables subject to linear inequalities*, cf. [**Koo**], 339–47.

Dav — Davie, A. M., *The approximation problem for Banach spaces*, Bulletin of the London Mathematical Society, **5**, (1973), 261–6.

Davi1 — Davis, M., *Computability and unsolvability*, McGraw-Hill Book Company, Inc., New York, 1958.

Davi2 — ————, *Hilbert's tenth problem is unsolvable*, Amer. Math. Monthly, **80**, (1973), 233–69.

Day — Day, M. M., *Normed linear spaces*, Third edition, Academic Press Inc. Publishers, New York, 1973.

Don — Donoghue, W. F., *Distributions and Fourier transforms*, Academic Press, New York, 1969.

Du — Dugundji, J., *Topology*, Allyn and Bacon, Boston, 1967.

Enf — Enflo, P., *A counterexample to the approximation problem in Banach spaces*, Acta Mathematica, **130**, (1973), 309–17.

Eng — Engelking, R., *Outline of general topology*, American Elsevier Publishing Company, Inc., New York, 1968.

FeT — Feit, W. and Thompson, J. G., *The solvability of groups of odd order*, Pac. J. Math., **13**, (1963), 775–1029.

FoAr — Fox, R. and Artin, E., *Some wild cells and spheres in three-dimensional space*, Ann. Math., **49**, No. 4, (1948), 979–90.

Fr — Freedman, M. H., *The topology of four-dimensional manifolds*, J. of Diff. Geom., **17**, (1982), 357–453.

Gar — Garnir, H. G., *Solovay's axiom and functional analysis*, Functional Analysis and its Applications, Lecture Notes in Mathematics, 399, Springer-Verlag, New York, 1974.

Ge1 — Gelbaum, B. R., *Expansions in Banach spaces*, Duke Mathematical Journal, **17**, (1950), 187–96.

Ge2 — ————, *A nonabsolute basis for Hilbert space*, Proc. Amer. Math. Soc., **2**, (1951), 720–1.

Ge3 — ————, *Notes on Banach spaces and bases*, An. Acad. Brasil. Ci., **30**, (1958), 29-36.

Ge4 — ————, *Free topological groups*, Proc. Amer. Math. Soc., **12**, (1961), 737–43.

Ge5 — ————, *On relatively free subsets of Lie Groups*, Proc. Amer. Math. Soc., **58**, (1976), 301–5.

Ge6 — ————, *Independence of events and of random variables*, Z. Wahrscheinlichkeitstheorie verw. Gebiete, **36**, (1976), 333–43.

Ge7 — ————, *Problems in analysis*, Springer-Verlag, New York, 1982.

Ge8 — ————, *Some theorems in probability theory*, Pac. J. Math., **118**, No. 2, (1985), 383–91.

Ge9 — ————, *Linear algebra,* Elsevier Science Publishing Company, Inc., New York, 1988.

GeO — Gelbaum, B. R. and Olmsted, J. M. H., *Counterexamples in analysis,* Holden-Day, San Francisco, 1964.

Gö1 — Gödel, K., *Die Vollständigkeit der Axiome der logischen Funktionenkalküls,* Monatshefte für Mathematik und Physik, **37**, (1930), 349–60.

Gö2 — ————, *Über formal unentscheidbare Sätze der Principia Mathematica und verwandter Systeme,* Monatshefte für Mathematik und Physik, **38**, (1931), 173–98; English translation: *On formally undecidable propositions of Principia Mathematica and related systems,* Oliver and Boyd, Edinburgh-London, 1962.

Gö3 — ————, *The consistency of the axiom of choice and of the generalized continuum-hypothesis with the axioms of set theory,* Princeton University Press, Princeton, 1940.

Goe — Goetze, E., *Continuous functions with dense set[s] of proper local extrema,* Journal of undergraduate mathematics, **16**, (1984), 29–31.

Gof — Goffman, C., *Real functions,* Rinehart & Company, Incorporated, New York, 1953.

Gor1 — Gorenstein, D., *Finite simple groups: An introduction to their classification,* Plenum Press, New York, 1982.

Gor2 — ————, *The classification of finite simple groups,* Plenum Press, New York, 1983.

Gr — Graves, L. M., *The theory of functions of real variables,* McGraw-Hill Book Company, Inc., New York, 1956.

GrooD — Groot, J. de and Dekker, T., *Free subgroups of the orthogonal group,* Compositio Math., **12**, (1954), 134–6.

Groth — Grothendieck, A., *Produits tensoriels topologiques et espaces nucléaires,* Mem. Amer. Math. Soc., **16**, 1955.

GuR — Gunning, R. and Rossi, H., *Analytic functions of several complex variables,* Prentice-Hall, Inc. Englewood Cliffs, N. J., 1965.

Hab — Haber, S., *On the nonomnipotence of regular summability,* Advances in mathematics, **28**, (1978), 231–2.

Hal — Hall, M., Jr., *The theory of groups,* The Macmillan Company, New York, 1959.

Halm — Halmos, P. R., *Measure theory,* D. van Nostrand Company, Inc., New York, 1950.

Har — Hardy, G. H., *Weierstraß's non-differentiable function,* Trans. Amer. Math. Soc., **17**, (1916), 301–25.

Hau — Hausdorff, F., *Grundzüge der Mengenlehre,* Von Veit, Leipzig, 1914.

HeSt — Hewitt, E. and Stromberg, K., *Real and abstract analysis,* Springer-Verlag New York, Inc., New York, 1965.

Hil — Hilbert, D., *Mathematical problems,* Bull. Amer. Math. Soc., **8**, (1902), 461–2.

Hi2 — ————, *Grundlagen der Geometrie,* 7. Auflage, B. G. Teubner, Leipzig, 1930.

HiB — Hilbert, D. and Bernays, P., *Grundlagen der Mathematik,* 2 vols., Verlag von Julius Springer, Berlin, 1939.

Hil — Hille, E., *Analytic function theory,* 2 vols., Ginn and Company, New York, 1962.

Hema — Hemasinha, R., I *The symmetric algebra of a Banach space;* II *Probability measures on Bergman space,* Dissertation, SUNY/Buffalo, 1983.

Hö — Hörmander, L., *Linear partial differential operators,* Third revised printing, Springer-Verlag, New York, 1969.

Hu — Hunt, R A., *On the convergence of Fourier series,* Proceedings of the Conference on Orthogonal Expansions and Their Continuous Analogues, Southern Illinois University Press, Carbondale, 1968.

J — Jacobson, N., *Basic algebra,* 2 vols., W. H. Freeman and Company, San Francisco, 1980.

Ja1 — James, R. C., *Bases and reflexivity of Banach spaces,* Ann. Math., **52**, (1950), 518–27.

Ja2 — ————, *A non-reflexive Banach space isometric with its second conjugate space,* Proc. Nat. Acad. Sci., **37**, (1951), 174–7.

Je — Jech, T. J., *Set theory,* Academic Press, New York, 1978.

Jo — Jones, J. P., *Recursive undecidability — an exposition,* Amer. Math. Monthly, **81**, (1974), 724–38.

KacSt — Kaczmarz, S. and Steinhaus, H., *Theorie der Orthogonalreihen,* Warsaw, 1935.

Kak1 — Kakutani, S., *Ueber die Metrisation der topologischen Gruppen,* Proc. Imp. Acad. Japan, **12**, (1936), 82.

Kak2 — ————, *Free topological groups and infinite direct product topological groups,* Proc. Imp. Acad. Japan, **20**, (1944), 595–98.

Kar1 — Karlin, S., *Unconditional convergence in Banach spaces,* Bull. Amer. Math. Soc., **54**, (1948), 148–52.

Kar2 — ————, *Bases in Banach spaces,* Duke Math. J., **15**, (1948), 971–85.

Karm — Karmarkar, N., *A new polynomial-time algorithm for linear programming,* Combinatorica, **4**, (4), (1984), 373–95.

Ke — Kelley, J. L., *General topology,* D. van Nostrand Company, Inc., New York, 1955; Springer-Verlag, New York, 1975.

KeS — Kemeny, J. G. and Snell, J. L., *Finite Markov chains,* D. van Nostrand Company, Inc., Princeton, 1960.

Kh — Khachiyan, L. G., *Polynomial algorithms in linear programming* (Russian), Zhurnal Vichislitel'noi Matematiki i Matematicheskoi Fiziki **20**, 1, (1980), 51–68.

KlM — Klee, V. and Minty, G. L., *How good is the simplex algorithm?*, Proceedings of the Third Symposium, UCLA, 1969, 159–175; Inequalities III, Academic Press, New York, 1972.

Kln — Kleene, S. C., *Introduction to metamathematics*, D. van Nostrand Company, Inc., Princeton, 1952.

KnKu — Knaster, B. and Kuratowski, C., *Sur les ensembles connexes*, Fund. Math. **2**, (1921), 206–55.

Kno — Knopp, K., *Infinite sequences and series*, Dover Publications, New York, 1956.

Ko1 — Kolmogorov, A. N., *Foundations of the theory of probability*, Second (English) edition, Chelsea Publishing Company, New York, 1956.

Ko2 — ————, *Grundebegriffe der Wahrscheinlichtsrechnung*, Chelsea Publishing Company, New York, 1956.

Ko3 — ————, *On the representation of continuous functions of several variables by superposition of continuous functions of one variable and addition* (Russian), Dokl. Akad. Nauk SSSR, **114**, (1957), 369–73.

Koo — Koopmans, T. C., *Activity analysis of production and allocation*, Cowles Commission for Research in Economics, Monograph 13, John Wiley & Sons, Inc., New York, 1951.

Kr — Krusemeyer, M., *Why does the Wronskian work?*, Amer. Math. Monthly, **95**, 1988, 46–9.

Kuo — Kuo, H., *Gaußian measures in Banach spaces*, Lecture Notes in Mathematics, **463**, Springer-Verlag, New York, 1975.

Kur — Kuratowski, C., *Topologie, I, II*, Quatrième édition, Hafner Publishing Company, New York, 1958.

Lam — Lambalgen, M. van, *Algorithmic information theory*, The Journal of Symbolic Logic, **54**, (1989), 1389–400.

Lan — Lang, S., *Algebra*, Addison-Wesley Publishing Company, Inc., Palo-Alto, 1965.

Law — Lawson, H. B., Jr., *The theory of gauge fields in four dimensions*, Regional Conference Series in Mathematics, Conference Board of the Mathematical Sciences, **58**, American Mathematical Society, Providence, 1983.

Le — Lewy, H., *An example of a smooth partial differential equation without solution*, Ann. Math. (2) **66**, (1957), 155–8.

Loe — Loève, M., *Probability theory*, D. van Nostrand Company, Inc., Princeton, 1955.

Loo — Loomis, L. H., *An introduction to abstract harmonic analysis*, D. van Nostrand Company, Inc., New York, 1953.

Lor — Lorentz, G. G., *Approximation of functions*, Holt, Rinehart and Winston, New York, 1966.

Löw — Löwenheim, L., *Über Möglichkeiten im Relativkalkül*, Math. Ann., **76**, (1915), 447–70.

M — Malliavin, P., *Impossibilité de la synthèse spectrale sur les groupes abéliens non compacts*, Publ. Math. Inst. Hautes Études Sci. Paris, **1**, (1959), 61–8.

Ma — Markov, A. A., *O svobodnich topologičeskich gruppach*, Izv. Akad. Nauk. SSSR. Ser. Mat., **9**, (1945), 3–64; Amer. Math. Soc. Translations Series, **1**, no. 30, (1950), 11–88.

MaS — Marshall, K. T. and Suurballe, J. W., *A note on cycling in the simplex method*, Naval Research Logistics Quarterly, **16**, (1969), 121–37.

Mat1 — Matijasevič, Ju. V., *Enumerable sets are Diophantine*, Dokl. Akad. Nauk SSSR, **191**, (1970), 279–82 (Russian); Soviet Mat. Dokl., **11**, (1970), 354–8 (English).

Mat2 — ———, *Diophantine representation of enumerable predicates*, Izvestija Akademii Nauk SSSR, Seria Matematičeskaja, **13**, (1971), 3–30; English translation: Mathematics of the USSR — Izvestija, **5**, (1971), 1–28.

Mau — Mauldin, R. D., editor, *The Scottish book, Mathematics from the Scottish Café*, Birkhäuser, Boston, 1981.

Me — Mendelson, E., *Introduction to mathematical logic*, D. van Nostrand Company, Inc., New York, 1964.

Mi — Milin, I. M., *On the coefficients of univalent functions*, Dokl. Akad. Nauk SSSR, **176**, (1967), 1015–8 (Russian); Soviet Math. Dokl., **8**, (1967), 1255–8 (English).

Miln1 — Milnor, J., *On manifolds homeomorphic to the 7-sphere*, Ann. Math., **64**, (1956), 399–405.

Miln2 — ———, *Some consequences of a theorem of Bott*, Ann. Math., **68**, (1958), 444–9.

MoH — Morse, M. and Hedlund, G. A., *Unending chess, symbolic dynamics and a problem in semigroups*, Duke Math. J., **11**, (1944), 1–7.

Mou — Moulton, F. R., *A simple non-desarguesian plane geometry*, Trans. Amer. Math. Soc., **3**, (1902), 192–5 .

Mu — Murray, F. J., *Linear transformations in Hilbert space*, Princeton University Press, Princeton, 1941.

Mur — Murty, K. G., *Linear programming*, John Wiley & Sons, New York, 1983.

My — Myerson, G., *First class functions*, Macquarie Mathematics Reports, 88-0026, September 1988.

N1 — Neumann, J. von, *Zur allgemeinen Theorie des Masses*, Fund. Math., **13**, (1929), 73–116.

N2 — ———, *Mathematische Grundlagen der Quantenmechanik*, Springer-Verlag, Berlin, 1932.

N3 — ———, *On rings of operators. Reduction theory*, Ann. Math., **50**, (1949), 401–85.

N4 — ———, *Mathematical foundations of quantum mechanics*, Princeton University Press, Princeton, 1955.

NM — Neumann, J. von and Morgenstern, O., *Theory of games and economic behavior,* Second edition, Princeton University Press, 1947.

New — Newman, M. H. A., *Elements of the topology of plane sets of points,* Cambridge University Press, Cambridge, 1939.

Nov — Novikov, P. S., *On the algorithmic unsolvability of the word problem for group theory,* Amer. Math. Soc. Translations, Series 2, **9**, 1–124.

NovA — Novikov, P. S. and Adian, S. I., *Defining relations and the word problem for free groups of odd exponent* (Russian), Izv. Akad. Nauk. SSSR, Ser. Mat., **32**, (1968), 971–9.

O1 — Olmsted, J. M. H., *Real variables,* Appleton-Century-Crofts, New York, 1956.

O2 — ————, *Advanced Calculus,* Appleton-Century-Crofts, New York, 1961.

O3 — ————, *Calculus with analytic geometry,* 2 vols., Appleton-Century-Crofts, New York, 1966.

Ox — Oxtoby, J. C., *Measure and category,* Springer-Verlag, Inc., New York, 1971.

PV — Posey, E. E. and Vaughan, J. E., *Functions with proper local maxima in each interval,* Amer. Math. Monthly, **90**, (1983), 281–2.

Rao — Rao, M. M., *Measure theory and integration,* John Wiley & Sons, New York, 1987.

Rin — Rinow, W., *Lehrbuch der Topologie,* VEB Deutscher Verlag der Wissenschaften, Berlin, 1975.

Rob — Robertson, M. S., *A remark on the odd schlicht functions,* Bull. Amer. Math. Soc., **42**, (1936), 366–70.

Robi — Robinson, R. M., *On the decomposition of spheres,* Fund. Math. **34**, (1947), 246–66.

Rog — Rogers, H., Jr., *Theory of recursive functions and effective computability,* McGraw-Hill Book Company, Inc., New York, 1967.

Ros — Rosenblatt, M., *Random processes,* Oxford University Press, New York, 1962.

Rosn — Rosenthal, P., *The remarkable theorem of Lévy and Steinitz,* Amer. Math. Monthly, **94**, No. 4, (1987), 342–51.

Ross — Rosser, J. B., *Extensions of some theorems of Gödel and Church,* Journal of Symbolic Logic, **1**, (1936), 87–91.

Rot — Rotman, J., *The theory of groups: An introduction,* Second edition, Allyn and Bacon, Boston, 1937.

Roy — Royden, H. L., *Real analysis,* Third edition, The Macmillan Company, New York, 1988.

Rub — Rubel, L. A., *A universal differential equation,* Bull. Amer. Math. Soc. **4**, (1981), 345–9.

Rud — Rudin, W., *Real and complex analysis,* Third edition, McGraw-Hill, Inc., New York, 1987.

Sc — Schaefer, H. H., *Topological vector spaces*, Springer-Verlag, New York, 1970.

Sch — Schwartz, L., *Radon measures on arbitrary topological spaces and cylindrical probabilities*, Oxford University Press, New York, 1973.

Si1 — Sierpinski, W., *Sur une propriété des séries qui ne sont pas absolument convergentes*, Bulletin International de l'Academie Polonaise des Sciences et des Lettres, Classe des Sciences Mathématiques et Naturelles, Cracovie [Cracow] **149**, (1911), 149–158.

Si2 — ———, *Sur la question de la mesurabilité de la base de M. Hamel*, Fund. Math., **1**, (1920), 105–11.

Si3 — ———, *Sur un problème concernant les ensembles mesurables superficiellement*, ibid., 112–5.

SiW — Singer, I. M. and Wermer, J., *Derivations on commutative Banach algebras*, Math. Ann., **129**, (1955), 260–4.

Sk — Skolem, T., *Logisch-kombinatorische Untersuchungen über die Erfüllbarkeit oder Beweisbarkeit mathematischer Sätze*, Skrifter Vidensk, Kristiana, **I**, (1919), 1–36.

Sm1 — Smale, S., *Generalized Poincaré's conjecture in dimension > 4*, Ann. Math., **74**, (1961), 391–466.

Sm2 — ———, *On the average speed of the simplex method of linear programming*, Technical report, Department of Mathematics, University of California, Berkeley, 1982.

Sm3 — ———, *On the average number of steps of the simplex method of linear programming*, Math. Programming, **27**, (1983), no. 3, 241–62.

Smi — Smith, K. T., *Primer of modern analysis*, Bogden & Quigley, Inc., Publishers, Tarrytown-on-Hudson, New York, 1971.

Sol — Solovay, R., *A model of set theory in which every set is Lebesgue-measurable*, Ann. Math., **92**, (1970), 1–56.

Sp — Sprecher, D. A., *On the structure of continuous functions of several variables*, Trans. Amer. Math. Soc., **115**, (1965), 340–55.

SS — Steen, L. A. and Seebach, J. A., Jr., *Counterexamples in topology*, Second edition, Springer-Verlag, New York, 1978.

St — Steinitz, E., *Bedingte konvergente Reihen und konvexe Systeme*, Jour. für Math. [Jour. für die reine und angewandte Math.], **143**, (1913), 128–75.

Sto1 — Stone, M. H., *Linear transformations in Hilbert space and their applications to analysis*, Amer. Math. Soc., New York, 1932.

Sto2 — ———, *Applications of the theory of Boolean rings to general topology*, Trans. Amer. Math. Soc., **41**, (1937), 375–481.

Sto3 — ———, *A generalized Weierstrass approximation theorem*, Studies in mathematics, Volume 1, 30–87, R. C. Buck, editor, Mathematical Association of America, Prentice-Hall, Inc., Englewood Cliffs, N. J., 1962.

Stoy — Stoyanov, J. M., *Counterexamples in probability*, John Wiley & Sons, New York, 1987.

Str — Stromberg, K., *The Banach-Tarski paradox*, Amer. Math. Monthly, **86**, (1979), 151–61.

Sz.-N — Sz.-Nagy, B., *Introduction to real functions and orthogonal expansions*, Oxford University Press, New York, 1965.

T — Tarski, A., *A decision method for elementary algebra and geometry*, Second edition, University of California Press, Berkeley, 1951.

TaMR — Tarski, A., Mostowski, A., and Robinson, R. M., *Undecidable theories*, North-Holland, Amsterdam, 1953.

Tay — Taylor, A. E., *Introduction to functional analysis*, John Wiley & Sons, Inc., New York, 1958.

To — Toeplitz, O., *Über allgemeine lineare Mittelbildungen*, Prace Matematyczne-Fizyczne, **22**, (1911), 113–9.

Tu1 — Tukey, J. W., *Convergence and uniformity in topology*, Princeton University Press, Princeton, 1940.

Tu2 — ———, *Some notes on the separation of convex sets*, Portugaliae Mathematica, **3**, (1942), 95–102.

Tur — Turing, A., *On computable numbers with an application to the Entscheidungsproblem*, Proc. Lon. Math. Soc. series 2, **42**, (1936–7), 230–65; corrections, ibid., **43**, (1937), 544–6.

V — Vesley, R., , Notices of the American Mathematical Society, **36**, (December, 1989), 1352.

Wa — Waerden, B. L. van der, *Algebra* (English translation of the Seventh edition), 2 vols., Frederick Ungar Publishing Co., New York, 1970.

Wag — Wagon, S., *The Banach-Tarski paradox*, Cambridge University Press, Cambridge, 1985.

Wi — Widder, D. V., *The Laplace transform*, Princeton University Press, Princeton, 1946.

ZA — Zukhovitskiy, S. I. and Avdeyeva, L. I., *Linear and convex programming*, W. B. Saunders Company, Philadelphia, 1966.

Zy — Zygmund, A., *Trigonometric series* 2 vols., Cambridge University Press, Cambridge, 1988.

SUPPLEMENTAL BIBLIOGRAPHY

The list below was compiled by Professor R. B. Burckel and it is with his kind permission that the items are included in this volume. The authors are grateful for his generosity and scholarship.

Bauer, W.R. and **Benner, R.H.** — *The nonexistence of a Banach space of countably infinite Hamel dimension,* Amer. Math. Monthly, **78**, (1971), 895–6.

Benedicks, M. — *On the Fourier transforms of functions supported on sets of finite Lebesgue measure,* Jour. Math., Anal. and Applications, **106**, (1985), 180–3.

Broadman, E. — *Universal covering series,* Amer. Math. Monthly **79**, (1972), 780–1.

Brown, A. — *An elementary example of a continuous singular function,* Amer. Math. Monthly, **76**, (1969), 295–7.

Bruckner, A. — *Some new simple proofs of old difficult theorems,* Real Analysis Exchange, **9**, (1984), 63–78.

Cantor, R., Eisenberg, M., and **Mandelbaum, E.M.** — *A theorem on Riemann integration,* Jour. Lon. Math. Soc., **37**, (1962), 285–6.

Cater, F.S. — *Most monotone functions are not singular,* Amer. Math. Monthly, **89**, (1982), 466–9.

———— — *Functions with prescribed local maximum points,* Rocky Mountain Jour. of Math., **15**, (1984), 215–7.

———— — *Differentiable nowhere analytic functions,* Amer. Math. Monthly, **91**, (1984), 618–24.

———— — *Equal integrals of functions,* Can. Math. Bull., **28**, (1985), 200–4.

———— — *Mappings into sets of measure zero,* Rocky Mountain Jour. of Math., **16**, (1986), 163–71.

———— — *An elementary proof of a theorem on unilateral derivatives,* Canadian Math. Bull., **29**, (1986), 341–3.

Conway, J. — *The inadequacy of sequences,* Amer. Math. Monthly, **76**, (1969), 68–9.

Darst, R. and **Goffman, C.** — *A Borel set which contains no rectangles,* Amer. Math. Monthly, **77**, (1970), 728–9.

De Guzmaán, M. — *Some paradoxical sets with applications in the geometric theory of real variables,* L'Enseignement de Math., (2), **29**, (1983), 1–14.

Donoghue, W. F. Jr. — *On the lifting property,* Proc. Amer. Math. Soc., **16**, (1965), 913–4.

Dressler, R.E. and **Kirk, R.B.** — *Non-measurable sets of reals whose measurable subsets are countable,* Israel Jour. of Math., **11**, (1972), 265–70.

Drobot, V. and **Morayne, M.** — *Continuous functions with a dense set of proper local maxima,* Amer. Math. Monthly, **92**, (1985), 209–11.

Dubuc, S. — *Courbes de von Koch et courbes d'Osgood,* C.R. Math. Rep. Acad. Sci. Canada, **5**, (1983), 173–8.

Edwards, D.A. — *On translates of L^∞-functions,* Jour. Lon. Math. Soc., **36**, (1961), 431–2.

Erdös, P. and **Stone, M.H.** — *On the sum of two Borel sets,* Proc. Amer. Math. Soc., **25**, (1970), 304–6.

Fremlin, D.H. — *Products of Radon measures: a counterexample,* Canadian Math. Bull., **19**, (1976), 285–9.

Gaudry, G.I. — *Sets of positive product measure in which every rectangle is null,* Amer. Math. Monthly, **81**, (1974), 889–90.

Gillis, J. — *Some combinatorial properties of measurable sets,* Quart. Jour. Math., **7**, (1936), 191–8.

———— *Note on a property of measurable sets,* Jour. Lon. Math. Soc., **11**, (1936), 139–41.

Goffman, C. — *A bounded derivative which is not Riemann integrable,* Amer. Math. Monthly, **84**, (1977), 205–6.

Goffman, C. and **Pedrick, G.** — *A proof of the homeomorphism of Lebesgue-Stieltjes measure with Lebesgue measure,* Proc. Amer. Math. Soc., **52**, (1975), 196–8.

Goldstein, A.S. — *A dense set in $L^1(-\infty, \infty)$,* Amer. Math. Monthly, **85**, (1978), 687–90.

Hanisch, H., Hirsch, W.M., and **Rényi, A.** — *Measure in denumerable spaces,* Amer. Math. Monthly, **76**, (1969), 494–502.

Hausdorff, F. — *Über halbstetige Funktionen und deren Verallgemeinerung,* Math. Zeit., **5**, (1919), 292–309.

Henle, J.M. — *Functions with arbitrarily small periods,* Amer. Math. Monthly, **87**, (1980), 816; **90**, (1983), 475.

Henle, J.M. and **Wagon, S.** — *A translation-invariant measure,* Amer. Math. Monthly, **90**, (1983), 62–3.

Hong, Y. and **Tong, J.** — *Decomposition of a function into measurable functions,* Amer. Math. Monthly, **90**, (1983), 573.

Jamison, R.E. — *A quick proof for a one-dimensional version of Liapounoff's theorem,* Amer. Math. Monthly, **81**, (1974), 507–8.

Johnson, B.E. — *Separate continuity and measurability,* Proc. Amer. Math. Soc., **20**, (1969), 420–2.

Johnson, G.W. — *An unsymmetric Fubini theorem,* Amer. Math. Monthly, **91**, (1984), 131–3.

Katznelson, Y. and **Stromberg, K.** — *Everywhere differentiable nowhere monotone functions,* Amer. Math. Monthly, **81**, (1974), 349–54.

Kaufman, R. and **Rickert, N.** — *An inequality concerning measures,* Bull. Amer. Math. Soc., **72**, (1966), 672–6.

Kirk, R.B. — *Sets which split families of measurable sets,* Amer. Math. Monthly, **79**, (1972), 884–6.

Knopp, K. — *Einheitliche Erzeugung und Darstellung der Kurven von Peano, Osgood und v. Koch,* Archiv der Math. u. Physik **26**, (1917), 103–14.

Leech, J. — *Filling an open set with squares of specified areas,* Amer. Math. Monthly, **87**, (1980), 755–6.

Leland, K.O. — *Finite dimensional translation invariant spaces,* Amer. Math. Monthly, **75**, (1968), 757–8.

Leonard, J.L. — *On nonmeasurable sets,* Amer. Math. Monthly, **76**, (1969), 551–2.

Lewin, J.W. — *A truly elementary approach to the bounded convergence theorem,* Amer. Math. Monthly, **93**, (1986), 395–7; **94**, (1987), 988–93.

Machara, R. — *On a connected dense proper subgroup of $\mathbb{R}^2$ whose complement is connected,* Proc. Amer. Math. Soc., **91**, (1986), 556–8.

Mattics, L.E. — *Singular monotonic functions,* Amer. Math. Monthly, **84**, (1977), 745–6.

Milcetich, J. — *Cartesian product measures,* Amer. Math. Monthly, **787**, (1971), 550–1.

Miller, A.D. and **Výborný, R.** — *Some remarks on functions with one-sided derivatives,* Amer. Math. Monthly, **93**,, (1986), 471–5.

Miller, W.A. — *Images of monotone functions,* Amer. Math. Monthly, **90**, (1983), 408–9.

Moran, W. — *Separate continuity and supports of measures,* Jour. Lon. Math. Soc., **44**, (1969), 320–4.

Mussman, D. and **Plachky, D.** — *Die Cantorsche Abbildung ist ein Borel-Isomorphismus,* Elemente der Math., **35**, (1980), 42–3.

Newman, D.J. — *Translates are always dense on the half line,* Proc. Amer. Math. Soc., **21**, (1969), 511–2.

Overdijk, D.A., **Simons, F.H.**, and **Thiemann, J.G.F.** — *A comment on unions of rings,* Indigationes Math., **41**, (1979), 439–41.

Pelling, M.J. — *Borel subsets of a product space,* Amer. Math. Monthly, **90**, (1983), 136–8.

Pettis, B.J. — *Sequence with arbitrarily slow convergence,* Amer. Math. Monthly, **68**, (1961), 302.

Randolph, J.F. — *Distances between points of the Cantor set,* Amer. Math. Monthly, **47**, (1940), 549–51.

Rao, B.V. — *Remarks on vector sums of Borel sets,* Colloq. Math., **25**, (1972), 103–4 and 64.

Rogers, C.A. — *Compact Borelian sets,* Jour. Lon. Math. Soc., **2,** (1970), 369–71.

Rosenthal, J. — *Nonmeasurable invariant sets,* Amer. Math. Monthly, **82,** (1975), 488–91.

Rubel, L.A. and **Siskakis, A.** — *A net of exponentials converging to a nonmeasurable function,* Amer. Math. Monthly, **90,** (1983), 394–6.

Rudin, W. — *An arithmetic property of Riemann sums,* Proc. Amer. Math. Soc., **15,** (1964), 321–4.

————— — *Well-distributed measurable sets,* Amer. Math. Monthly, **90,** (1983), 41–2.

Russell, A.M. — *Further comments on the variation function,* Amer. Math. Monthly, **86,** (1979), 480–2.

Šalát. T. — *Functions that are monotone on no interval,* Amer. Math. Monthly, **88,** (1981), 754–5.

Sinha, R. — *On the inclusion relations between $L^r(\mu)$ and $L^s(\mu)$,* Indian Jour. Pure and Appl. Math., **13,** (1982), 1046–8.

Stromberg, K. — *An elementary proof of Steinhaus's theorem,* Proc. Amer. Math. Soc., **36,** (1972), 308.

Takács, L. — *An increasing singular continuous function,* Amer. Math. Monthly, **85,** (1978), 35–7.

Thomas, R. — *A combinatorial construction of a nonmeasurable set,* Amer. Math. Monthly, **92,** (1985), 421–2.

Trautner, R. — *A covering principle in real analysis,* Quart. Jour. of Math., Oxford, **38,** (1987), 127–30.

Tsing, N.K. — *Infinite-dimensional Banach spaces must have uncountable basis – an elementary proof,* Amer. Math. Monthly, **91,** (1984), 505–6.

Villani, A. — *Another note on the inclusion $L^p(\mu) \subset L^q(\mu)$,* Amer. Math. Monthly, **92,** (1985), 485–7.

Walker, P.L. — *On Lebesgue integrable derivatives,* Amer. Math. Monthly, **84,** (1977), 287–8.

Walter, W. — *A counterexample in connection with Egorov's theorem,* Amer. Math. Monthly, **84,** (1977), 118–9.

Wesler, O. — *An infinite packing theorem for the sphere,* Proc. Amer. Math. Soc., **11,** (1960), 324–6.

Weston, J.D. — *A counterexample concerning Egoroff's theorem,* Jour. Lon. Math. Soc., **34,** (1959), 139–40; **35,** (1960), 366.

Wilker, J.B. — *Space curves that point almost everywhere,* Trans. Amer. Math. Soc., **250,** (1979), 263–74.

Young, R.M. — *An elementary proof of a trigonometric identity,* Amer. Math. Monthly, **86,** (1979), 296.

Zaanen, A.C. — *Continuity of measurable functions,* Amer. Math. Monthly, **93,** (1986), 128–30.

Zajiček, L. — *An elementary proof of the one-dimensional density theorem,* Amer. Math. Monthly, **86,** (1979), 297–8.

Zamfirescu, T. — *Most monotone functions are singular,* Amer. Math. Monthly, **88,** (1981), 47–8.

Zolezzi, T. — *On weak convergence in L^∞,* Indiana Univ. Math. Jour., **23,** (1974), 765–6.

The notation a.b.c. d indicates **Chapter** a, **Section** b, **Subsection** c, page d; similarly a.b. c indicates **Chapter** a, **Section** b, page c.

A 5.1. 224: the set of axioms of a formal logical system.

A° 2.1.1. 45: the interior of the set A.

$\mathbb{A}_E$ 2.3.4. 170: the set of algebraic complex numbers in the set E.

$\mathcal{A}_n$ 1.1.4. 8: the alternating group, i.e., the set of even permutations, on $\{1, \ldots, n\}$.

$\{A\}$ 1.2.2. 21: for an algebra A the associated algebra in which multiplication: $A \times A \ni (a, b) \mapsto ab$ is replaced by a new multiplication:

$$A \times A \ni (a, b) \mapsto a \circ b \overset{\text{def}}{=} ab - ba.$$

$(a_{ij})_{i,j=1}^{m,n}$ 1.3.1. 25

AA^{-1} 1.1.4. 5: for a set A in a group G, the set

$$\left\{ ab^{-1} \ : \ a, b \in A \right\}.$$

See also $X \circ Y$.

AC 2.1.2. 55, 2.1.2. 65: the set of Absolutely Continuous functions.

a.e. 1.1.4. 7: almost everywhere.

$\{A \ : \ P\}$ 1.1.4. 6: the set of all A for which P obtains.

$A(S)$ 2.2.1. 118: the (surface) area of the set S in $\mathbb{R}^3$.

B^{**} 2.3.2. 164

B_1 2.2.3. 144: the unit ball in $\mathbb{R}^n$.

BAP 2.3.1. 158: Bounded Approximation Property.

BES 2.2.1. 118: the BESicovitch sphere.

BV 2.1.2. 54: the set of functions of Bounded Variation.

$B_{(x)}$ 2.2.2. 142

$B^{(y)}$ 2.2.2. 142

$\mathcal{B}$ 2.2.1. 120: a homeomorphic image in $\mathbb{R}^3$ of B_1, the unit ball in $\mathbb{R}^3$; 2.3.1. 159: a (Schauder) basis for a Banach space; 2.3.3. 168: the category of Banach spaces and continuous homomorphisms.

B 2.3.2. 162: a biorthogonal set.

$\mathbf{c}_0$ 2.3.1. 159

$\mathbf{c}$ 2.3.2. 163

C 5.2. 229

$\mathbb{C}$ 1.1.5. 18

(C, α) 2.1.3. 76

CON 2.3.1. 156: Complete OrthoNormal Set.

Const 4.1. 212: the set of constant functions.

Cont(f) 2.1.1. 42

Conv 2.3.3. 168

C^∞ 2.1.1. 51, 2.1.2. 63

C ($\mathbb{T}, \mathbb{R}$) 2.3.3. 165

$C_b^{(k)}$ ($\mathbb{R}, \mathbb{C}$) 2.4.1. 172

C ($[0, 1], \mathbb{R}$) 2.1.1.0 51

C_0 1.1.4. 7: the Cantor set.

C_0 ($X, \mathbb{R}$) 2.3.2. 163

C_α 2.2.1. 115

C_p ($\mathbb{R}, \mathbb{R}$) 2.3.3. 165

$\mathcal{C}_0$ 2.1.2. 55

D 3.2.2. 200: the discrete topology.

$\mathcal{D}$ 1.3.1. 26: the set of diagonable matrices.

$\overset{\text{def}}{=}$ 1.1.2. 3: "(is) defined to be."

$\deg(P)$ 2.1.3. 94: the degree of polynomial P.

det 1.3.1. 26, 2.5.1. 177: determinant.

$\text{diam}(E)$: the diameter of the set E in a metric space (X, d).

Diff(f) 2.1.1. 42

Discont(f) 2.1.1. 42, 2.1.1. 49, 2.1.2. 64

$D(a, r)$ 2.6.1. 180

$D(a, r)^\circ$ 2.6.1. 180: $\{ z \ : \ z \in \mathbb{C}, \ |z - a| < r \}$.

$\mathcal{D}^{\mathbf{M}}$ 2.2.1. 112

$\mathcal{F}$ 2.3.1. 157: the Franklin system of orthogonal functions in C ($[0, 1], \mathbb{C}$);
    5.1. 223: a formal system of logic.

F 2.2.1. 105: the set of all closed subsets of a topological space.

f^{n*} 2.4.2. 176: the n-fold convolution of the function f with itself.

$f_{(x)}$ 2.2.2. 140

$f^{(y)}$ 2.2.2. 140

F_N 2.1.3. 74: Fejér's kernel.

F_σ 2.1.1. 43: the union of a countable set of closed sets.

$\mathcal{G}$ 1.1.4. 8: the category of groups.

GCD 1.2.3. 23: Greatest Common Divisor.

G_δ 2.1.1. 43: the intersection of a countable set of open sets.

$G : H$ 1.1.2. 3: the index of the subgroup H in the group G.

$\mathbb{H}$ 1.1.5. 13: the set of quaternions.

HEMIBES 2.2.1. 120: the BESicovitch HEMIsphere.

$H(\Omega)$ 2.6.1. 180: the set of functions holomorphic in the region Ω ($\subset \mathbb{C}$).

$\text{Hom}(A, B)$ 2.4.1. 172: the set of Homomorphisms: $h : A \mapsto B$.

$\mathcal{I}$ 1.3.2. 36: $[\mathcal{H}]_{inv}$

id 2.1.1. 52, 2.2.2. 143: the identity map.

$\mathbb{I}$ 2.1.3. 68: the set of Irrational (complex) numbers.

$\mathbb{I}_E$ 2.3.4. 170: the set of Irrational numbers in the subset E of $\mathbb{C}$.

iff 1.1.2. 3: if and only if.

im 1.1.3. 4

$\mathrm{Ind}(S)$ 4.1. 212: for a set S of random variables, the set of random variables f such that $S \cup f$ is independent.

$\mathbb{K}$ 1.1.4. 5, 2.2.1. 105: the set of compact subsets of a topological space.

$\mathbb{K}$ 1.2.2. 22: the generic notation for a field (German Körper).

$\mathbb{K}[x]$ 1.3.2. 31: for a field $\mathbb{K}$ (or, more generally, a ring R) and an "indeterminate" x, the set of all polynomials of the form

$$\sum_{m=0}^{n} a_m x^m, \ n \in \mathbb{N}, \ a_m \in \mathbb{K} \ (\text{or } a_m \in R).$$

Example 1. $\mathbb{C}[z]$ is the vector space of all polynomials in the (complex) variable z and with complex coefficients. **Example 2.** $\mathbb{Z}[x]$ is the ring consisting of all polynomials in the indeterminate x and with coefficients in $\mathbb{Z}$.

ker1.1.3. 4

$\mathcal{L}$ 1.2.2. 21: Lie algebra; 2.2.1. 104: the set of all Lebesgue measurable subsets of $\mathbb{R}^n$.

$\mathcal{LCG}$ 1.1.4. 9: the category of locally compact (topological) groups.

LCV 5.2. 229: locally convex vector space.

Lip α 2.5.2. 178: generalized Lipschitz condition: for positive α, f is in *Lip* α at a iff for some positive K (the Lipschitz constant) and some positive δ, $|x - a| < \delta \Rightarrow |f(x) - f(a)| \leq K|x - a|^\alpha$.

l^p 2.3.1. 159

$L^1(G)$ 2.4.2. 175

$L^p(X, \mathbb{C})$, $1 \leq p$ 2.3.4. 169

$L^2_{\mathbb{R}}(\Omega, \Pi)$ 4.2.1. 220: the set of $\mathbb{R}$-valued random variables defined on Ω and square integrable with respect to Π.

$\ell(\gamma)$ 2.2.1. 114, 2.2.1. 123: of the arc $\gamma : [0,1] \ni t \mapsto \mathbb{R}^n$, its length

$$\sup_{0=t_0 < \cdots < t_n=1} \sum_{i=1}^{n} \|\gamma(t_i) - \gamma(t_{i-1})\|, \ m \in \mathbb{N}.$$

$L(\gamma^*)$ 2.2.1. 114: of the arc-image γ^*, the infimum of the lengths $\ell(\gamma \circ s)$ for all parametric descriptions s.

l^∞ 2.3.2. 164

lim inf 2.1.3. 67: of a sequence $\{a_n\}_{n \in \mathbb{N}}$ of real numbers,

$$\inf_{n \in \mathbb{N}} \sup \{ a_m \ : \ m \geq n \};$$

of a sequence $\{A_n\}_{n \in \mathbb{N}}$ of sets,

$$\bigcup_{n=1}^{\infty} \left(\bigcap_{m=n}^{\infty} A_m \right),$$

i.e., the set of points belonging to all but finitely many A_n.

lim sup 2.1.3. 67: of a sequence $\{a_n\}_{n\in\mathbb{N}}$ of real numbers,

$$\sup_{n\in\mathbb{N}} \inf \{\, a_m \;:\; m \geq n \,\};$$

of a sequence $\{A_n\}_{n\in\mathbb{N}}$ of sets,

$$\bigcap_{n=1}^{\infty} \left(\bigcup_{m=n}^{\infty} A_m \right),$$

i.e., the set of points belonging to infinitely many A_n.

m 2.3.1. 159: the set of all bounded sequences of complex numbers.

m_A 1.3.1. 26

$m|n$ 1.1.2. 4: the integer n is an integral multiple of the integer m.

(m,n) 2.1.1. 48: the Greatest Common Divisor (GCD) of the integers m, n.

$M(m,n)$ 1.3.3. 38

$m \bmod k$ 2.1.3. 80: for k and m in $\mathbb{N}$, in $[0, k-1]$ the unique natural number r such that for some natural number q, $m = qk + r$.

Mat_{mn} 1.3.1. 29: the set of $m \times n$ matrices with entries from a field $\mathbb{K}$; more generally, $\mathrm{Mat}_{\Gamma\Lambda}$ denotes the set of all $\mathbb{K}$-valued functions on $\Gamma \times \Lambda$.

$\mathbb{N}$ 1.1.2. 3: the set of natural numbers.

NBG 5.2. 228: the axiom system proposed by von Neumann, Bernays and Gödel for set theory.

$\mathcal{N}$ 2.4.1. 173: the set of generalized nilpotents in a Banach algebra.

$\mathcal{ND}$ 2.3.3. 165: in $\mathbb{R}^{\mathbb{R}}$, the set of nowhere differentiable functions.

Nondiff(f) 2.1.1. 42, 2.1.2. 56

o 4.2. 219: Landau's "little o" notation: for some a,

$$f(t) = o(g(t)) \Leftrightarrow \lim_{t\to a} \frac{f(t)}{g(t)} = 0.$$

O 1.2.3. 25: in a vector space, the zero vector: $\mathbf{x} + \mathbf{O} = \mathbf{x}$.

$\mathcal{O}$ 2.2.1. 105: the set of open sets in a topological space X.

$|P|$ 2.1.2. 65: for a partition P of an interval in $\mathbb{R}$, the greatest of the lengths of the (finitely many) disjoint subintervals constituting P.

$\mathcal{P}(E)$ 2.2.1. 105: the power set of E, i.e., the set of all subsets of the set E.

$\mathcal{P}^n$ 2.2.1. 114: the parallelotope $[0,1]^n$.

PLPP 1.3.3. 37: the **P**rimal **L**inear **P**rogramming **P**roblem.

$\overline{q}(m,n)$ 1.3.3. 38

$\mathcal{Q}$ 1.1.5. 15: for a topological space X, in $(\mathbb{H}^*)^X$ the multiplicative group of functions f such that both f and $\frac{1}{f}$ are bounded and continuous; 1.3.3. 38: a (convex) polyhedron that is the intersection of a finite set of half-spaces in $\mathbb{R}^n$.

$\mathbb{Q}$ 1.1.4. 6: the set of rational numbers.

Q' 2.1.3. 67: the set of limit points of the set Q.

QL 1.1.4. 8: quotient lifting.

$\mathcal{R}$ 1.1.5. 16: for a topological space X, in $\left(\mathbb{R}^+\right)^X$ the set of functions f such that both f and $\frac{1}{f}$ are continuous and bounded; 2.2.3. 147: for a relation R in $N \overset{\text{def}}{=} \{1,\ldots,n\}$, the subset $\{\,(i,j)\ :\ iRj\,\}$ of $N \times N$; 2.4.1. 172: the radical of a (Banach) algebra.

R 5.1. 224: the set of rules of inference of a formal logical system.

$\mathbb{R}$ 1.1.4. 6: the set of real numbers.

$\mathbb{R}^+$ 1.1.5. 15: the multiplicative group of positive real numbers.

$\mathbb{R}^{(n,+)}$ 1.3.3. 37: the nonnegative orthant

$$\{\,(x_1,\ldots,x_n)\ :\ x_i \geq 0,\ 1 \leq i \leq n\,\}$$

of $\mathbb{R}^n$.

sgn 2.1.2.64: the signum function: $z \in \mathbb{C} \Rightarrow \overline{\mathrm{sgn}(z)}z = |z|$.

span(S) 2.3.3. 168: of a subset S in a vector space, the set of all (finite) linear combinations of vectors in S.

S_1 2.2.3. 144: in $\mathbb{R}^n$ the set $\{\,\mathbf{x}\ :\ \|\mathbf{x}\| = 1\,\}$.

S^n 3.3.1. 208: a homeomorphic image of S_1 in $\mathbb{R}^{n+1}$, q.v.; whereas S_R is used when the metric properties of the surface of a sphere are discussed, S^n is used when the topological properties of the surface of a sphere are studied.

$\mathcal{S}$ 2.2.1. 103: a σ-ring of (measurable) sets.

S_n 1.1.2. 3, 1.1.4. 8: the (symmetric) group of all permutations of $\{1,\ldots,n\}$.

$\mathcal{SD}$ 2.3.3. 166: in $\mathbb{R}^{\mathbb{R}}$, the set of somewhere differentiable functions.

$\mathcal{S}(\mathsf{A})$ 1.1.4. 5: the intersection of the set of all σ-rings containing the set A of sets, i.e., the σ-ring generated by the set A of sets.

$S(A)$ 2.1.3. 69: for a sequence $A \overset{\text{def}}{=} \{a_n\}_{n\in\mathbb{N}}$, the formal sum, i.e., without regard to convergence, $\sum_{n=1}^{\infty} a_n$.

$\mathbf{S}_\beta(A)$ 2.1.3. 69: for a subset β of $\mathbb{N}$, and a sequence $A \overset{\text{def}}{=} \{\mathbf{a}_n\}_{n\in\mathbb{N}}$ the formal sum $\sum_{n\in\beta} \mathbf{a}_n$.

$S_\pi(A)$ 2.1.3. 69: for a sequence $A \overset{\text{def}}{=} \{a_n\}_{n\in\mathbb{N}}$ and a permutation π in the set of Π of all permutations of $\mathbb{N}$, the formal sum $\sum_{n=1}^{\infty} a_{\pi(n)}$.

$S_{|\;|}(A)$ 2.1.3. 69: for a sequence $A \overset{\text{def}}{=} \{a_n\}_{n\in\mathbb{N}}$, the formal sum

$$\sum_{n=1}^{\infty} |a_n|\ (\leq \infty).$$

$SO(n)$ 2.2.3. 144: over $\mathbb{R}$, the set of all $n \times n$ orthogonal matrices M such that $\det(M) = 1$.

$S_R(f)$ 2.6.2. 181: on $\mathbb{T}_R$, the set of singularities of the function f.

supp(f): the support of the function f.

$\mathbb{T}$ 1.1.5. 18, 2.3.3. 165: the set of complex numbers of absolute value one; equivalently, the set $\mathbb{R}/\mathbb{Z}$.

$\mathbb{T}_R$: the set of complex numbers of absolute value R.

T^{-1} 1.3.1. 26: the inverse (if it exists) of the map T.

T_f 2.1.2. 65: the total variation of the function f.

T_{fP} 2.1.2. 65: for a function f in $\mathbb{R}^I$ and a partition P of the interval I, the total variation of f with respect to the partition P.

$\mathsf{Tr}_{\mathbb{R}}$ 2.3.4. 170: the set of Transcendental Real numbers.

U 1.2.3. 24: a uniform structure.

var 2.3.1. 161: the total variation of a function.

V^* 1.3.2. 34: the dual space of the vector space V.

$[V]$ 1.3.1. 25: the set of endomorphisms of the vector space V.

$[V]_{inv}$ 1.3.1. 26: the set of invertible elements in $[V]$.

$[V]_{sing}$ 1.3.1. 26: the set of singular (noninvertible) elements in $[V]$.

$[V, W]$ 1.3.1. 29: the set of morphisms (linear maps) of the vector space V into the vector space W.

wf 5.1. 224: well-formed formula.

WF 5.1. 223: the set of well-formed formulae.

$(\mathbf{x}, \mathbf{y})$ 1.3.2. 33: the scalar product of the vectors $\mathbf{x}$ and $\mathbf{y}$.

$\mathbf{x} \succ \mathbf{y}$ 1.3.3. 37

$\mathbf{x} \succeq \mathbf{y}$ 1.3.3. 37

x^{-1} 1.1.1. 1: the inverse of x.

(X, d) 2.1.1. 43: a metric space X with metric d.

$(X, \mathcal{S}, \mu)$ 2.2.1. 103: the measure situation consisting of the set X, the σ-ring $\mathcal{S}$ of (some) subsets of X, and the (countably additive) measure $\mu : \mathcal{S} \mapsto [0, \infty]$.

$X \circ Y$: for two sets X and Y in a set S where there is defined a map

$$S \times S \ni \{x, y\} \mapsto x \circ y \in S,$$

the set $\{\, x \circ y \; : \; x \in X, y \in Y \,\}$. **Example 1.** If S is a group, if $x \in S$, if $T \subset S$, and if $a \circ b \overset{\text{def}}{=} ab$ then

$$xT \overset{\text{def}}{=} \{\, xt \; : \; t \in T \,\}, \;\; Tx \overset{\text{def}}{=} \{\, tx \; : \; t \in T \,\}.$$

Example 2. If C_0 is the Cantor set in $\mathbb{R}$ then

$$C_0 + C_0 \overset{\text{def}}{=} \{\, x + y \; : \; x, y \in C_0 \,\} \; (= [0, 2]).$$

X^Y 1.1.4. 6: the set of all maps of the set Y into the set X.

$X \setminus Y$ 1.1.2. 4, 2.1.1. 44: $\{\, x \; : \; x \in X, \; x \notin Y \,\}$.

$\mathbb{Z}$ 1.1.5. 12: the set of integers.

$\mathbb{Z}_2, \mathbb{Z}/2$ 1.3.1. 28: the finite field $\{0, 1\}$.

ZF 5.2. 228: the Zermelo-Fraenkel formulation of the axioms of set theory.

ZFC 5.2. 229: ZF together with the Axiom of Choice.

ZFCCH 5.2. 229: ZFC together with the **C**ontinuum **H**ypothesis.

ZFCGCH 5.2. 229: ZFC together with the **G**eneralized **C**ontinuum **H**ypothesis.

ZFS 5.2. 229

$\hookrightarrow$ 1.1.4. 8, 2.2.2. 138: the inclusion map.

$\overset{\phi}{\mapsto}$ 1.1.3. 4: the morphism ϕ maps to.

γ 2.2.1. 113

γ^* 2.2.1. 113: for an arc $\gamma : [0,1] \ni t \mapsto \gamma(t) \in \mathbb{R}^n$ the set (image) $\gamma([0,1])$.

$\Gamma(A)$ 2.1.3. 69: for a sequence $A \overset{\text{def}}{=} \{a_n\}_{n \in \mathbb{N}}$ and the set Φ of all finite subsets ϕ of $\mathbb{N}$ the set $\{ S_\phi(A) \ : \ \phi \in \Phi \}$.

$\delta_{\lambda\mu}$ 1.3.2. 33: Kronecker's "delta function," i.e.,

$$\delta_{\lambda\mu} = \begin{cases} 1 & \text{if } \lambda = \mu \\ 0 & \text{otherwise.} \end{cases}$$

λ_n 2.2.1. 104: Lebesgue measure in $\mathbb{R}^n$.

λ_*, λ^* 2.2.1. 107: Lebesgue inner resp. Lebesgue outer measure in $\mathbb{R}^n$.

ρ_*, ρ^* 2.2.2. 138

$\sigma_T(A)$ $\left(\overset{\text{def}}{=} \lim_{m \to \infty} \sigma_{m,T}(A) \right)$ 2.1.3. 75

$\sigma_{m,T}(A)$ $\left(\overset{\text{def}}{=} \sum_{n=1}^\infty t_{mn} s_n(A) \right)$ 2.1.3. 75

$\sigma(T)$ 1.3.2. 31: the spectrum of the morphism T.

$\nu \ll \mu$ 2.2.2. 137: the measure ν is absolutely continuous with respect to the measure μ, i.e., $\mu(A) = 0 \Rightarrow \nu(A) = 0$.

Π 2.1.3. 69: the set of all permutations of $\mathbb{N}$.

ϕ 2.1.3. 69: a finite subset of $\mathbb{N}$.

Φ 2.1.3. 69: the set of all finite subsets of $\mathbb{N}$.

χ_A 1.1.4. 5: the characteristic function of the set A:

$$\chi_A(x) = \begin{cases} 1 & \text{if } x \in A \\ 0 & \text{otherwise.} \end{cases}$$

$\omega(f, x_0, \epsilon)$ 2.3.1. 161: the ϵ-modulus of continuity of f at x_0.

$\omega(f, \epsilon)$ 2.3.1. 161: the ϵ-uniform modulus of continuity of f.

$\#(S)$ 1.1.2. 2: the cardinality of the set S.

∂A 2.2.1. 121, 2.6.1. 180: the boundary of the set A in a topological space.

$\Rightarrow$ 1.1.1. 1: "implies."

$\mapsto$ 1.1.1. 1: "maps to."

$\Leftrightarrow$ 1.1.2. 4: "if and only if" ("iff").

$\in$ 1.1.2. 4: "is a member of."

$\subset$ 1.1.3. 4: "is contained in" ("is a subset of").

$\emptyset$ 1.1.2. 4: the empty set.

$\to$ 1.1.4. 6: "approaches," "converges (to);" 1.1.5. 17: "maps to" (in (commutative) diagrams); 5.1. 223: "implies" (in formal logic).

$\supset$ 1.1.4. 8: "contains."

$\cong$ 1.1.5. 10: in group theory, "is isomorphic to," 2.2.3. 145, 3.1.1. 186: in Euclidean space, "is congruent to," 3.3. 209: in topology, "is homotopic to."

$\subsetneqq$ 1.1.5. 10: "is a proper subset of."

$\downarrow$ 1.1.5. 17: "maps to" (in (commutative) diagrams); 2.1.1. 49: "approaches from above," "decreases monotonely (to)."

$\uparrow$ 2.1.1. 49: "approaches from below," "increases monotonely (to)."

$\overline{S}$ 1.1.5. 18: in topological contexts, the (topological) closure of the set S.

$\circ$ 1.2.2. 21: (binary operation); 2.1.1.42: (composition of functions).

$A \dot\cup B$ 1.2.3. 22: the union of the two *disjoint* sets A and B.

$\| \cdots \|$ 1.3.1. 26: the norm of the vector $\cdots$; 2.4.1. 172: the norm of the linear map $\cdots$.

$\|f\|_\infty$ 2.1.2. 64: for a measure situation $(X, \mathcal{S}, \mu)$ and a measurable function f in $\mathbb{C}^X$,

$$\inf \{ M \ : \ \mu (\{ x \ : \ |f(x)| \geq M \}) = 0 \} \ (\leq \infty).$$

$\|f\|_p$ 2.3.1. 159: for a measure situation $(X, \mathcal{S}, \mu)$, p in $\mathbb{R} \setminus \{0\}$, and a measurable function f in $\mathbb{C}^X$,

$$\left(\int_X |f(x)|^p \, d\mu \right)^{\frac{1}{p}} \ (\leq \infty).$$

$\succ$ 1.3.3. 37: partial order (strictly greater than).

$\succeq$ 1.3.3. 37, 2.3.4. 169: partial order (greater than or equal to).

$\overset{u}{\to}$ 2.1.2. 53: "converges uniformly (to)."

$\overset{a.e.}{\to}$ 2.1.3. 86, 2.2.2. 134: "converges almost everywhere (to)."

∇ 2.2.1. 122

$\overset{dom}{\to}$ 2.2.2. 135: "converges dominatedly (to)."

$\overset{meas}{\to}$ 2.2.2. 134: "converges in measure (to)."

$\overset{\| \ \|_p}{\to}$ 2.2.2. 134: "converges in p-norm (to)."

$\vee$ 2.3.4. 169: supremum (of a pair); 5.1. 223: logical "or."

$\wedge$ 2.3.4. 169: infimum (of a pair); 5.1. 223: logical "and."

$\neg$ 5.1. 223: logical "not."

$\forall$ 5.1. 223: logical "for all."

$\exists$ 5.1. 223: logical "there exist(s)."

$\dot\bigcup$ 1.1.2. 4, 2.1.2. 63, 2.2.1. 149: used instead of $\bigcup$ to signify the union of a set of pairwise *disjoint* sets.

The notation a.b.c. d indicates **Chapter** a, **Section** b, **Subsection** c, page d; similarly a.b. c indicates **Chapter** a, **Section** b, page c.

Abel, N. H. 2.1.3. 76
Abel summable 2.1.3. 76
Abel summation 2.1.3. 76
abelian 1.1.4. 8: of a group G, that the group operation is commutative.
abelianization 1.1.5. 15: for a group G and its commutator subgroup $Q(G)$, the (abelian) quotient group $G/Q(G)$.
absolutely continuous 2.1.2. 55, 2.1.2. 65, 2.1.3. 87: of a function f in $\mathbb{C}^{\mathbb{R}}$, that f' exists a.e. on $[a, b]$ and that for x in $[a, b]$,

$$f(x) = f(a) + \int_a^x f'(t)\, dt$$

cf. **Exercise 2.1.2.15. 65**; 2.2.2. 137: of a measure ν with respect to a measure μ, that every null set (μ) is also a null set (ν).
— convergent 2.1.3. 69: of a series $\sum_{n=1}^{\infty} a_n$, that $\sum_{n=1}^{\infty} |a_n| < \infty$.
ADIAN, S. I. 1.1.5. 11
adjacent 1.1.5. 10: in the context of free groups, of two words w_1 and w_2, that there is an x such that for some words u and v, $w_1 = ux^\epsilon x^{-\epsilon}v$ and $w_2 = uv$.
adjoint 1.3.1. 25: of a matrix $(a_{ij})_{i,j=1}^{m,n}$, the matrix $(b_{ij})_{i,j=1}^{n,m}$ in which $b_{ij} = \overline{a_{ji}}$; 1.3.2 34: of a linear transformation $T : V \mapsto W$ between vector spaces, the linear transformation $T^* : W^* \mapsto V^*$ between their duals and satisfying $\mathbf{w}^*\,(T\mathbf{v}) = T^*\mathbf{w}^*(\mathbf{v})$.
ALEXANDER, J. W. 3.2.2. 206
Alexander's horned sphere 3.2.2. 206
algebra 1.2.2. 21, 2.3.4. 169, 2.4.1. 172: a ring R that is a vector space over a field $\mathbb{K}$ and such that if $a \in \mathbb{K}$ and $\mathbf{x}, \mathbf{y} \in R$ then

$$a(\mathbf{xy}) = (a\mathbf{x})\mathbf{y} = \mathbf{x}(a\mathbf{y}).$$

algorithm 5.1. 227: a (computer) program for mapping $\mathbb{Z}$ into itself. (The preceding definition is a colloquial version of Church's thesis.)
almost every section, point, etc. 2.2.1. 110: every section outside a set of sections indexed by a null set, every point outside a null set, etc.
almost everywhere (a.e.): in the context of a measure situation, "except on a set of measure zero (a null set)."

alternating group 1.1.4. 8, 1.1.6. 19: the group $\mathcal{A}_n$ of even permutations of
    the set $\{1, 2, \ldots, n\}$.
— series theorem 2.1.3. 79: If $a_n \in \mathbb{R}$, $a_n = (-1)^n |a_n|$, and $|a_n| \downarrow 0$ then
    $\sum_{n=1}^{\infty} a_n$ converges.
alternative (division) algebra 1.2.2. 22: an algebra in which multiplication
    is neither necessarily commutative nor necessarily associative.
analytic continuation 2.6.2. 180: for a region Ω^* properly containing a region
    Ω in which a function f (in $\mathbb{C}^{\mathbb{C}}$) is analytic, the process of defining,a
    function f^* analytic in Ω^* and equal to f in Ω.
— function 1.3.1. 26, 2.1.1. 51: a function f in $\mathbb{C}^{\mathbb{C}}$ and such that f' exists
    (in some region Ω).
— set 5.1. 228: the continuous image of a Borel set.
antiderivative of order k 2.1.2. 62: for a function f, a function F such that
    $F^{(k)} = f$.
ANTOINE, L. 3.2.2. 207
Antoine's necklace 3.2.2. 207
APPEL, K. 3.1.2. 197
arc 2.2.1. 113: a continuous map of $\gamma : [0, 1] \mapsto X$ of $[0, 1]$ into a topological
    space X.
ARCHIMEDES 1.2.3. 23
Archimedean 1.2.3. 23: of an ordered field $\mathbb{K}$ that if $p, q \in \mathbb{K}$ and $0 < p < q$
    then, for some n in $\mathbb{N}$, $q < np$.
arc-image 2.2.1. 113: the range of an arc.
arcwise connected 1.3.2. 37: of a set S in a topological space X, that any
    two points of S are the endpoints of an arc in S.
area 2.1.2. 58: of a subset S of $\mathbb{R}^2$, the value of the Riemann (or, more
    generally, the Lebesgue) integral

$$\int \int_{\mathbb{R}^2} \chi_S(x, y) \, d\lambda_2(x, y)$$

(if it exists). The "problem of 'surface' area" for (images of) maps from
    $\mathbb{R}^m$ to $\mathbb{R}^n$, especially when $m < n$, is difficult. One of the difficulties,
    when $2 = m < n = 3$, is discernible from a reading of the discussion
    of **Example 2.2.1.11. 118, Remark 2.2.1.4. 121, Note 2.2.1.4.
    121,** and **Exercise 2.2.1.12. 123**. For extensive discussions of the
    topic cf. **[O1, O2, Smi]**.
ARTIN, E. 3.2.2. 206
ARZELÀ, C. 2.3.1 162
ASCOLI, G. 2.3.1. 162
Ascoli-Arzelà theorem 2.3.1. 162: A uniformly bounded set of equicontinu-
    ous $\mathbb{R}$-valued functions defined on a compact metric space contains a
    uniformly convergent subsequence.
associativity 1.1.1. 1: of a binary operation, that always $a(bc) = (ab)c$.
auteomorphism 3.1.2. 196: a homeomorphism of a topological space onto
    itself.

autojection 2.2.1. 114: a bijection of a set onto itself.

automorphism 1.2.2. 22: a bijective endomorphism (whence an "autojective" endomorphism).

average 2.1.3. 74: of a finite set $\{s_1, \ldots, s_n\}$ of numbers, the number

$$\frac{s_1 + \cdots + s_n}{n}.$$

axiom 1.1.1. 1, 5.1. 224

Axiom of Choice 1.1.4. 6, 2.3.3. 167, 5.2. 229: If $\{A_\lambda\}_{\lambda \in \Lambda}$ is a set of sets, there is a set A consisting of precisely one element from each A_λ.

— — Solovay 5.2. 229

Baire, R. 2.1.1. 43

Baire's (category) theorem 2.1.1. 43: The intersection of a countable set of dense open subsets of a complete metric space is a dense G_δ.

ball: see *closed ball, open ball.*

BANACH, S. 1.3.2. 37, 2.1.1. 51, 2.2.3. 145, 2.3.1. 156, 2.4.1. 172

Banach algebra 1.3.2. 37, 2.4.1. 172: an algebra A (over $\mathbb{R}$ or $\mathbb{C}$) that is a Banach space and such that for any scalar a and any vectors $\mathbf{x}$ and $\mathbf{y}$ the relations $\|a\mathbf{x}\mathbf{y}\| = |a| \|\mathbf{x}\mathbf{y}\| \le |a| \|\mathbf{x}\| \|\mathbf{y}\|$ obtain.

— space 1.3.2. 31, 2.1.1. 51, 2.3.1. 156: a complete normed vector space over $\mathbb{R}$ or $\mathbb{C}$.

— -Tarski paradox 2.2.3. 145

base 2.2.1. 112: for the topology of a space, a set S of open sets such that each open set of the topology is a union of sets in S.

basic neighborhood (in a Cartesian product) 2.2.1. 113: for a point

$$\mathbf{x} \overset{\text{def}}{=} \{x_\lambda\}_{\lambda \in \Lambda}$$

in a Cartesian product $\Pi_{\lambda \in \Lambda} X_\lambda$ of topological spaces, a set that is a Cartesian product in which finitely many factors, say those corresponding to the finite set $\{\lambda_1, \ldots, \lambda_n\}$, are neighborhoods U_{λ_i} of the components x_{λ_i}, $1 \le i \le n$, and in which the remaining factors are the full spaces $X_{\lambda'}$, $\lambda' \notin \{\lambda_1, \ldots, \lambda_n\}$.

— variables 1.3.3. 39: in linear programming, the variables constituting the complement of the set of free variables, q.v.

basis 2.3.1. 156: in a topological vector space, a Schauder basis.

BAUMSLAG, G. 1.1.5. 11

BESICOVITCH, A. S. 2.2.1. 118, 2.2.1. 123

BERNAYS, P. 5.2. 229

BESSEL, F. W. 2.3.1. 161

Bessel's inequality 2.3.1. 161: for an orthonormal system $\{\mathbf{x}_\lambda\}_{\lambda \in \Lambda}$ and any vector $\mathbf{x}$ in a Hilbert space, the relation: $\sum_{\lambda \in \Lambda} |(\mathbf{x}, \mathbf{x}_\lambda)|^2 \le \|\mathbf{x}\|^2$.

— approximation property 2.3.1. 158: of a normed vector space V, that there is in $[V]$ a sequence $\{F_n\}_{n\in\mathbb{N}}$ such that $\sup_{n\in\mathbb{N}}\|F_n\| < \infty$ and for every vector $\mathbf{x}$,

$$\lim_{n\to\infty}\|\mathbf{x} - F_n\mathbf{x}\| = 0.$$

— variation 2.1.2. 54, 2.1.3. 88, 3.1.2. 192: of a function f in $\mathbb{R}^{[a,b]}$, that

$$\sup_{a\leq x_1<\cdots<x_n\leq b}\sum_{k=2}^{n}|f(x_k) - f(x_{k-1}| < \infty,\ n \in \mathbb{N}.$$

BRANGES, L. DE 2.6.6. 184
bridging functions 2.1.2. 62, 2.2.2. 139
BRIESKORN, E. V. 3.3. 208
BRITTON, J. L. 1.1.5. 11
BROUWER, L. E. J. 3.1.2. 195, 3.1.2. 197
Brouwer's invariance of domain theorem 3.1.2. 197: If Ω is an open subset of $\mathbb{R}^2$ and if $f : \Omega \mapsto f(\Omega) \subset \mathbb{R}^2$ is a homeomorphism, then $f(\Omega)$ is open.
BURNSIDE, W. 1.1.5. 11
Burnside question 1.1.5. 11

canonical relation 2.2.3. 147
CANTOR, G. 1.1.4. 7, 2.1.2. 55, 2.1.3. 85
Cantor function 2.1.2. 55
— -Lebesgue theorem 2.1.3. 85
— set 1.1.4. 7
—'s theorem 5.1. 225
cardinality 1.1.2. 2: of a set, the bijection-equivalence class of a set; alternatively, the least ordinal among all bijectively equivalent ordinals.
CARLESON, L. 2.3.1. 158
Cartesian product 2.2.1. 112: of an indexed set $\{X_\lambda\}_{\lambda\in\Lambda}$, the set of all "vectors" $\{x_\lambda\}_{\lambda\in\Lambda}$ such that $x_\lambda \in X_\lambda$.
category 1.1.4. 8: a complex consisting of a) a class C of *objects*, $A, B, \ldots$; b) the class of pairwise disjoint sets $[A, B]$ (one for each pair A, B in $C \times C$) of *morphisms*; c) an associative law

$$\circ : [A, B] \times [B, C] \ni (f, g) \mapsto g \circ f \in [A, C] \stackrel{\text{def}}{=} [A, B] \circ [B, C];$$

of composition of morphisms; d) for each A in C a morphism 1_A in $[A, A]$ and such that if $f \in [A, B]$ and $g \in [C, A]$ then $f \circ 1_A = f$ and $1_A \circ g = g$. **Example 1.** The class of all groups is a category $\mathcal{G}$ if $[A, B]$ is the set of homomorphisms $\phi : A \mapsto B$, if $\phi \circ \psi$ denotes the composition of ϕ and ψ, and 1_A is the identity map. **Example 2.** A partially ordered set $\{S, \prec\}$ is a category $\mathcal{S}$ if the objects are the elements $A, B, \ldots$ of S and if the set of morphisms associated with the pair (A, B) is

$$[A, B] \stackrel{\text{def}}{=} \begin{cases} (\text{A,B}) & \text{if } A \preceq B \stackrel{\text{def}}{=} A \prec B \vee A = B \\ \emptyset & \text{otherwise.} \end{cases}$$

Since $\preceq$ is transitive, i.e.,

$$A \preceq B \wedge B \preceq C \Rightarrow A \preceq C,$$

the composition of morphisms is well-defined:

$$[A, B] \circ [B, C] = \begin{cases} [A, C] & \text{if } [A, B] \neq \emptyset, \ [B, C] \neq \emptyset \\ \emptyset & \text{otherwise.} \end{cases}$$

Since $\preceq$ is reflexive, i.e., $A \preceq A$, for each A in S, the identity 1_A exists and is $[A, A]$. In particular, the set of morphisms need not be a set of maps. (See also *first category* and *second category*.)

CAUCHY, A. L. DE 1.2.3. 24

Cauchy complete 1.2.3. 24: of a uniform space X, that every Cauchy net in X has a limit in X.

— completion 1.2.3. 24: for a uniform space X, the set of equivalence classes of Cauchy nets.

— criterion 2.1.3. 71: for a complete uniform space X, the statement that a net converges iff it is a Cauchy net.

— -Kowalewski theorem 2.5.2. 179: For a function f in $\mathbb{R}^{\mathbb{R}^n}$ and analytic in a neighborhood of $\mathbf{O}$, let the partial differential equation

$$\sum_{0 \leq k_1 + \cdots + k_n \leq N} a_{k_1 \ldots k_n} \left(\frac{\partial}{\partial x_1} \right)^{k_1} \cdots \left(\frac{\partial}{\partial x_n} \right)^{k_n} u = f$$

have coefficients analytic near $\mathbf{O}$ and let the coefficient of

$$\frac{\partial^N u}{\partial^N x_n}$$

be different from zero at $\mathbf{O}$. Then if ϕ is analytic near $\mathbf{O}$ the differential equation has a a unique solution u, analytic near $\mathbf{O}$ and satisfying

$$\left(\frac{\partial}{\partial x_1} \right)^{k_1} \cdots \left(\frac{\partial}{\partial x_n} \right)^{k_n} (u - \phi) = 0$$

$$\text{when } x_1 = \cdots = x_n = 0 \text{ and } a_1 + \cdots + a_n < N$$

[Hö].

— net 1.2.3. 24: in a uniform space X with uniformity U, a net $\{x_\lambda\}_{\lambda \in \Lambda}$ such that for each U in U there is a λ_0 such that $(x_\lambda, x_{\lambda'}) \in U$ if $\lambda, \lambda' \succeq \lambda_0$.

— product 2.1.3. 82: of two series $\sum_{n=0}^{\infty} a_n$ and $\sum_{m=0}^{\infty} b_m$, the series

$$\sum_{n=0}^{\infty} \left(\sum_{p+q=n} a_p b_q \right).$$

CAYLEY, A. 1.2.2. 22

Cayley algebra 1.2.2. 22

central limit theorem 4.1. 214: If $\{f_n\}_{n \in \mathbb{N}}$ is a sequence of independent random variables such that

$$\int_X f_n(\omega) \, d\omega = 0$$

$$\int_X f_n(\omega)^2 \, d\omega = 1$$

and such that for every positive ϵ,

$$\lim_{n \to \infty} \frac{1}{\sqrt{n}} \int_{\{\omega \,:\, |f_n(\omega)| \geq \epsilon \sqrt{n}\}} f(\omega)^2 \, d\omega = 0,$$

then

$$\lim_{n \to \infty} g_n \overset{\text{def}}{=} \lim_{n \to \infty} \frac{f_1 + \cdots + f_n}{\sqrt{n}} \overset{\text{def}}{=} g$$

exists almost everywhere and

$$P\{\omega \,:\, g(\omega) \leq x\} = \frac{1}{\sqrt{2\pi}} \int_{-\infty}^{x} \exp(-\frac{t^2}{2}) \, dt.$$

CESÀRO, E. 2.1.3. 76

Cesàro summation 2.1.3. 76

chain 2.5.2. 179: a sequence $\{F_J\}$ of functions, indexed by and defined over intervals J of a partition of an interval $[a, b]$, and such that the F_J and all their derivatives coincide at common endpoints of the intervals J.

characteristic function 1.1.4. 5, 2.1.1. 46, 2.2.1. 104, 2.2.2. 133: of a set A, the function χ_A such that

$$\chi_A(x) \overset{\text{def}}{=} \begin{cases} 1 & \text{if } x \in A \\ 0 & \text{if otherwise.} \end{cases}$$

CHARNES, A. 1.3.3. 40

CHURCH, A. 5.1. 227

conjugate 1.1.5. 13: of a quaternion $\mathbf{q} \overset{\text{def}}{=} a\mathbf{1} + b\mathbf{i} + c\mathbf{j} + d\mathbf{k}$, the quaternion $\overline{\mathbf{q}} \overset{\text{def}}{=} a\mathbf{1} - b\mathbf{i} - c\mathbf{j} - d\mathbf{k}$.

— linear 1.3.2. 33: of a map $T : V \mapsto \mathbb{C}$ of a vector space V over $\mathbb{C}$, that $a \in \mathbb{C} \Rightarrow T(a\mathbf{x}) = \overline{a}T(\mathbf{x})$.

— symmetric 1.3.2. 33

connected 2.1.1. 43: of a topological space X, that it is not the union of two nonempty and disjoint open sets.

consequence 5.1. 224

constant (in formal logic) 5.1. 223

contain a constant 2.2.3. 151: of a relation R on $\{1, \ldots, n\}$, that for some k, (iRk), $1 \leq i \leq n$; by abuse of language, R contains the constant k.

content: see *n-dimensional content*.

continuous 1.1.4. 5, 2.1.1. 42, 3.2.2. 203

Continuum Hypothesis 5.2. 229: The cardinality of the set S of all well-ordered finite or countably infinite sets no two of which are in order-preserving bijective correspondence is the cardinality of $\mathbb{R}$: $\aleph_1 = 2^{\aleph_0}$.

contradictory 5.1. 224: of two wfs, that each is the negation of the other.

convergent (converges) 1.2.3. 24: A net in a topological space is convergent (converges) iff the net has a limit.

converges absolutely 2.1.3. 82: of a series $\sum_{n=1}^{\infty} a_n$ consisting of real or complex terms, that $\sum_{n=1}^{\infty} |a_n| < \infty$.

— uniformly 2.1.2. 56, 2.1.3. 84: of a sequence $\{f_n\}_{n=0}^{\infty}$ of functions defined on a set S that for each positive ϵ there is an $N(\epsilon)$ such that for all x in S, $|f_n(x) - f_0(x)| < \epsilon$ if $n > N(\epsilon)$.

convex 1.1.4. 7: for an open set U in $\mathbb{R}$ and of a function f on $\mathbb{R}^U$, that if $t \in [0, 1]$ and $x, y, tx+(1-t)y \in U$ then $f(tx+(1-t)y) \leq tf(x)+(1-t)f(y)$: "the curve lies below the chord;" 2.3.3. 167: of a set S in a vector space V, that $(\mathbf{u}, \mathbf{v} \in S) \wedge (t \in [0, 1]) \Rightarrow t\mathbf{u} + (1 - t)\mathbf{v} \in S$.

— hull 2.3.3. 168: of a set S in a vector space, the intersection of the set of all convex sets containing S.

— polyhedron 1.3.3. 38: in a vector space, the intersection of a finite number of half-spaces.

convolution 2.1.3. 74, 2.3.4. 170, 2.4.2. 176: of two functions f and g defined on a locally compact group G (with Haar measure μ) and in $L^1(G, \mathbb{C})$, the function

$$f * g : G \ni x \mapsto \int_G f(t^{-1}x)g(t)\, d\mu(t).$$

corner 4.2. 216: for $(x_1, \ldots, x_n)$ in $\mathbb{R}^n$, the set

$$\{(a_1, \ldots, a_n) \ : \ (a_1, \ldots, a_n) \in \mathbb{R}^n, a_i \leq x_i, \ 1 \leq i \leq n\}.$$

coset 1.1.2. 3: of a subgroup H of a group G, for some x in G, a set of the form xH or Hx.

countable 1.1.4. 5: of a set, that its cardinality is that of $\mathbb{N}$.

countably additive 2.2.1. 104: of a $\mathbb{C}$-valued set function Φ, that if $\{A_n\}_{n\in\mathbb{N}}$ is a sequence of pairwise disjoint sets in the domain of Φ then

$$\Phi\left(\bigcup\nolimits_{n\in\mathbb{N}}A_n\right) = \sum_{n=1}^{\infty}\Phi(A_n).$$

— subadditive, (see *subadditive*): of a nonnegative set function Φ, that

$$\Phi\left(\bigcup_{n\in\mathbb{N}}A_n\right) \le \sum_{n=1}^{\infty}\Phi(A_n).$$

counting measure 2.2.2. 134, 2.2.2. 141: for a measure situation $(X, 2^X, \mu)$, the measure μ such that for every subset A of X,

$$\mu(A) = \begin{cases} \#(A) & \text{if } A \text{ is finite} \\ \infty & \text{otherwise.} \end{cases}$$

cycling 1.3.3. 38: in linear programming by the simplex algorithm, the phenomenon in which a finite set of vertices is recurrently visited without the conclusion that any one of them is optimal: the algorithm *cycles*.

cylinder set 4.2. 216: in a Cartesian product, a set determined by conditions on finitely many vector components; 4.2. 217: in a vector space V, a set $Z_{\mathbf{x}_1^*,\ldots,\mathbf{x}_n^*;A}$ defined by a finite subset $\{\mathbf{x}_1^*,\ldots,\mathbf{x}_n^*\}$ of the dual space V^* and a Borel subset A of $\mathbb{R}^n$:

$$Z_{\mathbf{x}_1^*,\ldots,\mathbf{x}_n^*;A} \stackrel{\text{def}}{=} \{\mathbf{x} : \mathbf{x} \in V, (\mathbf{x}_1^*(\mathbf{x}),\ldots,\mathbf{x}_n^*(\mathbf{x})) \in A\}.$$

(C,α)-summable 2.1.3. 76

Daniell, P. J. 2.2.1. 104

Daniell integral 2.2.1. 104: a linear functional I defined on a linear lattice L of extended $\mathbb{R}$-valued functions and such that:

$$f \in L \wedge f \succeq 0 \Rightarrow I(f) \ge 0;$$
$$f_n \downarrow 0 \Rightarrow I(f_n) \downarrow 0.$$

DANTZIG, G. 1.3.3. 38

DAVIE, A. M. 2.3.1. 157

decreasing 1.2.3. 24: for ordered sets $(X_i, \succeq_i)$, $i = 1, 2$, and of an f in Y^X, that $x_1 \succeq_1 y_1 \Rightarrow f(x_1) \succeq_2 f(y_1)$.

DEDEKIND, R. 5.1. 224

dense 1.1.4. 5, 2.1.1. 48: of a subset A of a topological space X, that the closure $\overline{A}$ of A is X; equivalently, A meets every nonempty open subset of X.

derivation 2.4.1. 173: in an algebra A, a linear endomorphism D such that for $\mathbf{x}, \mathbf{y} \in A$, $D(\mathbf{xy}) = D(\mathbf{x})\mathbf{y} + \mathbf{x}D(\mathbf{y})$.

derived set 3.2.2. 200: the set of limit points of a set.

DESARGUE, G. 3.1.1. 187

Desargue's theorem 3.1.1. 187

diagonable 1.3.1. 26: of a SQUARE matrix A, that there is an invertible matrix P such that $P^{-1}AP$ is a diagonal matrix.

diagonal (matrix) 1.3.1. 26: a SQUARE matrix $(a_{ij})_{i,j=1}^{n,n}$ such that $a_{ij} = 0$ if $i \neq j$.

diameter 2.2.1. 126: of a set S in a metric space (X, d), $\sup_{x,y \in S} d(x, y)$.

DIAMOND, H. 2.2.1. 110

diffeomorphism 3.3. 208: a C^{∞} surjective homeomorphism $D : X \mapsto Y$ between differential geometric structures X and Y.

difference set 2.2.1. 109: for two subsets A, B of a group resp. abelian group, the set

$$\left\{ ab^{-1} \ : \ a \in A, b \in B \right\} \text{ resp. } \left\{ a - b \ : \ a \in A, b \in B \right\}.$$

differentiable (at $\mathbf{x}_0$) 1.1.4. 7: of a vector function

$$\mathbf{f} \overset{\text{def}}{=} (f_1, \ldots, f_n) : \mathbb{R}^m \ni \mathbf{x} \overset{\text{def}}{=} \begin{pmatrix} x_1 \\ \vdots \\ x_m \end{pmatrix} \mapsto \begin{pmatrix} f_1(\mathbf{x}) \\ \vdots \\ f_n(\mathbf{x}) \end{pmatrix} \in \mathbb{R}^n,$$

that there is in $[\mathbb{R}^m, \mathbb{R}^n]$ a T such that

$$\lim_{\mathbf{h} \neq \mathbf{O}, \|\mathbf{h}\| \to 0} \frac{\| (\mathbf{f}(\mathbf{x}_0 + \mathbf{h}) - \mathbf{f}(\mathbf{x}_0) - T(\mathbf{x}_0)\mathbf{h}) \|}{\|\mathbf{h}\|} = 0.$$

The vector $T(\mathbf{x}_0)$ is the *derivative* of $\mathbf{f}$ at $\mathbf{x}_0$. If $\mathbf{f}$ is differentiable at each point of the domain R of $\mathbf{f}$ then $\mathbf{f}$ is *differentiable* on R.

When, for a choice of bases, X for $\mathbb{R}^m$ and Y for $\mathbb{R}^n$, T is realized as an $m \times n$ matrix

$$T_{XY} \overset{\text{def}}{=} \left(\frac{\partial f_j(\mathbf{x})}{\partial \mathbf{x}_i} \right) \Bigg|_{\mathbf{x}=\mathbf{x}_0, \ 1 \leq i \leq m, \ 1 \leq j \leq n}$$

then T_{XY} is the *Jacobian matrix*

$$\frac{\partial \mathbf{f}(\mathbf{x})}{\partial \mathbf{x}} \Bigg|_{\mathbf{x}=\mathbf{x}_0} \overset{\text{def}}{=} \frac{\partial (f_1, \ldots, f_n)}{\partial (x_1, \ldots, x_m)} \Bigg|_{x_1 = x_{10}, \ldots, x_m = x_{m0}}.$$

If $m = n$ then $\det(T_{XY})$ is the *Jacobian determinant* for $\mathbf{f}$. If $m = n = 1$ and $X = Y = (1)$ and $\mathbf{f}$ resp. $\mathbf{x}_0$ is written f resp. x_0 then

$$T_{XY}(x_0) = \frac{df(x)}{dx} \Bigg|_{x=x_0}$$

$$= \det [T_{XY}(x_0)].$$

differential geometric structure 3.3. 208: a Hausdorff space X, an open covering $\mathcal{U}$ of X, and a set $\Phi \overset{\text{def}}{=} \{\phi_U\}_{U \in \mathcal{U}}$ of homeomorphisms

$$\phi_U : U \mapsto \mathbb{R}^n$$

such that if $U \cap U' \overset{\text{def}}{=} W \neq \emptyset$ then

$$\phi_{U'} \circ \phi_U^{-1} \in C^\infty \left(\phi_U(U) \cap \phi_{U'}(U'), \phi_U(U) \cap \phi_{U'}(U') \right).$$

The structure is an *n-dimensional differentiable manifold*.

DINI, U. 2.1.3. 93

Dini's theorem 2.1.3. 93

DIOPHANTUS 5.1. 228

Diophantine 5.1. 228: of a set of polynomial equations, that their coefficients are in $\mathbb{Z}$ and that their solutions are to be sought in $\mathbb{Z}$.

directed 1.2.3. 24: of a set, that it is partially ordered and that every pair in the set has an upper bound.

direct product 1.1.3. 5: for a set of algebraic structures, their Cartesian product endowed with component-wise operations.

discontinuity 1.1.4. 7: for a map of a topological space into a topological space, a point where the map is not continuous.

discrete topology 2.3.2. 163, 3.2.2. 200: the topology in which every set is open.

diset 1.2.3. 24: a directed set.

distribution 2.5.2. 179: for the set $W \overset{\text{def}}{=} C^\infty(\mathbb{R}^n, \mathbb{R})$ endowed with a suitable locally convex topology, a continuous linear functional on W, i.e., an element of the dual space W^* of W.

— function 4.2. 216: for a set $\{f_k\}_{k=1}^n$ of random variables, the function

$$\mathbb{R}^n \ni (x_1, \ldots, x_n) \mapsto P \left(\{ \omega \; : \; f_1(\omega) \leq x_1, \ldots, f_n(\omega) \leq x_n \} \right).$$

divergent 2.1.3. 70: of a series, that it fails to converge.

division algebra (division ring) 1.1.5. 13, 1.2.1. 19: an algebraic object governed by all the axioms for a field, save the axiom of commutativity for multiplication.

domain 2.1.2. 55, 2.4.1. 174: for a map $T : X \mapsto Y$, the set X.

dominate 2.1.3. 72

— absolutely 2.1.3. 72

dual space 1.3.2. 34, 2.3.1. 159, 2.3.2. 163: for a (topological) vector space V over a (topological) field $\mathbb{K}$, the set V^* of (continuous) linear maps of V into $\mathbb{K}$.

dyadic space 2.2.1. 112

edge 1.3.3. 38, 2.2.1. 123: of a polyhedron Π in $\mathbb{R}^n$, the intersection of Π with $n - 1$ of the hyperplanes determining Π.

$$\cdot : \mathbb{K} \times \mathbb{K} \ni (a, b) \mapsto a \cdot b \stackrel{\text{def}}{=} ab \in \mathbb{K}$$
$$+ : \mathbb{K} \times \mathbb{K} \in (a, b) \mapsto a + b \in \mathbb{K}.$$

Furthermore both $\cdot$ and $+$ are commutative and associative and

$$a(b + c) = ab + ac.$$

With respect to $+$, $\mathbb{K}$ is a group (with identity 0) and with respect to
$\cdot$, $\mathbb{K} \setminus \{0\}$ is a group (with identity 1).

1. The (countably many) *symbols* of the system are:
   a. the connectives $\neg$ ("negation"), $\rightarrow$ ("implies");
   b. the quantifier $\forall$ ("for all");
   c. derived connectives $\vee$ ("or"), $\wedge$ ("and"), and the derived quantifier
      $\exists$ ("there exists");
   d. predicate letters A, $\ldots$, function letters f, $\ldots$, variables x, $\ldots$,
      and constants a, $\ldots$.

2. An *expression* is a finite sequence of symbols. An expression E is a *term* of the system iff E is:
   a. a variable, a constant; or
   b. for terms $t_1, \ldots, t_n$ and a function letter f, $f(t_1, \ldots, t_n)$.
3. If A is a predicate letter and $t_1, \ldots, t_n$ are terms, $A(t_1, \ldots, t_n)$ is an *atomic formula*. An expression $\mathcal{E}$ is a *well-formed formula (wf)* iff $\mathcal{E}$ is:
   a. an atomic formula; or
   b. $\neg\mathcal{A}$ for some wf $\mathcal{A}$;or
   c. $\mathcal{A} \to \mathcal{B}$ for wfs $\mathcal{A}$ and $\mathcal{B}$;or
   d. for some variable y and wf $\mathcal{A}$, $\forall y \mathcal{A}$.
4. The *axioms* of the system deal with the relations among wfs $\mathcal{A}, \ldots,$ and are:
   a. $\mathcal{A} \to (\mathcal{B} \to \mathcal{A})$;
   b. $(\mathcal{A} \to (\mathcal{B} \to \mathcal{C})) \to ((\mathcal{A} \to \mathcal{B}) \to (\mathcal{A} \to \mathcal{C}))$;
   c. $(\neg\mathcal{B} \to \neg\mathcal{A}) \to ((\neg\mathcal{B} \to \mathcal{A}) \to \mathcal{B})$;
   d. if t is a term *free* for x in $\mathcal{A}(x)$, i.e., in t there is no occurrence of $(\forall x)$ or $(\exists x)$ then: $(\forall x)\,\mathcal{A}(x) \to \mathcal{A}(t)$;
   e. if x is *not* free in $\mathcal{A}$, i.e., every occurrence of x is *bound* by a quantifier then: $(\forall x)$ or $(\exists x)$: $(\forall x)\,(\mathcal{A} \to \mathcal{B}) \to (\mathcal{A} \to (\forall x)\,\mathcal{B})$;
4. the *rules of inference* are:
   a. modus ponens: $\mathcal{B}$ follows from $\mathcal{A}$ and $\mathcal{A} \to \mathcal{B}$;
   b. generalization: $(\forall x)\,\mathcal{A}$ follows from $\mathcal{A}$.
5. Proper axioms, e.g., the axioms of group theory.

The descriptor *first-order* indicates that the system does not deal with predicates or functions as arguments of predicates nor as arguments of quantifiers.

The items 1.–5. permit the chaining together of axioms to form *proofs* of *theorems*.

$$c_n = \int_{-\pi}^{\pi} f(\theta)\frac{e^{-in\theta}}{\sqrt{2\pi}}\, d\theta, \; n \in \mathbb{Z}.$$

— transform 2.1.3. 88: for an f Lebesgue integrable on $\mathbb{R}$, the function

$$\hat{f} : \mathbb{R} \ni t \mapsto \frac{1}{\sqrt{2\pi}} \int_{\mathbb{R}} f(x)e^{-itx}\,dx.$$

If, for $(X, \mathcal{S}, \mu)$ and $(Y, \mathcal{T}, \nu)$, $f : X \times Y \mapsto \mathbb{C}$ is $\mathcal{S} \times \mathcal{T}$-measurable then for almost every fixed x resp. y

$$f(x,y) \left(\overset{\text{def}}{=} f_{(x)}(y) \right) \text{ resp. } f(x,y) \left(\overset{\text{def}}{=} f^{(y)}(x) \right)$$

is $\mathcal{T}$-measurable resp. $\mathcal{S}$-measurable. If

$$\int_{X \times Y} |f(x,y)|\,d(\mu \times \nu) < \infty$$

then all the integrals in

$$\int_Y \left(\int_X f^{(y)}(x)\,d\mu \right) d\nu \text{ and } \int_X \left(\int_Y f_{(x)}(y)\,d\nu \right) d\mu$$

exist and both iterated integrals are equal to

$$\int_{X \times Y} f(x,y)\,d(\mu \times \nu).$$

for a power series $\sum_{n=0}^{\infty} a_n z^n$, a set $K \overset{\text{def}}{=} \{k, k+1, \ldots, k+l\}$ such that $a_n = 0$, $n \in K$.

GARNIR, H. G. 5.2. 230

GAUSS, C. F. 1.3.1. 29, 1.3.3. 38, 4.2. 220

Gaußian elimination 1.3.3. 38: the reduction via an invertible $m \times m$ matrix $\mathcal{G}$ of an $m \times n$ matrix A to an $m \times n$ matrix $B \overset{\text{def}}{=} (b_{ij})_{i,j=1}^{m,n} \overset{\text{def}}{=} \mathcal{G}A$ such that $b_{ij} = 0$ if $i > j$.

— measure (on the algebra Z of cylinder sets in Hilbert space) 4.2. 220

Gauß-Seidel algorithm 1.3.1. 29: a recursive method for finding approximate solutions of the matrix-vector equation $A\mathbf{x} = \mathbf{b}$; when A is SQUARE and for an invertible P, $A = P - Q$ and the spectral radius of $P^{-1}Q$ is less than 1, for an arbitrary $\mathbf{x}_0$, the algorithm generates the sequence

$$\left\{ x_{n+1} \overset{\text{def}}{=} (P^{-1}Q)^{n+1}\mathbf{x}_0 + \sum_{k=0}^{n}(P^{-1}Q)^k P^{-1}\mathbf{b} \right\}_{n \in \mathbb{N}},$$

which converges to a solution of the equation.

GELLÉS, G. 2.2.1. 110

Generalized Continuum Hypothesis 5.2. 229: For any α, $\aleph_{\alpha+1} = 2^{\aleph_\alpha}$.

generalized nilpotent 2.4.1. 172: in a Banach algebra, an element $\mathbf{x}$ such that

$$\lim_{n \to \infty} \|\mathbf{x}^n\|^{\frac{1}{n}} = 0.$$

— ratio test (for an infinite series) 2.1.3. 81

— root test (for an infinite series) 2.1.3. 81

generate(s) (a σ-ring) 1.1.4. 5, (a free group) 1.1.5. 11, (the commutator subgroup) 1.1.5. 15, (a group of Lie type) 1.2.2. 22, (a σ-ring) 2.2.1. 105, 2.2.3. 144, (a σ-ring) 4.2. 217: A set X generates an object A in a category $\mathcal{C}$ iff A is the intersection of the class of all $\mathcal{C}$-objects containing X; the elements of X are called the *generators* of A.

generator(s) 1.1.5. 11: see *generate(s)*.

GÖDEL, K. 5.1. 224, 5.1. 225

Gödel's completeness theorem 5.1. 224

— undecidability (incompleteness) theorem 5.1. 225

good (topological vector space) 5.2. 229

GORENSTEIN, D. 1.1.5. 19

gradient 1.3.3. 38: in the context of linear programming, the directional derivative of the *Cost* function; more generally, of a function f in $\mathbb{R}^{\mathbb{R}^n}$, the vector

$$\mathbf{grad}(f) \overset{\text{def}}{=} (f_{x_1}, \ldots, f_{x_n})$$

of partial derivatives.

GRAM, J. P. 2.3.1. 162

Gram-Schmidt orthonormalization 2.3.1. 162: for a linearly independent set $\{\mathbf{x}_n\}_{n \in \mathbb{N}}$ in a Hilbert space, the algorithm leading to the sequence

$$\mathbf{y}_1 \overset{\text{def}}{=} \frac{\mathbf{x}_1}{\|\mathbf{x}_1\|}$$

$$\mathbf{y}_{n+1} \overset{\text{def}}{=} \frac{\mathbf{x}_{n+1} - \sum_{k=1}^{n} (\mathbf{x}_{n+1}, \mathbf{y}_k) \mathbf{y}_k}{\left\| \mathbf{x}_{n+1} - \sum_{k=1}^{n} (\mathbf{x}_{n+1}, \mathbf{y}_k) \mathbf{y}_k \right\|}, \quad n > 1.$$

graph 1.3.2. 34, 2.2.2. 142: of a function $f : X \ni x \mapsto f(x) \in Y$, in $X \times Y$ the set

$$\{ (x, f(x)) \ : \ x \in X \}.$$

greatest common divisor: for a pair $\{m, n\}$ of natural numbers, the greatest natural number that is a factor of each; more generally, in a commutative ring R without divisors of zero, for a pair a, b of ring elements, an element c that is a factor of both a and b and such that if d is a factor of both a and b then d is also a factor of c.

GROTHENDIECK, A. 2.3.1. 157

group 1.1.1. 1

— algebra 2.4.2. 175: over a locally compact group G with Haar measure μ, the set $L^1(G)$ with multiplication defined by convolution:

$$* : L^1(G) \times L^1(G) \ni (f, g) \mapsto \left[f * g : G \ni x \mapsto \int_G f(y^{-1}x) g(y) \, d\mu \right].$$

group-invariant (measure) 2.2.3. 143

group of Lie type 1.2.2. 22

Haar, A. 1.1.4. 5, 2.2.1. 104

— measure 1.1.4. 5, 2.2.1. 104: on a locally compact group G, a Borel measure μ that is a) translation invariant, i.e., if E is measurable and $x \in G$ then $\mu(xE) = \mu(E)$, b) positive for each nonempty open set, and c) finite for every compact set.

HADAMARD, J. 2.6.2. 181

Hadamard's gap theorem 2.6.2. 181: If the series $S \overset{\text{def}}{=} \sum_{n=0}^{\infty} a_n z^n$ has a finite and positive radius of convergence R, if $n_1 < n_2 < \cdots$,

$$\frac{n_{k+1}}{n_k} \geq \lambda > 1,$$

and $a_n = 0$ if $n \notin \{n_k\}_{k \in \mathbb{N}}$, then S represents a function f holomorphic in $\{ z \ : \ |z| < R \}$ and $\{ z \ : \ |z| = R \}$ is a natural boundary for f.

HAHN, H. 2.3.1. 162

Hahn-Banach theorem 2.3.1. 162, 2.3.2. 162: If V is a vector space over $\mathbb{C}$ and p is a seminorm defined on V then a linear functional f, defined on a subspace W of V and such that $|f(\mathbf{w})| \leq p(\mathbf{w})$ on W, may be extended to a linear functional F defined on V and satisfying the inequality

$$|F(\mathbf{v})| \le p(\mathbf{v})$$

on V.

HAKEN, W. 3.1.2. 198

half-space: in a vector space V and for a linear functional $f : V \mapsto \mathbb{R}$, a set of the form $\{ \mathbf{x} : \mathbf{x} \in V, f(\mathbf{x}) \le r \}$.

halting problem 5.1. 227: that of determining whether a program applied to a data-set stops (*halts*) after finitely many operations.

HAMEL, G. 1.1.4 6

Hamel basis 1.1.4. 6: in a vector space V, a maximal linearly independent subset.

HARDY, G. H. 2.1.1. 51

HAUSDORFF, F. 2.2.1. 112; 2.2.3. 144

Hausdorff 2.2.1. 112: of (the topology of) a space, that any two points lie in disjoint neighborhoods.

— dimension 2.2.1. 122: of a subset S of a metric space (X, d) and for a positive p,

$$\sup_{r>0} \rho_p^r(S) \overset{\text{def}}{=} \sup_{r>0} \left[\inf_{\text{diam}(A_n)<r} \left\{ \sum_{n=1}^{\infty} (\text{diam}(A_n))^p : S \subset \bigcup_{n=1}^{\infty} A_n \right\} \right].$$

— topology 2.4.1. 174: a topology in which any two points lie in disjoint neighborhoods.

— -Young theorem 2.3.4. 169: If

$$1 < p \le 2, \quad \frac{1}{p} + \frac{1}{p'} = 1,$$

if $f \in L^p([-\pi, \pi], \mathbb{C})$, and if $\{c_n\}_{n \in \mathbb{Z}}$ is the set of Fourier coefficients of f then $\{c_n\}_{n \in \mathbb{Z}} \in l^{p'}$; if $\{c_n\}_{n \in \mathbb{Z}} \in l^p$ then $\sum_{n=-\infty}^{\infty} c_n e^{inx}$ norm-converges in $L^{p'}([-\pi, \pi], \mathbb{C})$ (to a function f in $L^{p'}([-\pi, \pi], \mathbb{C})$).

HEDLUND, G. A. 1.1.5. 11

HELLY, E. 2.3.1. 161

Helly selection theorem 2.3.1. 161: If $S \overset{\text{def}}{=} \{F_n\}_{n \in \mathbb{N}}$ is a sequence of functions from $[0, 1]$ to $\mathbb{R}$ and such that for some finite M,

$$\sup_{n \in \mathbb{N}} \text{var}(F_n), \ \sup_{n \in \mathbb{N}} |F_n(0)| < \infty$$

then S contains a subsequence converging everywhere on $[0, 1]$ to a function F, also of bounded variation.

HEMASINHA, R. 4.2. 218, 4.2. 220

HERBRAND, J. 5.1. 227

HILBERT, D. 2.1.4. 101, 2.3.1. 156, 2.3.1. 159, 3.1.1. 186

— space 1.3.2. 36, 2.3.1. 156, 2.3.1. 159: a vector space $\mathcal{H}$ over $\mathbb{C}$ and on which there is defined a positive definite, conjugate symmetric, conjugate bilinear functional; (If $\mathcal{H}$ is not finite-dimensional then it is assumed that $\mathcal{H}$ is complete with respect to the metric associated to the bilinear functional.)

—'s problems 2.1.4. 101

HIRZEBRUCH, F. 3.3. 208

HÖLDER, O. 2.3.4. 169

Hölder inequality 2.3.4. 169: For the measure situation $(X, \mathcal{S}, \mu)$, if

$$f \in L^p(X, \mathbb{C}), \ g \in L^{p'}(X, \mathbb{C}), \ 1 < p, \ \frac{1}{p} + \frac{1}{p'} = 1$$

then $fg \in L^1(X, \mathbb{C})$ and $\|fg\|_1 \le \|f\|_p \|g\|_{p'}$.

holomorphic 2.1.2. 61, 2.6.1. 180, 2.6.3. 182, 2.6.4. 183: of a $\mathbb{C}$-valued function f defined in a region Ω of $\mathbb{C}$, that $f'(z)$ exists for all z in Ω.

homogeneous linear differential equation 2.5.2. 177

homomorphism 1.1.3. 4: for algebraic structures A and B, a mapping

$$h : A \mapsto B$$

that, $*$ representing the generic k-ary operations, respects them, e.g., $h(a * b) = h(a) * h(b)$; 1.1.4. 5: if h is between topological structures, it is assumed to be continuous.

homotopic 1.3.2. 37, 3.2.2. 209: see *homotopy*.

homotopically equivalent 3.3. 209

homotopy 1.3.2. 36: a continuous map

$$h : [0, 1] \times X \ni (t, x) \mapsto h_t(x)$$

whereby two maps $f : X \mapsto Y$ and $g : X \mapsto Y$ are homotopic:

$$h_0(x) \equiv f(x), \ h_1(x) \equiv g(x).$$

HÖRMANDER, L. 2.5.2. 179

HUNT, R. A. 2.3.1. 158

hull 2.4.2. 175: of an ideal I in a Banach algebra, the set of regular maximal ideals containing I; see also *(closed) convex hull*.

hyperplane 4.3. 222: in a vector space V, a translate of the kernel of a linear functional, i.e., for a linear functional ϕ and a constant a or a vector $\mathbf{y}$ such that $\phi(\mathbf{y}) = a$, a set of the form

$$\{\mathbf{x} \ : \ \phi(\mathbf{x}) = a\} \ \left(= \phi^{-1}(0) + \mathbf{y}\right).$$

ideal 2.4.2. 174: in a ring R, a *proper* subring I such that $x \in R \Rightarrow xI \cup Ix \subset I$.

identities 1.1.5. 11
identity (of a group) 1.1.1. 1
— relation 2.2.3. 147
Identity Theorem 1.3.1. 26: If f and g are analytic in a region Ω, if $S \subset \Omega$
and S has a limit point in Ω, and if $f = g$ on S then $f = g$ on Ω.
image measure 2.2.2. 137
— — catastrophe 2.2.2. 138
implies 5.1. 223, 5.1. 224
inbreeding 2.1.2. 59, 2.2.1. 115: the process of repeating an operation, performed on the whole of a structure, on similar parts of the structure,
e.g., as in **Example 2.1.2.2. 58**
inclusion 5.2. 229
— map 2.2.2. 138: for a subset B of a set A, the restriction to B of the
identity map $A \ni x \mapsto x \in A$, i.e., $B \hookrightarrow A$.
indeterminates 1.3.1. 28: with respect to a ring R, symbols $x_1, \ldots, x_n$ used
to form polynomials

$$\sum_{0 \leq i_k \leq K,\ 1 \leq k \leq n} a_{i_1 i_2 \ldots i_k} x_1^{i_1} \cdots x_k^{i_k}$$

in which the coefficients are in R.
index 1.1.2. 3: for a subgroup H of a finite group G, the quotient

$$\frac{\#(G)}{\#(H)} \overset{\text{def}}{=} G : H.$$

— (indices) 1.3.2. 37: for a rectifiable closed curve γ in $\mathbb{C}$ and with respect
to a point a in $\mathbb{C} \setminus \gamma^*$, the integer

$$\mathrm{Ind}_\gamma(a) \overset{\text{def}}{=} \frac{1}{2\pi i} \int_0^1 \frac{d\gamma(t)}{\gamma(t) - a}.$$

When γ is absolutely continuous $\mathrm{Ind}_\gamma(a)$ is written

$$\frac{1}{2\pi i} \int_0^1 \frac{\gamma'(t)}{\gamma(t) - a}\, dt \overset{\text{def}}{=} \frac{1}{2\pi i} \int_\gamma \frac{dz}{z - a}.$$

infinitely differentiable 2.1.1. 51, 2.1.2. 61: of a function f in $\mathbb{C}^{\mathbb{R}^n}$, that f
has derivatives of all orders, see *differentiable*.
initial segment 2.2.1. 130: in a well-ordered set S, for some b in $S \setminus \inf(S)$
a subset of the form

$$\{c\ :\ c \in S,\ c \prec b\}.$$

injection 1.1.3. 5: a one-one map (not necessarily a surjection).
injective 1.1.5. 10: of a map, that it is an injection.

isoperimetric inequality 2.2.1. 121: if C is a rectifiable Jordan curve in $\mathbb{R}^2$ and if S is the bounded component of $\mathbb{R}^2 \setminus C$, then $A(S) \leq \frac{\ell(C)^2}{4\pi}$.

Jacobi, C. G. J. 1.2.2. 21
Jacobi identity 1.2.2. 21: in the set Mat_{nn} of $n \times n$ matrices, $A \circ B$ denoting $AB - BA$, the identity:

$$A \circ (B \circ C) + C \circ (A \circ B) + B \circ (C \circ A) = O.$$

JAMES. R. D. 2.3.2. 165
JORDAN C. 2.2.1. 113
Jordan block 4.3. 222: a SQUARE matrix of the form

$$\begin{pmatrix} \lambda & * & & & \\ & \lambda & * & & \\ & & \ddots & \ddots & \\ & & & \lambda & * \\ & & & & \lambda \end{pmatrix}, \text{ each } * \text{ is 1 or each } * \text{ is } 0.$$

— contour 2.6.5. 184: a Jordan curve-image.
— curve 2.2.1. 113: a homeomorphism $\gamma : \mathbb{T} \mapsto \mathbb{C}$.
— — -image 2.2.1. 113
— "curve" theorem: Let J in $\mathbb{R}^n$ be the homeomorphic image of the boundary ∂B of the unit ball

$$B_1 \overset{\text{def}}{=} \{\, \mathbf{x} \; : \; \mathbf{x} \in \mathbb{R}^n, \|\mathbf{x}\| \leq 1 \,\}.$$

Then $\mathbb{R}^n \setminus J$ is the union of two disjoint regions and J is the boundary of each of them. When $n = 2$ the statement given is known as the *Jordan curve theorem*.
— normal form 1.3.1. 27: for a SQUARE matrix A, a similar matrix J, i.e., for some invertible matrix P, $J = P^{-1}AP$, consisting of Jordan blocks situated on its diagonal.
— region 3.1.2. 198: a region R for which the boundary ∂R is a Jordan curve-image.

Kakeya, S. 2.2.1. 124
Kakeya problem 2.2.1. 124
KARLIN, S. 2.3.1. 160
KARMARKAR, N. 1.3.3. 38
k-ary marker, representation 2.1.1. 52
kernel 1.1.3. 4: of a homomorphism ϕ of an algebraic structure, the inverse image of the identity, e.g., if ϕ is a group homomorphism, the kernel of ϕ

is $\phi^{-1}(e)$; if ϕ is an algebra homomorphism, the kernel of ϕ is $\phi^{-1}(0)$;
    2.4.1. 172: of a set of regular maximal ideals in a Banach algebra,
    their intersection. The intersection of all regular maximal ideals in a
    commutative Banach algebra A is the *radical* of A.

kernel (hull(I)) 2.4.2. 175

KLEE, V. L. 1.3.3. 38

KLEENE, S. 5.1. 227

KNASTER, B. 3.1.2. 194

KOCH, J. 3.1.2. 198

KOLMOGOROV, A. N. 2.1.3. 89, 2.1.4. 101, 4.2. 216

— criteria 4.2. 216

KOWALEWSKI, S. 2.5.2. 179

KREIN, M. 2.3.2 162

Krein-Milman theorem 2.3.2. 162: A compact convex set K in a topological
    vector space V is the closed convex hull of the set of the extreme points
    of K.

KURATOWSKI, C. 3.1.2. 194

lattice 2.3.4. 169: a partially ordered set in which each pair of elements
    has both a least upper bound and a greatest lower bound.

least upper bound 1.2.3. 23: for a subset S of an ordered set X, in X an
    element x such that $s \in S \Rightarrow s \preceq x$ and such that if $s \in S \Rightarrow s \preceq y$
    then $y \not\prec x$.

LEBESGUE, H. 1.1.4. 6, 2.1.2. 63, 2.1.3. 87, 2.2.1. 104

— integrable 2.1.2. 63, 2.1.3. 87: of a function f in $\mathbb{R}^{\mathbb{R}^n}$ and with respect
    to the measure situation $(\mathbb{R}^n, \mathcal{L}, \lambda)$, that f is measurable and that

$$\int_{\mathbb{R}^n} |f(x)| \, d\lambda_n < \infty.$$

— measurable 1.1.4. 6: of a map f in $\mathbb{R}^{\mathbb{R}^n}$, that for every open set U in $\mathbb{R}$,
    $f^{-1}(U)$ is a Lebesgue measurable subset of $\mathbb{R}^n$; 1.1.4. 6: of set S and
    two Borel null sets N_1 and N_2 in $\mathbb{R}^n$, that there is in $\mathbb{R}^n$ a Borel set A
    such that

$$S = (A \setminus N_1) \cup N_2.$$

left identity 1.1.1. 1

— inverse 1.1.1. 1

LEIBNIZ, G. W. von 2.4.1. 173

Leibniz's rule 2.4.1. 173: if D is a derivation defined on a Banach algebra
    then

$$D^n(\mathbf{xy}) = \sum_{k=0}^{n} \binom{n}{k} D^k(\mathbf{x}) D^{n-k}(\mathbf{y}).$$

length (of an arc) 2.2.1. 114, 2.2.1. 123

$$\begin{pmatrix} t_{11} & \cdots & t_{1n} \\ \vdots & \ddots & \vdots \\ t_{m1} & \cdots & t_{mn} \end{pmatrix}$$

of elements of a ring R; in more general terms, for a pair Γ, Λ of sets, an element of $\mathrm{Mat}_{\Gamma\Lambda} \overset{\mathrm{def}}{=} R^{\Gamma \times \Lambda}$ (frequently it is assumed that a matrix is a function that is zero at all but finitely many points of $\Gamma \times \Lambda$, i.e., that a matrix is a function with compact support when $\Gamma \times \Lambda$ is viewed as a space with discrete topology); in the language of linear algebra, if V resp. W are vector spaces over $\mathbb{K}$ and with Hamel bases

$$V \overset{\mathrm{def}}{=} \{\mathbf{v}_\gamma\}_{\gamma \in \Gamma} \text{ resp. } W \overset{\mathrm{def}}{=} \{\mathbf{w}_\lambda\}_{\lambda \in \Lambda}$$

and if $T \in [V, W]$ then $T\mathbf{v}_\gamma \overset{\mathrm{def}}{=} \sum_{\lambda \in \Lambda} t_{\gamma\lambda}\mathbf{w}_\lambda$; the matrix

$$T_{VW} \overset{\mathrm{def}}{=} (t_{\gamma\lambda})_{\gamma \in \Gamma, \lambda \in \Lambda} \in \mathbb{K}^{\Gamma \times \Lambda}$$

is associated with T and the pair $\{V, W\}$ of bases; T_{VW} is a function with compact support; if $S \in [W, U]$ corresponds to the matrix $S_{WU} \overset{\mathrm{def}}{=}$

$(s_{\lambda\delta})$ for the Hamel bases $\mathcal{W}$ and $\mathcal{U} \stackrel{\text{def}}{=} \{\mathbf{u}_\delta\}_{\delta\in\Delta}$, then the composition $S \circ T$ is associated for the Hamel bases $\mathcal{V}$ and $\mathcal{U}$ with the matrix product

$$(S \circ T)_{\mathcal{V}\mathcal{U}} \stackrel{\text{def}}{=} \left(\sum_{\lambda\in\Lambda} t_{\gamma\lambda} s_{\lambda\delta}\right)_{\gamma\in\Gamma,\delta\in\Delta} \stackrel{\text{def}}{=} T_{\mathcal{V}\mathcal{W}} S_{\mathcal{W}\mathcal{U}};$$

if $X = Y$ and $\mathcal{V} = \mathcal{W}$, then the association $[V] \leftrightarrow \text{Mat}_{\Lambda\Lambda}$ of elements of $[V]$ with their correspondents as matrices is an *anti*-isomorphism of the algebra $[V]$ of endomorphisms of V and the algebra of their associated matrices: compositions of endomorphisms are mapped into products, *in reversed order*, of their associated matrices.

metric 1.3.2. 33: of a topological space X, that there is a map (a metric)

$$d : X \times X \ni (a, b) \mapsto [0, \infty)$$

such that a) $d(a, b) = 0 \Leftrightarrow a = b$, b) $d(b, c) \leq d(a, b) + d(a, c)$, and c) the set of all open balls is a neighborhood base for the topology of X.

— density theorem 4.1. 213: If E is a Lebesgue measurable subset of $\mathbb{R}$ then for almost every x in E

$$\lim_{\delta \to 0} \frac{\lambda \left(E \cap (x - \delta, x + \delta) \right)}{2\delta} = 1.$$

midpoint-convex 1.1.4. 7: of a function f in $\mathbb{R}^{\mathbb{R}}$, that always

$$f \left(\frac{x + y}{2} \right) \leq \frac{f(x) + f(y)}{2}.$$

$$P_n(z) \overset{\text{def}}{=} \sum_{k=1}^{m_n} \frac{c_{-k}}{(z - a_n)^k}$$

then in a neighborhood $D(a_n, r)^{\circ}$ of a_n,

$$f(z) = P_n(z) + \sum_{k=1}^{\infty} c_k (z - a_n)^k.$$

Define a sequence $\{a_n\}_{n \in \mathbb{N}}$ as follows:

$$a_{n+1} \overset{\text{def}}{=} a_n - \frac{f(a_n)}{f'(a_n)}$$

so long as $f'(a_n) \neq 0$. The algorithm is occasionally successful in generating a sequence $\{a_n\}_{n \in \mathbb{N}}$ such that

$$\lim_{n \to \infty} a_n \overset{\text{def}}{=} a$$

exists. Furthermore, in some instances, $f(a) = 0$.

— (of a quaternion) 1.1.5. 13

— -separable 2.3.2. 164: of a normed vector space V, that it is separable in its norm-induced topology.

normal distribution (function) 4.2. 218: for a random variable f, the distribution function

$$P\left(\{\,\omega\;:\;f(\omega)\le x\,\}\right)\overset{\text{def}}{=}\frac{1}{\sqrt{2\pi}}\int_{-\infty}^{x}\exp\left(-\frac{t^2}{2}\right)dt.$$

— operator (see *spectral theorem*): for a Hilbert space $\mathcal{H}$, an endomorphism N such that $NN^* = N^*N$.

— subgroup 1.1.2. 3: in a group G, a subgroup H such that for all x in G, $xH = Hx$.

normalized (measure) 2.2.3. 143

normally distributed 4.2. 218: of a random variable, that its distribution function is the normal distribution.

normed vector space 3.2.1. 199: a vector space endowed with a norm.

NOVIKOV, P. S. 1.1.5. 11

nowhere dense 2.1.1. 43, 2.2.1. 107: of a set E in a topological space X, that $\overline{X \setminus \overline{E}} = X$; alternatively, that in every neighborhood of every point of X there is a nonempty open subset that does not meet E.

null set 2.1.2. 56: in a measure situation $(X, \mathcal{S}, \mu)$, a set of measure zero.

odd 2.1.3. 87: of a map $f : V \mapsto W$ between vector spaces, that $f(-\mathbf{x}) = -f(\mathbf{x})$; (also, of a permutation, that it is not even).

one-one (see *injection*): of a map f in Y^X, that

$$a \ne b \Rightarrow f(a) \ne f(b).$$

open 1.1.4. 5, 3.2.2. 203: of a map $f : X \mapsto Y$ between topological spaces, that the images of open sets are open; 1.1.4. 5, 2.1.1. 43: of a set U in a topological space X, that U is one of the sets defining the topology of X.

— arc-image 2.2.1. 113: in a topological space X, the image $\gamma\left((0,1)\right)$ for a γ in $C\left((0,1), X\right)$.

— ball 2.1.3. 67, 3.2.1. 200: for a point P in a metric space (X, d) and a positive r, the set $\{\,Q\;:\;Q \in X,\; d(P,Q) < r\,\}$.

optimal vertex 1.3.3. 38

— vector 1.3.3. 38

orbit 2.2.3. 147: in a set X on which a group G acts, for some P in X, a set of the form $\{\,g(P)\;:\;g \in G\,\}$.

order 1.1.2. 2: of a group G, its cardinality $\#(G)$; 1.1.2. 4: of an element a of a group, the least natural number m such that $a^m = e$; 2.3.4. 168: in a set S, a binary relation $\succ$ such that if $a, b \in S$ then at most one of

$$\int_I f(x_1, \ldots, x_n) \, d\mathbf{x}$$

exists, i.e., that f is bounded and that $\mathrm{Discont}(f)$ is a null set.

$$c_n \overset{\mathrm{def}}{=} \int_{-\pi}^{\pi} f(x) \frac{e^{-inx}}{\sqrt{2\pi}} \, dx, \ n \in \mathbb{Z}$$

then $\lim_{|n| \to \infty} c_n = 0$; 2.1.3. 88: If $f \in L^1(\mathbb{R}, \mathbb{C})$ and

$$\hat{f}(t) \overset{\mathrm{def}}{=} \frac{1}{\sqrt{2\pi}} \mathrm{int}_{\mathbb{R}} f(x) e^{-itx} \, dx$$

then $\lim_{|t| \to \infty} \hat{f}(t) = 0$. (A natural generalization of the Riemann-Lebesgue theorem is valid for a locally compact abelian group G endowed with Haar measure μ defined on the σ-ring $\mathcal{S}(\mathsf{K})$ generated by the set K of compact subsets of G: If $\mathbb{T}$ is regarded as an abelian group with respect to multiplication of complex numbers, if

$$\hat{G} \overset{\mathrm{def}}{=} \{\alpha \ : \ \alpha \text{ a homomorphism of } G \text{ into } \mathbb{T}, \}$$

and if $f \in L^1(G, \mathbb{C})$ then the (Gelfand-) Fourier transform

$$\hat{f} : \hat{G} \ni \alpha \mapsto \int_G f(x) \overline{\alpha(x)} \, d\mu(x) \overset{\mathrm{def}}{=} \hat{f}(\alpha)$$

vanishes at infinity, i.e., if $\epsilon > 0$ there is in $\hat{G}$ a compact set $K(\epsilon)$ such that $|\hat{f}(\alpha)| < \epsilon$ if $\alpha \notin K$.)

$$\alpha, \beta, \sigma_i, \rho_j > 0, \ 1 \leq i \leq m, \ 1 \leq j \leq n$$

$$A \overset{\mathrm{def}}{=} (a_{ij})_{i,j=1}^{m,n}, \ \mathbf{x} \overset{\mathrm{def}}{=} (x_1, \ldots, x_n), \ \mathbf{X} \overset{\mathrm{def}}{=} (X_1, \ldots, X_m) \overset{\mathrm{def}}{=} A\mathbf{x}^t$$

$$M_{\alpha,\beta} \overset{\mathrm{def}}{=} \sup_{\mathbf{x} \neq \mathbf{O}} \frac{\left(\sum_{i=1}^m \sigma_i |X_i|^{\frac{1}{\beta}}\right)^\beta}{\left(\sum_{j=1}^n \rho_j |x_j|^{\frac{1}{\alpha}}\right)^\alpha}.$$

Then on every line in $\Delta\{(\alpha,\beta) : 0 < \alpha \leq 1, 0 < \beta \leq \alpha\}$ $M_{\alpha,\beta}$ is a multiplicatively convex function of its variables, i.e., if

$$(\alpha_1, \beta_1), (\alpha_2, \beta_2) \in \Delta,$$
$$t \in (0,1), \text{ and}$$
$$(\alpha, \beta) = t(\alpha_1, \beta_1) + (1-t)(\alpha_2, \beta_2)$$

then

$$M_{\alpha,\beta} \leq M_{\alpha_1,\beta_1}^t M_{\alpha_2,\beta_2}^{1-t}.$$

If one carefully interprets the formulae in the limiting case in which $\alpha = \beta = 0$ the result remains valid.

section 2.2.1. 110: in a Cartesian product $X_{\lambda \in \Lambda} X_\lambda$, for λ_0 in Λ, $X_{\lambda \neq \lambda_0} X_\lambda$.

segment 2.3.3. 167: in a vector space V, the convex hull of two vectors.

self-adjoint 1.3.2. 35: of an endomorphism T of a Hilbert space $\mathcal{H}$, that for all $\mathbf{x}$ and $\mathbf{y}$ in $\mathcal{H}$,

$$(T\mathbf{x}, \mathbf{y}) = (\mathbf{x}, T\mathbf{y});$$

of a matrix $(a_{ij})_{i,j=1}^{n,n}$ that $a_{ij} = \overline{a_{ji}}$.

semicontinuous (lower resp. upper) 2.1.1. 48, 2.2.2. 138, 2.3.4. 171: of a function f defined in the neighborhood of a point a in a topological space X, that

$$f(a) = \liminf_{x \to a} f(x) \text{ resp. } f(a) = \limsup_{x \to a} f(x);$$

of a function f defined throughout X, that f is lower resp. upper semicontinuous at each point of X; equivalently that for every t in $\mathbb{R}$,

$$f^{-1}[(t, \infty)] \text{ resp. } f^{-1}[(-\infty, t)]$$

is open. Occasionally the qualifiers *lower, upper* are omitted and *semicontinuous* is used alone.

semigroup 1.1.5. 11: a set S on which there is defined a binary operation subject to the sole requirement of associativity.

seminorm 2.4.1. 174, 5.2. 229: for a vector space V, a subadditive map

$$p : V \ni \mathbf{x} \mapsto [0, \infty).$$

semisimple 2.4.1. 172: of a commutative Banach algebra B, that its radical is $\{\mathbf{O}\}$.

separable 2.2.1. 117, 3.2.2. 201: of a topological space X, that it has a countable base of neighborhoods; 2.1.3. 66: of a metric space (X, d), (equivalently) that it contains a countable dense subset.

separate, separating 2.1.4. 102, 5.2. 229: A set S of functions in Y^X *separates* or *is separating* if whenever $x_1 \neq x_2 \in X$ there is in S an f such that $f(x_1) \neq f(x_2)$.

separated locally convex vector space 5.2. 229: a locally convex Hausdorff vector space.

sequence (of groups) 1.1.3. 4

set 5.2. 229

sfield 1.1.5. 13, 1.2.1. 19

shift operator 1.3.2. 32

short exact sequence of groups 1.1.4. 8

SIERPINSKI, W. 2.1.3. 70, 2.2.1. 104, 2.2.1. 130, 2.2.3. 146

σ-finite(ness) 2.2.2. 137, 2.3.1. 162: of a measure situation $(X, \mathcal{S}, \mu)$, that every element of $\mathcal{S}$ is the countable union of sets of finite measure.

— -ring 1.1.4. 5, 2.2.1. 103: a set $\mathcal{S}$ of sets closed with respect to the formation of set differences and countable unions.

additive; b) if

$$A_i \in \mathcal{S}, \ i = 1, 2$$

$$M_i \overset{\text{def}}{=} E(A_i)\mathcal{H}, \ i = 1, 2$$

$$M_\vee \overset{\text{def}}{=} \overline{\text{span}(M_1 \cup M_2)}$$

$$M_\wedge \overset{\text{def}}{=} \overline{\text{span}(M_1 \cap M_2)}$$

and if $E_\vee$ resp. $E_\wedge$ is the orthogonal projection on $M_\vee$ resp. $M_\wedge$ then

$$E(A_1 \cup A_2) = E_\vee \text{ and } E(A_1 \cap A_2) = E_\wedge;$$

c) $E(\emptyset) = O$ and $E(X) = I$.
— radius 1.3.1. 29: of an element (an $n \times n$ matrix M) of $[\mathbb{R}^n]$, the number ρ_M equal to the maximum of the set of absolute values of the eigenvalues of M; 2.4.1. 172: of an element $\mathbf{x}$ of a Banach algebra B,

$$\sup\{|\lambda| \ : \ \lambda \in \sigma(\mathbf{x})\} = \lim_{n \to \infty} \|\mathbf{x}^n\|^{\frac{1}{n}}.$$

(If $[\mathbb{R}^n]$ is regarded as a Banach algebra the two definitions just given are equivalent.)
— theorem 1.3.2. 36: If N is a normal operator in $[\mathcal{H}]$, if $\mathcal{P}$ is the set of self-adjoint projections in $[\mathcal{H}]$, and if $\mathcal{B}_2$ is the set of Borel sets in $\mathbb{C}$, there exists a spectral measure $E : \mathcal{B}_2 \mapsto \mathcal{P}$ such that

$$N = \int_{\mathbb{C}} z \, dE(z).$$

spectrum 1.3.2. 31: for an element $\mathbf{x}$ in a Banach algebra B with identity $\mathbf{e}$, the set $\sigma(\mathbf{x})$ of complex numbers λ such that $\mathbf{x} - \lambda\mathbf{e}$ is not invertible.
splits 1.1.3. 5: said of an exact sequence

$$G \overset{\phi}{\mapsto} H \overset{\psi}{\mapsto} K$$

if H and $G \times K$ are isomorphic.
SPRECHER, D. A. 2.1.4. 101
STEINITZ, E. 2.1.3. 70
stereographic projection 2.2.3. 145: the map

$$\mathbb{R}^2 \ni (x, y) \mapsto \left(\frac{x}{2(x^2 + y^2 + .25)}, \frac{y}{2(x^2 + y^2 + .25)}, \frac{x^2 + y^2 - .25}{2(x^2 + y^2 + .25)} \right) \in S_{\frac{1}{2}}.$$

STIRLING, J. 2.1.1. 52, 2.5.1. 176
Stirling's formula 2.1.1. 52, 2.5.1. 176:

$$\lim_{n \to \infty} \frac{n!}{\left(\frac{n}{e}\right)^n \sqrt{2\pi n}} = 1.$$

tamely embedded sphere 3.2.2. 205

tame sphere 3.2.2. 205

TARSKI, A. 1.3.2. 22, 2.2.3. 145, 5.1. 224

ternary marker 2.1.1. 52

the abelianization of a group G 1.1.5. 15: the quotient group of G by its commutator subgroup $Q(G)$.

— free group 1.1.5. 10

— k-ary representation (of a number) 2.1.1. 52: for a number x in $\mathbb{R}$,

  if the k-ary representation of x is unique, the k-ary representation of x;

  if there are two k-ary representations, the one in which, for some N in $\mathbb{N}$, all k-ary markers x_n are $k-1$ if $n \geq N$.

theorem 5.1. 224

thick 2.2.1. 107, 4.2. 218: of a set E in a measure situation $(X, \mathcal{S}, \mu)$, that the outer measure of $X \setminus E$ is zero.

THOMPSON, J. 1.1.6. 18

THORIN, G. O. 2.3.4. 169

TOEPLITZ, O. 2.1.3. 75

Toeplitz matrix 2.1.3. 75

TONELLI, L. 2.2.2. 140

Tonelli's theorem 2.2.2. 141

topological division algebra 1.2.3. 25: a topological space $\mathcal{A}$ that is a division algebra such that

$$\mathcal{A} \times \mathcal{A} \ni (a, b) \mapsto a - b \in \mathcal{A}$$

and

$$\mathcal{A} \times (\mathcal{A} \setminus \{0\}) \ni (a, b) \mapsto ab^{-1} \in \mathcal{A}$$

are continuous.

— group 1.1.4. 5: a topological space G that is a group such that $G \times G \ni (x, y) \mapsto xy^{-1} \in G$ is continuous.

— vector space 2.3.4. 168: a vector space V for which the additive structure is a topological group and that is defined over a topological field $\mathbb{K}$ so that the map

$$\mathbb{K} \times V \ni (a, \mathbf{x}) \mapsto a\mathbf{x} \in V$$

is continuous.

topology 2.1.1. 42: for a set X, in the power set 2^X a subset $\mathcal{O}$ containing $\emptyset$ and X and closed with respect to the formation of finite intersections and arbitrary unions of the elements of $\mathcal{O}$.

topopathology 3.2.2. 200: a topological phenomenon that goes counter to the expectations of many mathematicians; the study of such phenomena.

— continuous 1.1.4. 6: of a map $f : X \mapsto Y$ and for uniform structures
U for X and V for Y, that if $V \in \mathsf{V}$ there is in U a U such that
$(a, b) \in U \Rightarrow (f(a), f(b)) \in V$.

unit ball 2.3.4. 171: in a metric space (X, d), a set of the form

$$\{ x \ : \ d(x, a) \leq 1 \}.$$

unitary 1.3.2. 36: of an automorphism U of a Euclidean space $\mathcal{H}$, that for
all $\mathbf{x}$ and $\mathbf{y}$ in $\mathcal{H}$, $(U\mathbf{x}, U\mathbf{y}) = (\mathbf{x}, \mathbf{y})$.

univalent 2.6.6. 184: of a holomorphic function, that it is injective.

universal comparison test 2.1.3. 72

vanishes at infinity 2.1.3. 88, 2.3.2. 163: of a $\mathbb{C}$-valued function f defined
on a topological space X, that for every positive ϵ there is in X a
compact set $K(\epsilon)$ such that $x \notin K(\epsilon) \Rightarrow |f(x)| < \epsilon$.

variable 5.1. 223

variation 2.3.1. 160: See *total variation*.

— lattice: See *linear lattice*.

— space 1.1.4. 6, 1.3.1. 25: an abelian group V that is a module over a field
$\mathbb{K}$, i.e., there is a map

$$\mathbb{K} \times V \ni (a, \mathbf{x}) \mapsto a \cdot \mathbf{x} \in V$$

such that a) $a \cdot (b \cdot \mathbf{x}) = ab \cdot \mathbf{x}$ and b) $a \cdot (\mathbf{x} + \mathbf{y}) = a \cdot \mathbf{x} + a \cdot \mathbf{y}$. The
elements of V are *vectors*.

VESLEY, R. 5.1. 228

vicinity 1.2.3. 24: an element U of a uniform structure U.

Walsh, J. L. 4.1. 215

Walsh function 4.1. 215: for a set $\{r_{k_1}, \ldots, r_{k_m}\}$ of Rademacher functions,
the function

$$\prod_{i=1}^{m} r_{k_i}.$$

weak 2.5.2. 179: of a solution of a differential equation, that it is a distri-
bution; 3.2.2. 202: of a topology O for a space X and a set $\{f_\lambda\}_{\lambda \in \Lambda}$ of
maps from X into a topological space Y, that O is generated by the
set $\{ f_\lambda^{-1}(V) \ : \ V \text{ open in } Y \}$.

weaker 3.2.2. 201: of a topology O, that it is a subset of another topology
O'.

weakest 2.3.2. 162: of a topology O, that it is weaker than each topology
of a set of topologies.

WEIERSTRASS, K. 2.2.2. 139, 2.6.4. 183

Zermelo, E. 5.2. 228

Zermelo-Fraenkel 5.2. 228: of the set of axioms provided by Zermelo and
    Fraenkel as the foundation for set theory.

zero homomorphism 2.4.2. 175: the homomorphism mapping each element
    of an algebra into 0.

ZORN, M. 1.1.4. 6

Zorn's lemma 1.1.4. 6: If $(S, \prec)$ is a partially ordered set in which each
    linearly ordered subset has an upper bound, then S has a maximal
    element, i.e., there is in S an s such that for any s' in S, either $s' \preceq s$
    or s' and s are not comparable, i.e., *never* $s \prec s'$.

Problem Books in Mathematics *(continued)*

Problem-Solving Through Problems
by *Loren C. Larson*

A Problem Seminar
by *Donald J. Newman*

Exercises in Number Theory
by *D.P. Parent*